Mathematical Decision Making:
Predictive Models
and Optimization

Scott P. Stevens, Ph.D.

THE
GREAT
COURSES®

PUBLISHED BY:

THE GREAT COURSES
Corporate Headquarters
4840 Westfields Boulevard, Suite 500
Chantilly, Virginia 20151-2299
Phone: 1-800-832-2412
Fax: 703-378-3819
www.thegreatcourses.com

Scott P. Stevens, Ph.D.

Professor of Computer Information Systems
and Business Analytics
James Madison University

Professor Scott P. Stevens is a Professor of Computer Information Systems and Business Analytics at James Madison University (JMU) in Harrisonburg, Virginia. In 1979, he received B.S. degrees in both Mathematics and Physics from The Pennsylvania State University, where he was first in his graduating class in the College of Science. Between completing his undergraduate work and entering a doctoral program, Professor Stevens worked for Burroughs Corporation (now Unisys) in the Advanced Development Organization. Among other projects, he contributed to a proposal to NASA for the Numerical Aerodynamic Simulation Facility, a computerized wind tunnel that could be used to test aeronautical designs without building physical models and to create atmospheric weather models better than those available at the time.

In 1987, Professor Stevens received his Ph.D. in Mathematics from The Pennsylvania State University, working under the direction of Torrence Parsons and, later, George E. Andrews, the world's leading expert in the study of integer partitions.

Professor Stevens's research interests include analytics, combinatorics, graph theory, game theory, statistics, and the teaching of quantitative material. In collaboration with his JMU colleagues, he has published articles on a wide range of topics, including neural network prediction of survival in blunt-injured trauma patients; the effect of private school competition on public schools; standards of ethical computer usage in different countries; automatic data collection in business; the teaching of statistics and linear programming; and optimization of the purchase, transportation, and deliverability of natural gas from the Gulf of Mexico. His publications have appeared in a number of conference proceedings, as well as in the *European*

Journal of Operational Research; the *International Journal of Operations & Production Management*; *Political Research Quarterly*; *Omega: The International Journal of Management Science*; *Neural Computing & Applications*; *INFORMS Transactions on Education*; and the *Decision Sciences Journal of Innovative Education*.

Professor Stevens has acted as a consultant for a number of firms, including Corning Incorporated, C&P Telephone, and Globaltec. He is a member of the Institute for Operations Research and the Management Sciences and the Alpha Kappa Psi business fraternity.

Professor Stevens's primary professional focus since joining JMU in 1985 has been his deep commitment to excellence in teaching. He was the 1999 recipient of the Carl Harter Distinguished Teacher Award, JMU's highest teaching award. He also has been recognized as an outstanding teacher five times in the university's undergraduate business program and once in its M.B.A. program. His teaching interests are wide and include analytics, statistics, game theory, physics, calculus, and the history of science. Much of his recent research focuses on the more effective delivery of mathematical concepts to students.

Professor Stevens's previous Great Course is *Games People Play: Game Theory in Life, Business, and Beyond.* ■

Table of Contents

Table of Contents

Table of Contents

Mathematical Decision Making:
Predictive Models and Optimization

Scope:

People have an excellent track record for solving problems that are small and familiar, but today's world includes an ever-increasing number of situations that are complicated and unfamiliar. How can decision makers—individuals, organizations in the public or private sectors, or nations—grapple with these often-crucial concerns? In many cases, the tools they're choosing are mathematical ones. Mathematical decision making is a collection of quantitative techniques that is intended to cut through irrelevant information to the heart of a problem, and then it uses powerful tools to investigate that problem in detail, leading to a good or even optimal solution.

Such a problem-solving approach used to be the province only of the mathematician, the statistician, or the operations research professional. All of this changed with two technological breakthroughs, both in the field of computing: automatic data collection and cheap, readily available computing power. Automatic data collection (and the subsequent storage of that data) often provides the analyst with the raw information that he or she needs. The universality of cheap computing power means that analytical techniques can be practically applied to much larger problems than was the case in the past. Even more importantly, many powerful mathematical techniques can now be executed much more easily in a computer environment—even a personal computer environment—and are usable by those who lack a professional's knowledge of their intricacies. The intelligent amateur, with a bit of guidance, can now use mathematical techniques to address many more of the complicated or unfamiliar problems faced by organizations large and small. It is with this goal that this course was created.

The purpose of this course is to introduce you to the most important prediction and optimization techniques—which include some aspects of statistics and data mining—especially those arising in operations research (or operational research). We begin each topic by developing a clear intuition of the purpose of the technique and the way it works. Then, we apply it to a problem in a step-by-step approach. When this involves using a computer, as often it does, we keep it accessible. Our work can be done in a spreadsheet environment, such as OpenOffice's Calc (which is freely distributable) or Microsoft Office's Excel. This has two advantages. First, it allows you to see our progress each step of the way. Second, it gives you easy access to an environment where you can try out what we're examining on your own. Along the way, we explore many real-world situations where various prediction and optimization techniques have been applied—by individuals, by companies, by agencies in the public sector, and by nations all over the world.

Just as there are many kinds of problems to be solved, there are many techniques for addressing them. These tools can broadly be divided into predictive models and mathematical optimization.

Predictive models allow us to take what we already know about the behavior of a system and use it to predict how that system will behave in new circumstances. Regression, for example, allows us to explore the nature of the interdependence of related quantities, identifying those ones that are most useful in predicting the one that particularly holds our interest, such as profit.

Sometimes, what we know about a system comes from its historical behavior, and we want to extrapolate from that. Time series forecasting allows us to take historical data as a guide, using it to predict what will happen next and informing us how much we can trust that prediction.

The flood of data readily available to the modern investigator generates a new kind of challenge: how to sift through those gigabytes of raw information and identify the meaningful patterns hidden within them. This is the province of data mining, a hot topic with broad applications—from online searches to advertising strategies and from recognizing spam to identifying deadly genes in DNA.

But making informed predictions is only half of mathematical decision making. We also look closely at optimization problems, where the goal is to find a best answer to a given problem. Success in this regard depends crucially on creating a model of the situation in a mathematical form, and we'll spend considerable time on this important step. As we'll discover, some optimization problems are amazingly easy to solve while others are much more challenging, even for a computer. We'll determine what makes the difference and how we can address the obstacles. Because our input data isn't always perfect, we'll also analyze how sensitive our answers are to changes in those inputs.

But uncertainty can extend beyond unreliable inputs. Much of life involves unpredictable events, so we develop a variety of techniques intended to help us make good decisions in the face of that uncertainty. Decision trees allow us to analyze events that unfold sequentially through time and evaluate future scenarios, which often involve uncertainty. Bayesian analysis allows us to update our probabilities of upcoming events in light of more recent information. Markov analysis allows us to model the evolution of a chance process over time. Queuing theory analyzes the behavior of waiting lines— not only for customers, but also for products, services, and Internet data packets. Monte Carlo simulation allows us to create a realistic model of an environment and then use a computer to create thousands of possible futures for it, giving us insights on how we can expect things to unfold. Finally, stochastic optimization brings optimization techniques to bear even in the face of uncertainty, in effect uniting the entire toolkit of deterministic and probabilistic approaches to mathematical decision making presented in this course.

Mathematical decision making goes under many different names, depending on the application: operations research, mathematical optimization, analytics, business intelligence, management science, and others. But no matter what you call it, the result is a set of tools to understand any organization's problems more clearly, to approach their solutions more sensibly, and to find good answers to them more consistently. This course will teach you how some fairly simple math and a little bit of typing in a spreadsheet can be parlayed into a surprising amount of problem-solving power. ∎

The Operations Research Superhighway
Lecture 1

This course is all about the confluence of mathematical tools and computational power. Taken as a whole, the discipline of mathematical decision making has a variety of names, including operational research, operations research, management science, quantitative management, and analytics. But its purpose is singular: to apply quantitative methods to help people, businesses, governments, public services, military organizations, event organizers, and financial investors find ways to do what they do better. In this lecture, you will be introduced to the topic of operations research.

What Is Operations Research?

- **Operations research** is an umbrella term that encompasses many powerful techniques. Operations research applies a variety of mathematical techniques to real-world problems. It leverages those techniques by taking advantage of today's computational power. And, if successful, it comes up with an implementation strategy to make the situation better. This course is about some of the most important and most widely applicable ways that that gets done: through predictive **models** and mathematical **optimization**.

- In broad terms, predictive models allow us to take what we already know about the behavior of a system and use it to predict how that system will behave in new circumstances. Often, what we know about a system comes from its historical behavior, and we want to extrapolate from that.

- Sometimes, it's not history that allows us to make predictions but, instead, what we know about how the pieces of the system fit together. Complex behavior can emerge from the interaction of even simple parts. From there, we can investigate the possibilities—and probabilities.

- But making informed predictions is only half of what this course is about. We'll also be looking closely at optimization and the tools to accomplish it. Optimization means finding the best answer possible to a problem. And the situation can change before the best answer that you found has to be scrapped. There are a variety of optimization techniques, and some optimization questions are much harder to solve than others.

- Mathematical decision making offers a different way of thinking about problems. This way of looking at problems goes all the way back to the rise of the scientific approach—in particular, investigating the world not only qualitatively but quantitatively. That change turned alchemy into chemistry, natural philosophy into physics and biology, astrology into astronomy, and folk remedies into medicine.

- It took a lot longer for this mindset to make its way from science and engineering into other fields, such as business and public policy. In the 1830s, Charles Babbage, the pioneer in early computing machines, expounded what today is called the Babbage principle—namely, the idea that highly skilled, high-cost laborers should not be "wasting" their time on work that lower-skilled, lower-cost laborers could be doing.

- In the 1880s, this idea became part of Fredrick Taylor's scientific management, which attempted to apply the principles of science to manufacturing workflow. His approach focused on such matters as efficiency, knowledge transfer, analysis, and mass production. Tools of statistical analysis began to be applied to business.

- Then, Henry Ford took the idea of mass production, coupled it with interchangeable parts, and developed the assembly line system at his Ford Motor Company. The result was a company that, in the early 20th century, paid high wages to its workers and still sold an affordable automobile.

- But most historians set the real start of operations research in Britain in 1937 during the perilous days leading up to World War II—specifically, the Bawdsey Research Station near Suffolk. It was the center of radar research and development in Britain at the time. It was also the location of the first radar tower in what became Britain's essential early-warning system against the German Luftwaffe.

- A. P. Rowe was the station superintendent in 1937, and he wanted to investigate how the system might be improved. Rowe not only assessed the equipment, but he also studied the behavior of the operators of the equipment, who were, after all, soldiers acting as technicians. The results allowed Britain to improve the performance of both men and machines. Rowe's work also identified some previously unnoticed weaknesses in the system.

- This analytical approach was dubbed "operational research" by the British, and it quickly spread to other branches of their military and to the armed forces of other allied countries.

Computing Power

- Operational research—or, as it came to be known in the United States, operations research—was useful throughout the war. It doubled the on-target bomb rate for B-29s attacking Japan. It increased U-boat hunting kill rates by about a factor of 10. Most of this, and other work, was classified during the war years. So, it wasn't until after the war that people started turning a serious eye toward what operational research could do in other areas. And the real move in that direction started in the 1950s, with the introduction of the electronic computer.

- Until the advent of the modern computer, even if we knew how to solve a problem from a practical standpoint, it was often just too much work. Weather forecasting, for example, had some mathematical techniques available from the 1920s, but it was impossible to reasonably compute the predictions of the models before the actual weather occurred.

- Computers changed that in a big way. And the opportunities have only accelerated in more recent decades. Gordon E. Moore, cofounder of Intel, first suggested in 1965 what has since come to be known as **Moore's law**: that transistor chip count on an integrated circuit doubles about every two years. Many things that we care about, such as processor speed and memory capacity, grow along with it. Over more than 50 years, the law has continued to be remarkably accurate.

- It's hard to get a grip on how much growth that kind of doubling implies. Moore's law accurately predicted that the number of chips on an integrated circuit in 2011 was about 8 million times as high as it was in 1965. That's roughly the difference between taking a single step and walking from Albany, Maine, to Seattle, Washington, by way of Houston and Los Angeles. All of that power was now available to individuals and companies at an affordable price.

Mathematical Decision-Making Techniques

- Once we have the complicated and important problems, like it or not, along with the computing power, the last piece of the puzzle is the mathematical decision-making techniques that allow us to better understand the problem and put all that computational power to work.

- To do this, first you have to decide what you're trying to accomplish. Then, you have to get the data that's relevant to the problem at hand. Data collection and cleansing can always be a challenge, but the computer age makes it easier than ever before. So much information is automatically collected, and much of it can be retrieved with a few keystrokes.

- But then comes what is perhaps the key step. The problem lives in the real world, but in order to use the powerful synergy of mathematics and computers, it has to be transported into a new, more abstract world. The problem is translated from the English that we use to describe it to each other into the language of mathematics. Mathematical language isn't suited to describe everything, but what it can capture it does with unparalleled precision and stunning economy.

- Once you've succeeded in creating your translation—once you have modeled the problem—you look for patterns. You try to see how this new problem is like ones you've seen before and then apply your experience with them to it.

- But when an operations researcher thinks about what other problems are similar to the current one, he or she is thinking about, most of all, the mathematical formulation, not the real-world context. In daily life, you might have useful categories like business, medicine, or engineering, but relying on these categories in operations research is as sensible as thinking that if you know how to buy a car, then you know how to make one, because both tasks deal with cars.

- In operations research, the categorization of a problem depends on the mathematical character of the problem. The industry from which it comes only matters in helping to specify the mathematical character of the problem correctly.

Modeling and Formulation
- The translation of a problem from English to math involves modeling and formulation. An important way that we can classify problems is as either **stochastic** or **deterministic**. Stochastic problems involve random elements; deterministic problems don't.

- Many problems ultimately have both deterministic and stochastic elements, so it's helpful to begin this course with some statistics and data mining to get a sense of that combination. Both topics are fields in their own right that often play important roles in operations research.

- Many deterministic operations research problems focus on optimization. For problems that are simple or on a small scale, the optimal solution may be obvious. But as the scale or complexity of the problem increases, the number of possible courses of action tends to explode. And experience shows that seat-of-the-pants decision making can often result in terrible strategies.

- But once the problem is translated into mathematics, we can apply the full power of that discipline to finding its best answer. In a real sense, these problems can often be thought of as finding the highest or lowest point in some mathematical landscape. And how we do this is going to depend on the topography of that landscape. It's easier to navigate a pasture than a glacial moraine. It's also easier to find your way through open countryside than through a landscape crisscrossed by fences.

- Calculus helps with finding highest and lowest points, at least when the landscape is rolling hills and the fences are well behaved, or non-existent. But in calculus, we tend to have complicated functions and simple boundary conditions. For many of the practical problems we'll explore in this course through linear programming, we have exactly the opposite: simple functions but complicated boundary conditions.

- In fact, calculus tends to be useless and irrelevant for linear functions, both because the **derivatives** involved are all constants and because the **optimum** of a linear function is always on the boundary of its domain, never where the derivative is zero. So, we're going to focus on other ways of approaching optimization problems—ways that don't require a considerable background in calculus and that are better at handling problems with cliffs and fences.

- These deterministic techniques often allow companies to use computer power to solve in minutes problems that would take hours or days to sort out on our own. But what about more sizeable uncertainty? As soon as the situation that you're facing involves a random process, you're probably not going to be able to guarantee that you'll find the best answer to the situation—at least not a "best answer" in the sense that we mean it for deterministic problems.

- For example, given the opportunity to buy a lottery ticket, the best strategy is to buy it if it's a winning ticket and don't buy it if it's not. But, of course, you don't know whether it's a winner or a loser at the time you're deciding on the purchase. So, we have to come up with a different way to measure the quality of our decisions when we're dealing with random processes. And we'll need different techniques, including probability, Bayesian statistics, Markov analysis, and simulation.

Important Terms

derivative: The derivative of a function is itself a function, one that essentially specifies the slope of the original function at each point at which it is defined. For functions of more than one variable, the concept of a derivative is captured by the vector quantity of the gradient.

deterministic: Involving no random elements. For a deterministic problem, the same inputs always generate the same outputs. Contrast to **stochastic**.

model: A simplified representation of a situation that captures the key elements of the situation and the relationships among those elements.

Moore's law: Formulated by Intel founder Gordon Moore in 1965, it is the prediction that the number of transistors on an integrated circuit doubles roughly every two years. To date, it's been remarkably accurate.

operations research: The general term for the application of quantitative techniques to find good or optimal solutions to real-world problems. Often called operational research in the United Kingdom. When applied to business problems, it may be referred to as management science, business analytics, or quantitative management.

optimization: Finding the best answer to a given problem. The best answer is termed "optimal."

optimum: The best answer. The best answer among all possible solutions is a global optimum. An answer that is the best of all points in its immediate vicinity is a local optimum. Thus, in considering the heights of points in a mountain range, each mountain peak is a local maximum, but the top of the tallest mountain is the global maximum.

stochastic: Involving random elements. Identical inputs may generate differing outputs. Contrast to **deterministic**.

Suggested Reading

Budiansky, *Blackett's War*.

Gass and Arjang, *An Annotated Timeline of Operations Research*.

Horner and List, "Armed with O.R."

Yu, Argüello, Song, McCowan, and White, "A New Era for Crew Recovery at Continental Airlines."

Questions and Comments

1. Suppose that you decide to do your holiday shopping online. You have a complete list of the presents desired by your friends and family as well as access to the inventory, prices, and shipping costs for each online site. How could you characterize your task as a deterministic optimization problem? What real-world complications may turn your problem from a deterministic problem into a stochastic one?

    Answer:

    The most obvious goal is to minimize total money spent, but it is by no means the only possibility. If you are feeling generous, you might wish to maximize number of presents bought, maximize number of people for whom you give presents, and so on. You'll face some constraints. Perhaps you are on a limited budget. Maybe you have to buy at least one present for each person on your list. You might have a lower limit on the money spent on a site (to get free shipping). You also can't buy more of an item than the merchant has. In this environment, you're going to try to determine the number of items of each type that you buy from each merchant.

The problem could become stochastic if there were a chance that a merchant might sell out of an item, or that deliveries are delayed, or that you may or may not need presents for certain people.

2. Politicians will often make statements like the following: "We are going to provide the best-possible health care at the lowest-possible cost." While on its face this sounds like a laudable optimization problem, as stated this goal is actually nonsensical. Why? What would be a more accurate way to state the intended goal?

Answer:

It's two goals. Assuming that we can't have negative health-care costs, the lowest-possible cost is zero. But the best-possible health care is not going to cost zero. A more accurate way to state the goal would be to provide the best balance of health-care quality and cost. The trouble, of course, is that this immediately raises the question of who decides what that balance is, and how. This is exactly the kind of question that the politician might want not to address.

The Operations Research Superhighway
Lecture 1—Transcript

We live on a planet of seven billion people, organized into communities and companies and nations. And every day, all of those people, and the communities and companies and nations that they make up, have to make decisions—decisions that affect themselves, usually, the lives of others as well.

This isn't a new state of affairs, and as long as the situation faced is a familiar one, or if it's new but not too complicated, then it's relatively easy for them to find a good course of action. Experience informs intuition, maybe the personal experience of the individual, maybe the experience of the community, codified in its traditions and rules.

All in all, it's a pretty good system, and we've been using it to figure out how to hunt, or how to build a fire, or when to plant and harvest crops for a long time. Preserving and passing on this kind of knowledge is the foundation of any society. But cultures and traditions develop over a long period of time, and those eventual good answers have usually been found after a long period of trial and error.

But many of the problems faced today are anything but traditional. They're new. They're complicated. And some of them come with huge, even global consequences for bad decisions. But the Brave New World of these complicated problems also contains new possibilities for finding answers, including two that will be especially important to us in this course, the development of mathematical tools and techniques to access, and the access to extensive and affordable computer power.

The confluence of mathematical tools and computational power has led us to where we are now and to what this course is all about. Taken as a whole, the discipline has a variety of names, occupational research, operations research, management science, quantitative management, analytics. But its purpose is single minded—to apply quantitative methods to help people, businesses, governments, public services, military organizations, event organizers, financial investors, to find ways to do what they do—better.

Let me give you some examples. Medicine: University of Pittsburgh researchers looked at how liver transplants are assigned and livers delivered in the U.S. It turns out that there are 11 distributional zones that evolved over time in a way that had more to do with history than getting livers in time to people who need them. So, researchers revisited the problem. They looked at factors such as geography and population density, and they anticipated supply and demand of future liver transplants. Using an optimization technique called integer programming, they analyzed more than a trillion alternatives and came up with new map consisting of six, rather than 11, distribution zones. They estimated that implementing this new system could mean that 14% more people got liver transplants; that's over 800 people each year getting an organ that they can't live without.

Transportation: The Netherlands is a small country, but one that relies heavily on its railway system. On average, a Dutch citizen travels about 1000 miles per year by rail. About 1.1 million people use the system on an average workday. That kind of a load often leads to substantial train delays. The Dutch could build more track, but, that's expensive, and frankly, land in the country is at a premium. So, in 2006, Netherlands Railways threw out their old passenger timetable and created a completely new one. No new track, just a new timetable. The new scheduling was done using operations research techniques, including a network flow model. The result? A new system with fewer delays, more trains in operation, and a happier public. Netherlands Railways saw a $60 million increase in its annual profit, a quarter of which came from increased passenger usage, a win for the company, a win for the public, and a win for the rest of us with the reduction in greenhouse gases that accompanied the shift from private vehicles to rail transport.

Let's go global. After the Cold War, both the Soviet Union and the United States had huge stockpiles of plutonium from that they had to have decommissioned from the warheads, about 34 tons of it, in fact. So what to do with it? Plutonium is highly toxic, even minute quantities, and the fact that it could be stolen by terrorists presented a clear and present danger to both national and global security. Further, the terms of the treaty between the two superpowers required that their stockpiles be depleted in tandem.

Both countries applied operations research to the problem in the form of a multi-attribute utility model. They created a hierarchy of objectives, nonproliferation of nuclear weapons, operational effectiveness, attention to the environment, health and safety issues, and so on. The ordering of the goals differed between the two countries, but in the U.S., 37 criteria emerged for assessing the relative desirability of 13 different disposal alternatives. The one that was selected, interestingly enough, was the same one that was adopted by the Soviets, to mix the plutonium into an oxide fuel for irradiation in commercial nuclear reactors.

Well, these examples are all very large scale, liver-transplant networks, national rail systems, nuclear builddown. You might be getting the impression that quantitative analytics is only useful, and only used, on problems of that magnitude. Not so. The tools are increasingly applied in small- and mid-sized companies as well. Let me give you two examples from the aviation industry. Let's start with a small one: Bombardier Flexjet. They had a fleet of 25 aircraft and operated, essentially, an airline without a flight schedule. Flexjet offered planes with 24-hour-a-day availability flying to a destination of your choice. You could literally fly almost anywhere in the world with 48 hours–notice, or less.

Think of the logistics of this. Planes are expensive, and letting them sit on the ground or flying them from place to place with no customer quickly eats up the profits. Likewise, crew need rest, and federal regulations are quite strict on how long a crew can be active before getting time off. A crew that's not where it's needed has to be deadheaded, traveling as passengers to their next point of departure. And Flexjet didn't even know what its schedule was going to be! They did quite well for themselves in spite of this, and within a four-year period, Flexjet grew from a company with 25 aircraft to manage to one with 80. And the logistical headaches became much, much worse.

So they turned to analytics, including nonlinear programming. The final result was a decision support system consisting of three pieces, each handling a different aspect of Flexjet's scheduling needs. The first determined the annual leaves for the crews. The second worked out the basic monthly schedules. The third had the huge job of working out the daily and weekly scheduling. The support system took three years to implement, but the results were worth it: a 20% decrease in the flight-crew size per flight, a 10% increase in flight hours available per craft, and no reduction in the fraction of flights that went off perfectly. A great solution for a small company.

On the other end of the scale, we have Continental Airlines. Long ago, Continental decided where it would fly, what its flight schedules would be, where its transportation hubs would be located, and so on. But every airline faces unexpected disruptions in its schedule, from weather to mechanical problems to crew availability. Like Flexjet, Continental turned to operations research (OR) for a way to prepare for what they couldn't predict. The result was CrewSolver, a decision support system that quickly determined the best, or nearly the best, way to reassign crews to craft in the event of a disruption. It takes into account not only the flights that needed to be covered and the regulations that must be obeyed, but also factors in quality of life issues for the crews being scheduled. The system was in place before the 9/11 terrorist attacks, and Continental led the industry in the rapidity of its recovery from the disruptions of that day. Continental estimates that CrewSolver saved the company $40 million in major disruptions in 2001 alone. Since then, many other airlines have followed Continental's lead. And the list of applications goes on, and on, and on.

Kimberly Clark uses OR to schedule its factory floors in Latin America, letting it manufacture its dozens of products quickly and inexpensively on the same few machines. Electric companies use OR to determine schedules for buying, selling, and generating electricity, and then delivering it cheaply and reliably to wherever it's needed. Hospital emergency rooms use OR to reduce waiting times, cut costs, improve quality of service. OR has helped Travelocity to regain its market share and double its revenue over a three-year period. It's been used to develop an expert system to assist patients suffering from prostate cancer to make intelligent, informed decisions based on their personal situations and preferences. In short, OR is everywhere.

But what is it? What are we doing when we are applying OR to a problem? Because, to some extent, the success stories that I've told you about so far have been black boxes. The only thing that they seemed to have in common was that there was a problem. Along comes operations research. Whamo! Things got better. Well, operations research is an umbrella term, and a lot of powerful techniques fit under that umbrella.

To characterize what OR is, the Institute for Operations Research and Management Science has coined the phrase "the science of better," and that's an excellent way to put it. OR applies a variety of mathematical techniques to real-world problems. It leverages those techniques by taking advantage of today's computational power. And, if successful, it comes up with an implementation strategy to make the situation better. This course is about some of the most important and most widely applicable ways that that gets done, through predictive models optimization.

Speaking broadly, predictive models let us take what we already know about the behavior of a system and use it to predict how that system will behave in new circumstances. Often, what we know about a system comes from its historical behavior, and we want to extrapolate from that. What's going to happen in the use of solar power? In the speed and power of computer chips? Can we look at the emails that you've received in the past and build a better spam filter? How does Amazon make good guesses about the other products that you might be interested in? How does Pandora pick the next song to play on your personal radio station? We'll look at all of these questions and techniques that can address them.

Sometimes, it's not history that lets us make predictions, but what we already know about how the pieces in the system fit together. Complex behavior can emerge from the interaction of even simple parts. From there, we can investigate the possibilities and probabilities. How can search parties that have failed to find a downed plane turn that negative information into guidance on how to carry on? What can a direct marketing firm expect to happen if it changes its catalog shipping practices? What's going to happen to the waiting times of customers if the service facility's capacity is cut by 10%? How much of a safety margin is needed in bidding a job in the face of uncertainty if you don't want to lose your shirt? We'll explore all of these questions, too, and the tools needed to address them.

But making informed predictions is only half of what we'll be talking about; we'll also be looking closely at optimization and the tools to accomplish it. Optimization means finding the best answer to a problem. How should an airline schedule its crews? Its flights? Select its hub cities? How should a railroad get its empty freight cars from where they are now to where they're needed, cheaply and quickly? How should hospitals schedule their staff while keeping costs down? How should you divide your hard-earned money among various investments in light of some expenses that you know you have coming up? And how much can the situation change in any of these matters before the best answer that you just found has to be scrapped? Looking at these examples will lead us to a variety of optimization techniques and give us the insights into why some optimization questions are so much harder to answer and solve.

In creating this course, I've aimed for the middle ground in terms of rigor. My guiding principle has been to cover the material that has proven useful, and to do so in the context of interesting cases. Because of that, you'll see neither mathematical proofs nor drill work. What you will see are lots and lots of real-world examples, along with the development of the OR tools necessary to address them. And, the actual solution process applied to a simplified version of the situation. In some cases, where a computer is capable of, or even essential for, conducting the needed grunt work, I'll show you how to use a simple spreadsheet program to get that part of the job done. But in every case, I've tried to combine the actual solution process with a useful intuition about what's really going on. So if you're looking for a roll-up-your-sleeves course, there'll be plenty to do. By the time you finish it, you'll have a basic competence in operations research, and a solid foundation, if you want to go further.

But, just getting the sense of what it's all about can also be very empowering, and you can come back later and improve specific skills, if your intellectual curiosity or professional interests take you in that direction; you don't need technical fluency in order to see how mathematical decision making can apply to a problem that you may face. Maybe you call in the experts, or maybe that becomes your own cue to engage with this material more fully.

Regardless, mathematical decision making offers a different way of thinking about problems, a perspective that, I believe, can inspire anyone to seek and find new solutions [to] all sorts of problems.

It may be helpful say something about how all this all got started. This way of looking at problems goes all the way back to the rise of the scientific approach, in particular, investigating the world not only qualitatively, but quantitatively. That change turned alchemy into chemistry, natural philosophy into physics and biology, astrology into astronomy, folk remedies into medicine.

It took a lot longer for this mindset to make its way [from] science and engineering to other fields, such as business and public policy. Least squares regression, for example, was pioneered in astronomy in the early 19th century, but it didn't get the name regression until the late 19th century, and it didn't get applied to business and economic problems until the 20th century. In the 1830s, Charles Babbage pioneered early computing machines, expounded what's today called the Babbage Principle, namely, the idea that highly skilled, high-cost laborers should not be wasting their time working on things that lower-skilled, lower-cost laborers could be doing. In the 1880s, this idea became part of Fredrick Taylor's scientific management, which attempted to apply the principles of science to the manufacturing workflow. His approach focused on such matters as efficiency, knowledge transfer, analysis, and mass production. Tools of statistical analysis began to be applied to business. Then Henry Ford took the idea of mass production, coupled it with interchangeable parts, and developed the assembly line system at his Ford Motor Company. The result was a company that, in the early 20th century, paid high wages to its workers and still sold an affordable automobile.

But most historians set the real start of operations research in Britain, in 1937, during the perilous days leading up to World War II—specifically, at the Bawdsey Research Station near Suffolk. It was the center of radar research and development in Britain at the time. It was also the location of the first radar tower in what became Britain's essential early-warning system against the German Luftwaffe. A. P. Rowe was the station superintendent in 1937, and he wanted to investigate how the system might be improved. Rowe not only assessed the equipment, he also studied the behavior of the operators of the equipment, who were, after all, soldiers acting as technicians. The results allowed Britain to improve the performance of both men and machines. Rowe's work also identified some previously unnoticed weaknesses in the system. This analytical approach was dubbed operational research by the British at the time. It quickly spread to other branches of their military and to the armed forces of other allied countries.

Let me give you an idea of what I mean by focusing on one man—P. M. S. Blackett. Blackett was a physicist and a veteran of World War I. In the 1940s, he became the scientific advisor to the commander in chief of the British anti-aircraft command. Many years later, Blackett would go on to win the Nobel Prize for his work with cosmic rays, but in the early 1940s, he and his team were looking at the problem of how to more effectively shoot down German planes attacking Britain. Early on during the Battle of Britain in the summer and fall of 1940, it took an average of 20,000 rounds of anti-aircraft ammunition to bring down just one plane. A year later, with Blackett's team on the job, that dropped from 20,000 rounds to only 4,000!

From there, Blackett went to work for the Royal Navy. Commercial ships were being targeted and sunk by German U-boats with frightening frequency. They needed to travel in convoys, escorted by military vessels. A convoy can travel no faster than its slowest ship, so some military leaders wanted lots and lots of small convoys. Others favored fewer, large convoys. The idea here was that a large convoy can have more escort warships assigned to it— better protection. Blackett reviewed the data and found that the losses that the convoy suffers are, for the most part, dependent on the density of escorts on the convoy's perimeter. And it follows from this that the larger convoys were the way to go. Let's see why.

Imagine for a moment that the convoy's geometry is a square; the same analysis applies to any other shape. The area of the square determines how many ships the convoy can contain; the perimeter is the border that the escorting warships have to secure. In the schematic picture that I'm using, we have four escorts for one cargo ship icon. You can see that doubling each side of the square gives you an area that can hold four times as many ships.

But the perimeter has only increased by a factor of two. That means that combining four smaller convoys into one larger one would enable the British to double the density of escorts on the perimeter.

Once an analytic approach starts to take hold, you sometimes see things that you really should have seen earlier. During the war, the bottoms of the bomber aircraft were painted black, since they were often on night missions. But, was this the best choice when bombers were being used to hunt subs during the daytime? Experiment showed that painting the belly of the plane white made it blend in better with the gray skies of the North Atlantic. A white-bellied bomber could get 20% closer to its target before being sighted, on average. And that meant a projected 30% increase in kill rate, a no-brainer.

On the other hand, what sometimes seemed like a no-brainer really wasn't. For example, if our planes have enough engine power to carry more armor, where should we put it? To answer that question, people looked at the planes that limped home after suffering German anti-aircraft fire. Operational research is data-driven. The damage [tended] to fall in certain areas of the plane, so naturally, they suggested that those areas receive extra armor. Except Blackett's team came to exactly the opposite conclusion. The planes that were being seen were the ones that made it back, which meant, the areas where they took damage were not vital to being able to return from a mission. His team recommended reinforcing those parts of the plane that were conspicuously not damaged on the returning aircraft.

Operational research, or as it came to be known in the U.S., operations research, was useful throughout the war. It doubled the on-target bomb rate for B-29s attacking Japan; it increased U-boat hunting kills by a factor of about 10. Most of this and other work was classified during the war years, so it wasn't until after the war that people started turning a serious eye toward what operational research could do in other areas. And the real move in that direction started in the 1950s with the introduction of the electronic computer.

Until the advent of the modern computer, even if we knew how to solve a problem, from a practical standpoint, it was often just too much darned work. Weather forecasting, for example, had some mathematical techniques available from the 1920s, but it was impossible to reasonably compute the predictions of the model before the actual weather occurred. Computers changed that in a big way.

And the opportunities have only accelerated more in recent decades. Gordon E. Moore, cofounder of Intel, first suggested in 1965 what's since come to be known as Moore's Law, that transistor chip count on an integrated circuit doubles about every two years. A lot of things that we care about, like processor speed and memory capacity, grow along with it. Over more than 50 years, the law has continued to remain remarkably accurate. It's hard to get a grip on how much growth that kind of doubling implies. Moore's Law predicted, accurately, that the number of chips on an integrated circuit in 2011 was about 8 million times as high as it was in 1965. Still hard to grasp? That's the difference between taking a single step and walking from Augusta, Maine to Seattle, Washington, by way of Houston and L.A., and that power now available to individuals and companies at an affordable price.

So, we've got complicated and important problems, like it or not. We've got computing power. The last piece of the puzzle is the mathematical decision-making techniques that allow us to better understand the problem and put all that computational power to work. How does that work? Well, let me answer that first very broadly, and then sharpen the picture. First, you've got to decide what you're trying to accomplish. Then, you have to get the data that's relevant to the problem at hand. Data collection and cleansing can always be a challenge, but our computer age makes that easier than ever before. So much information is automatically collected, and much of it can be retrieved with a few keystrokes.

But then comes perhaps the key step. The problem lives in the real world, but in order to use the powerful synergy of mathematics and computers, it has to be transported into a new, more abstract world. The way that it always feels to me is not so much the problem is transported, as translated, from the English that we use to describe it to one another into "mathematese," if you like. The language of math isn't suited to describe everything; the beauty of a sunset is, and I suspect always will be, beyond mathematics. But what it can capture, it does with unparalleled precision and stunning economy.

I'll come back to this step in a few minutes. But for now, let's imagine that you've succeeded in creating your translation. You've modeled the problem. What do you do with this shiny new formulation? You do what humans always do. You look for patterns. You try to see how this new problem is like the ones that you've seen before, and then apply experience with them to it, just like the ancient wisdom that we discussed at the beginning of this lecture, that experience may be personal, or it may be hard won by others and passed on to you.

But, when an operations researcher thinks about what other problems are similar to the current one, he or she is thinking about, most of all, the mathematical formulation, not the real-world context. In daily life, you might have useful categories, like business, or medicine, or engineering, but relying on these categories in operations research is as sensible as thinking that if you know how to buy a car, then you know how to make one, since both tasks deal with cars.

In OR, the categorization of the problem depends on the mathematical character of the problem. The industry in which it comes from matters only in helping to specify the mathematical character of that problem correctly. So we'll be introducing a wide variety of useful mathematical categories in this course. Each one is a category, because some collection of very bright people figured out an approach that works for everything in that category. That's lovely. The approach is like a superhighway to your answer, very fast, very smooth, and, once you're on the right road, quite easy to follow.

But almost every trip is going to have an initial segment, taking you from home to the highway by secondary roads. Without that part, the superhighway isn't going to do you much good. That initial part of the journey is the translation of the problem from English to math that we discussed earlier—modeling, formulation. So, we're also going to spend quite a bit of time making sure that you have the expertise that will let you reach the highway for a lot of problem types. And we'll give you a feel, an intuition, for each problem type or technique, to help you find your way.

One important way that we can classify problems is either stochastic or deterministic. Stochastic problems involve random elements. Deterministic problems don't. Kimberly Clark knew what their machines could make, what quantities of each product they needed, how long it would take to reconfigure a machine between one product and the next—a deterministic problem. Determining a corporate response to a potential public-relations nightmare? That's stochastic. Many problems ultimately have both deterministic and stochastic elements, so it's helpful to begin our course with some statistics and data mining to get a sense of what that combination looks like right up front. Both topics are fields in their own right that often play important roles in operations research.

A lot of the deterministic problems in OR focus on optimization. For problems that are simple or on a small scale, the optimal solution may be obvious. But as the scale or complexity of the problem increases, the number of possible courses of action tends to explode. Think of guessing someone's one-character password, as opposed to guessing their 16-character one. And experience shows that seat-of-the-pants decision making can often result in terrible strategies. But once the problem is translated into mathematics, we can apply the full power of that discipline to finding its best answer.

We're going to see that, in a real sense, these problems can often be thought of as finding the highest or lowest point in some mathematical landscape. How we do this is going to depend on the topography of that landscape. It's easier to navigate a pasture than a glacial moraine. It's also easier to find your way through open countryside than through a landscape crisscrossed with fences.

If you've taken calculus, you know that it's great at finding highest and lowest points, at least when the landscape is rolling hills and the fences are well behaved, or, non-existent. But in calculus, we tend to have complicated functions and simple boundary conditions. For many of the practical problems we'll explore in this course through linear programming, we have exactly the opposite, simple functions but complicated boundary conditions.

In fact, calculus tends to be useless and irrelevant for linear functions, both because the derivatives involved are all constants, and because the optimum of a linear function is always on the boundary of its domain, never where the derivative is zero. So, I'm going to focus on other ways of approaching optimization problems, ways that don't require a considerable background in calculus and that are better at handling problems with cliffs and fences.

These deterministic techniques often allow companies like Flexjet and Continental to use computer power to solve, in minutes, problems that would take hours or days to sort out on our own. When Flexjet finally knows what its routing requirements are, or when Continental becomes aware of the disruptions to its normal operations, their models can compute solutions almost instantaneously, not only prescriptions of what should be done, but reports on how sensitive the solutions are to the minor errors in the data that were used to create the model in the first place.

How about more sizeable uncertainty? As soon as the situation that you're facing involves a random process, you're probably not going to be able to guarantee that you'll find the best answer to the situation, at least not a best answer in the sense that we meant it for deterministic problems. For example, given the opportunity to buy a lottery ticket, the best strategy is: Buy it if it's a winning ticket, and don't buy it if it's not. Lovely strategy. But of course, we don't know whether it's a winner or a loser at the time that you're deciding to make the purchase. So, we have to come up with a different way to measure the quality of our decisions when we're dealing with random processes. And we'll need different techniques, which we'll develop in the last section of the course—probability, Bayesian statistics, Markov analysis, simulation. One thing's for sure: In a stuff-happens kind of situation, we're going to want to be able to get a handle on the range of possible results of a strategy. How good can things end up? How bad? And how likely is "how bad"?

So, there's a lot to talk about, predictive models, optimization, stochastic models, a lot of powerful tools, a lot of practical applications, and some surprises along the way.

Forecasting with Simple Linear Regression
Lecture 2

Ⅰn this lecture, you will learn about linear regression, a forecasting technique with considerable power in describing connections between related quantities in many disciplines. Its underlying idea is easy to grasp and easy to communicate to others. The technique is important because it can—and does—yield useful results in an astounding number of applications. But it's also worth understanding how it works, because if applied carelessly, linear regression can give you a crisp mathematical prediction that has nothing to do with reality.

Making Predictions from Data

- Beneath Yellowstone National Park in Wyoming is the largest active volcano on the continent. It is the reason that the park contains half of the world's geothermal features and more than half of its geysers. The most famous of these is Old Faithful, which is not the biggest geyser, nor the most regular, but it is the biggest regular geyser in the park—or is it? There's a popular belief that the geyser erupts once an hour, like clockwork.

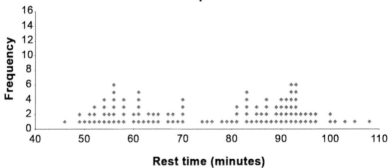

Time between Eruptions of Old Faithful

Figure 2.1

- In **Figure 2.1**, a dot plot tracks the rest time between one eruption and the next for a series of 112 eruptions. Each rest period is shown as one dot. Rests of the same length are stacked on top of one another. The plot tells us that the shortest rest time is just over 45 minutes, while the longest is almost 110 minutes. There seems to be a **cluster** of short rest times of about 55 minutes and another cluster of long rest times in the 92-minute region.

- Based on the information we have so far, when tourists ask about the next eruption, the best that the park service can say is that it will probably be somewhere from 45 minutes to 2 hours after the last eruption—which isn't very satisfactory. Can we use predictive modeling to do a better job of predicting Old Faithful's next eruption time? We might be able to do that if we could find something that we already know that could be used to predict the rest periods.

- A rough guess would be that water fills a chamber in the earth and heats up. When it gets hot enough, it boils out to the surface, and then the geyser needs to rest while more water enters the chamber and is heated to boiling. If this model of a geyser is roughly right, we could imagine that a long eruption uses up more of the water in the chamber, and then the next refill/reheat/erupt cycle would take longer. We can make a scatterplot with eruption duration on the horizontal axis and the length of the following rest period on the vertical.

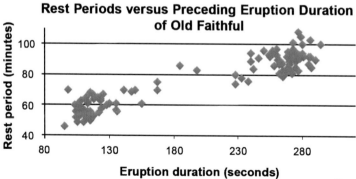

Rest Periods versus Preceding Eruption Duration of Old Faithful

Figure 2.2

- When you're dealing with bivariate data (two variables) and they're both quantitative (numerical), then a scatterplot is usually the first thing you're going to want to look at. It's a wonderful tool for exploratory data analysis.

- Each eruption gets one dot, but that one dot tells you two things: the *x*-coordinate (the left and right position of the dot) tells you how long that eruption lasted, and the *y*-coordinate (the up and down position of the same dot) tells you the duration of the subsequent rest period.

- We have short eruptions followed by short rests clustered in the lower left of the plot and a group of long eruptions followed by long rests in the upper right. There seems to be a relationship between eruption duration and the length of the subsequent rest. We can get a reasonable approximation to what we're seeing in the plot by drawing a straight line that passes through the middle of the data, as in **Figure 2.3**.

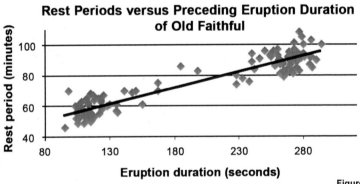

Figure 2.3

- This line is chosen according to a specific mathematical prescription. We want the line to be a good fit to the data; we want to minimize the distance of the dots from the line. We measure this distance vertically, and this distance tells us how much our prediction of rest time was off for each particular point. This is called the **residual** for that point. A residual is basically an **error**.

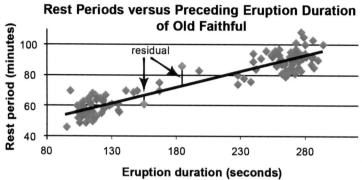

Figure 2.4

- The graph has 112 points, so we could find their 112 residuals—how well the line predicts each point. We want to combine these residuals into a single number that gives us a sense of how tightly the dots cluster around the line, to give us a sense of how well the line predicts all of the points.

- You might think about averaging all of the distances between the dots and the line, but for the predictive work that we're doing, it's more useful to combine these error terms by squaring each residual before we average them together. The result is called the mean squared error (MSE). The idea is that each residual tells you how much of an error the line makes in predicting the height of a particular point—and then we're going to square each of these errors, and then average those squares.

- A small mean squared error means that the points are clustering tightly around the line, which in turn means that the line is a decent approximation to what the data is really doing. The straight line drawn in the Old Faithful scatterplot is the one that has the lowest MSE of any straight line you can possibly draw. The proper name for this prediction line is the **regression line**, or the least squares line.

- Finding and using this line is called **linear regression**. More precisely, it's simple linear regression. The "simple" means that we only have one input variable in our model. In this case, that's the duration of the last eruption.

- If you know some calculus, you can use the definition of the regression line—the line that minimizes MSE—to work out the equation of the regression line, but the work is time consuming and tedious. Fortunately, any statistical software package or any decent spreadsheet, such as **Excel** or OpenOffice's **Calc**, can find it for you. In those spreadsheets, the easiest way to get it is to right-click on a point in your scatterplot and click on "add trendline." For the cost of a few more clicks, it'll tell you the equation of the line.

- For the eruption data, the equation of the line is about $y = 0.21x + 34.5$, where x is the eruption duration and y is the subsequent rest. So, the equation says that if you want to know how long a rest to expect, on average, after an eruption, start with 34.5 minutes, and then add an extra 0.21 minutes for every additional second of eruption.

- Any software package will also give you another useful number, the r^2 value, which is also called the **coefficient of determination**, because it tells you how much the line determines, or explains, the data. For the Old Faithful data, the spreadsheet reports the r^2 value as about 0.87. Roughly, that means that 87% of the variation in the height of the dots can be explained in terms of the line. In other words, the model explains 87% of the variation in rest times in terms of the length of the previous eruption.

Linear Regression

- Linear regression assumes that your data is following a straight line, apart from "errors" that randomly bump a data point up or down from that line. If that model's not close to true, then linear regression is going to give you nonsense. We'll expect data to follow a straight line when a unit change in the input variable can be expected to cause a uniform change in the output variable. For Old Faithful, each additional second of eruption adds about 12 seconds of rest.

- If r^2 is low, we're on shaky ground—and that's one thing everyone learns quite early about linear regression. But linear regression is so easy to do (at least with a statistical calculator or computer) that you'll often see people becoming overconfident with it and getting themselves into trouble.

- The problem is that linear regressions aren't always as trustworthy as they seem. For example, using a small data set is a very bad way to make predictions. Even though you could draw a straight line between two data points and get an r^2 of 1—a perfect straight-line fit—the line that you find might be a long way from the true line that you want, the one that gives the true underlying relationship between your two variables.

- Not only might the line that you find differ significantly from the true line, but the farther you get to the left or right of the middle of your data, the larger the gap between the true line and your line can be. This echoes the intuitive idea that the farther you are from your observed data, the less you can trust your prediction.

- It's a general principle of statistics that you get better answers from more data, and that principle applies to **regression**, too. But if so, how much data is enough? How much can we trust our answers? Any software that can find the regression equation for you can probably also give you some insights into the answer to these questions. In Excel, it can be done by using the program's regression report generator, part of its data analysis add-in. You put in your x and y values, and it generates an extensive report.

- The software isn't guaranteeing that the real intercept lies in the range it provides, but it's making what is known as **confidence interval** predictions based on some often-reasonable assumptions about how the residuals are distributed. It's giving a range that is 95% likely to contain the real intercept.

- The uncertainties in the slope and intercept translate into uncertainties in what the correct line would predict. And any inaccuracy of the line gets magnified as we move farther from the center of our data. The calculations for this are a bit messy, but if your data set is large and you don't go too far from the majority of your **sample**, the divergence isn't going to be too much.

- Suppose that we want to be 95% confident about the value of one variable, given only the value of the second variable. There's a complicated formula for this **prediction interval**, but if your data set is large, there's a rule of thumb that will give you quite a good working approximation. Find one number in your regression report: It's usually called either the **standard error** or standard error of the regression. Take that number and double it. About 95% of the time, the value of a randomly selected point is going to be within this number's range of what the regression line said.

- So, if you're talking about what happens on average, the regression line is what you want. If you're talking about an individual case, you want this prediction interval.

Important Terms

Calc: The OpenOffice suite's equivalent to Excel. It's freely downloadable but lacks some of the features of Excel.

cluster: A collection of points considered together because of their proximity to one another.

coefficient of determination: See r^2.

confidence interval: An interval of values generated from a sample that hopefully contains the actual value of the population parameter of interest. See **confidence (statistics)**.

error: In a forecasting model, the component of the model that captures the variation in output value not captured by the rest of the model. For regression, this means the difference between the actual output value and the value forecast by the true regression line.

Excel: The Microsoft Office suite's spreadsheet program.

linear regression: A method of finding the best linear relationship between a set of input variables and a single continuous output variable. If there is only one input variable, the technique is called simple; with more than one, it is called multiple.

prediction interval: The prediction interval is an interval with a specified probability of containing the value of the output variable that will be observed, given a specified set of inputs. Compare to **confidence interval**.

r^2: The coefficient of determination, a measure of how well a forecasting model explains the variation in the output variable in terms of the model's inputs. Intuitively, it reports what fraction of the total variation in the output variable is explained by the model.

regression: A mathematical technique that posits the form of a function connecting inputs to outputs and then estimates the coefficients of that function from data. The regression is linear if the hypothesized relation is linear, polynomial if the hypothesized relation is polynomial, etc.

regression line: The true regression line is the linear relationship posited to exist between the values of the input variables and the mean value of the output variable for that set of inputs. The estimated regression line is the approximation to this line found by considering only the points in the available sample.

residual: Given a data point in a forecasting problem, the amount by which the actual output for that data point exceeds its predicted value. Compare to **error**.

sample: A subset of a population.

standard error: Not an "error" in the traditional sense. The standard error is the estimated value of the standard deviation of a statistic. For example, the standard error of the mean for samples of size 50 would be found by generating every sample of size 50 from the population, finding the mean of each sample, and then computing the standard deviation of all of those sample means.

Suggested Reading

Hyndman and Athanasopoulos, *Forecasting*.

Miller and Hayden, *Statistical Analysis with the General Linear Model*.

Ragsdale, *Spreadsheet Modeling & Decision Analysis*.

Questions and Comments

1. Imagine that we set a group of students on a task, such as throwing 20 darts and trying to hit a target. We let them try, record their number of successes, and then let them try again. When we record their results in a scatterplot, we are quite likely to get something similar to the following graph. The slope of the line is less than 1, the students who did the best on the first try tend to do worse on the second, and the students who did worst on the first try tend to improve on the second. If we praised the students who did well on the first try and punished those who did poorly, we might take these results as evidence that punishment works and praise is counterproductive. In fact, it is just an example of regression toward the mean. (See **Figure 2.5**.)

   Assume that a student's performance is a combination of a skill factor and a luck factor and that the skill factor for a student is unchanged from trial to trial. Explain why you would expect behavior like that suggested by the graph without any effects of punishment or praise.

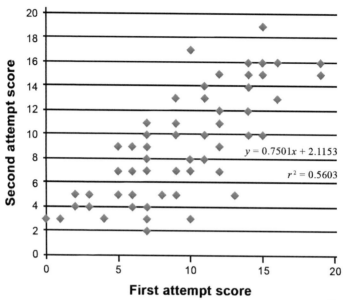

y = 0.7501x + 2.1153

r² = 0.5603

First attempt score

Figure 2.5

Answer:

Consider the highest scorers in the original round. Their excellence is probably due to the happy coincidence of considerable skill and considerable luck. When this student repeats the exercise, we can expect the skill factor to be essentially unchanged, but the luck factor is quite likely to decrease from the unusually high value it had in the first round. The result is that the performance of those best in round 1 is likely to decrease in round 2. On the low end, we have a mirror of this situation. The worst performers probably couple low skill with bad luck in round 1. That rotten luck is likely to improve in round 2—it can hardly get worse!

This effect is seen in a lot of real-life data. For example, the children of the tallest parents are usually shorter than their parents, while the children of the shortest parents are usually taller than their parents.

2. Suppose that you are given a sack that you know contains 19 black marbles and 1 white marble of identical size. You reach into the bag, close your hand around a marble, and withdraw it from the bag. It is correct to say that you are 95% confident that the marble in your hand is black, and it is in this sense that the term "confidence" is used in statistics. Consider each of the statements below, and find the one that is equivalent to your "95% confidence" statement.

a) This particular marble is 95% black and 5% white. (Maybe it has white spots!)

b) This particular marble is black 95% of the time and white 5% of the time. (Perhaps it flickers!)

c) This particular marble doesn't have a single color, only a probability. Its probability of being black is 95%.

d) The process by which I got this particular marble can be repeated. If it were repeated many, many times, the resulting marble would be black in about 95% of those trials.

Answer:

The answer is d), but the point of the question is that answers a) through c) correspond roughly to statements that are often made by people when interpreting confidence. For example, given a 95% confidence interval for mean income as $40,000 to $50,000, people will often think that 95% of the population makes money between these bounds. Others will say that the mean is in this range 95% of the time. (The mean of the population is a single, fixed number, so it is either in the interval or it is not.) When we declare confidence, we are speaking of confidence in a process giving an interval that manages to capture the population parameter of interest.

Forecasting with Simple Linear Regression
Lecture 2—Transcript

Knowing what's going to happen, or even having a pretty reliable estimate of it, is power. If you know the future, you can capitalize off it. I'm often amazed how much of the behavior of the physical universe can be captured by relatively simple mathematical models. And of course, they are models. A good model is one that shows close agreement with the facts, so that we can use to make predictions that might not otherwise be possible.

I want to talk with you today about a forecasting technique with considerable power in describing connections between related quantities in many disciplines—linear regression. Its underlying idea is easy to grasp and easy to communicate to others. The technique is important because it can, and does, yield useful results in an astounding number of applications. But it's also worth understanding how it works, because, if applied carelessly, linear regression can give you a crisp mathematical prediction that has nothing whatsoever to deal with reality.

To develop these ideas, it's best to start easy. In fact, let's all take it easy. Let's go on vacation to Wyoming, to Yellowstone National Park. Impressive place, Yellowstone. Beneath the park is an active super volcano, the largest on the continent, which is the reason the park contains half of the world's geothermal features, also more than half of its geysers. The most famous of these, of course, is Old Faithful. It's not the biggest geyser, nor the most regular, but it's the biggest regular geyser in the park. Or is it? There's a popular belief that the geyser erupts once an hour, like clockwork. Let's look at some data and see.

This graph is a dot plot, tracking the rest time between one eruption and the next for a series of 112 eruptions. Each rest period is shown as one dot. Rests of the same length are stacked up on top of one another. And what the plot tells us is you shouldn't set your watch by Old Faithful. The shortest rest time that we saw, over on the left, is just over 45 minutes. The longest one, way over on the right, is almost 110 minutes. If you wanted to be generous, you could call the geyser Old Semi-Faithful, since there seems to be a cluster of short rest times of about 55 minutes, and another cluster of long rest times in the 92-minute range.

So, based on what we've got right now, when tourists ask about the next eruption, the best the park service can say is, "Folks, it'll probably be somewhere between 45 minutes to 2 hours after the last eruption," which isn't very satisfactory. Can we take a predictive modeling approach to do this better? Let's see.

We want to do a better job of predicting Old Faithful's next eruption time. We might be able to do that if we could find something that we already know that could be used to predict the rest periods. But what? We might reason like this; "I don't know the details of how a geyser works, but my rough guess is that water fills a chamber in the earth and heats up. When it gets hot enough, it boils out to the surface, and then the geyser needs to rest while more water enters the chamber and is heated to boiling." It doesn't sound like a bad guess, but an analyst would also want to take advantage of domain knowledge, if possible, talking to people who know and understand geysers. It's an important rule of thumb for any project. When you can, pick the brains of the experts.

Still, if our model of a geyser is roughly right, then we could imagine that a long eruption uses up more water in the chamber, and then the next refill-reheat-erupt cycle would take longer. We can look into this, making a scatter plot with eruption duration on the horizontal axis, and the length of the following rest period on the vertical.

Here's what we get; when you're looking at bivariate data, two variables, and they're both quantitative, numerical, then a scatter plot is usually the first thing you're going to want to look at. It's a wonderful tool for exploratory data analysis. Each eruption gets one dot, but that dot tells you two things. The x-coordinate, the left-right position, tells you how long the eruption lasted. The y-coordinate, the up-and-down position on that same dot, tells you the duration of the subsequent rest period. For example, look at this point. It represents an eruption that lasted about 190 seconds, followed by a rest of about 86 minutes.

So, we have short eruptions followed by short rest periods clustered in the lower left. And then we have a group of long eruptions followed by long rest periods in the upper right. It looks like we're on to something! There does, indeed, seem to be a relationship between the eruption duration and the length of the subsequent rest.

OK. We have a prediction. Can we capture this relationship quantitatively? Well, we can get a reasonable approximation to what we're seeing here by drawing a straight line that passes through the middle of the data, like this. If I asked you to sketch a straight line through the middle of this data, it's likely that the line you'd have drawn would have been close to the one that I'm showing here. So, for an eruption of 180 seconds, this line would let us predict a rest period of about 72 minutes. Great. We've predicted to within a minute how long to wait for the next eruption. But how reliable is that prediction? Before we can answer that, we have to talk a bit more about the line itself.

You see, the line that I drew really isn't just one that I eyeballed from the graph; it's chosen according to a specific mathematical prescription. Not surprisingly, we want the line to be a good fit to the data. We want to minimize the distance of the dots from the line. If you've got a line like this, and a dot like that, you might think that's the way you'd measure the distances along the diagonal, like that, but we don't. We measure it vertically, like that. That distance tells us how much our prediction of rest time was off for that particular case. It's also called the residual of that point. A residual is basically an error.

So, this dot has a positive residual of about +13; it's 13 minutes above the line. This one has a residual of about −5; it's 5 minutes below the line. Remember that for this problem, the vertical axis is rest time in minutes. So, our graph has 112 points, and we could, therefore, find 112 residuals—how well the line predicts each data point. We want to combine these residuals into a single number that gives us a sense of how tightly the dotted data is clustered around the line, to give us a sense of how well the line predicts all of the points.

You might just think about averaging the distances between the dots and the line, but for the predictive work that we're doing here, it's more useful to combine these errors in a different way, by squaring each residual before we average them together. The result is called the mean squared error, or MSE. The idea is that the residual tells you how much of an error the line makes in predicting the height of that particular point, and then we're going to square each of those errors and then average all those squares. A small mean squared error means that the points are clustering tightly around the line, which in turn means that the line is a decent approximation to what the data is really doing. The straight line that I drew on our Old Faithful scatterplot is the one that has the lowest MSE of any straight line that you can possibly draw. The proper name for the prediction line is the regression line, or the least squares line.

The decision to square all of the residuals before averaging them seems an odd thing to do, so it's probably worth justifying this decision a bit. In several senses, it's really a matter of practicality. One aspect of practicality is simplicity, and that seems to favor just averaging the residuals directly. But a residual can be positive or negative, depending on whether the point is above or below the line. Average +5 and −5, and you get 0. But the error for those two points is obviously not zero. If you shoot two arrows at an archery target, one five feet to the left and one five feet to the right, you can't claim two bulls eyes!

Okay, then. We don't use the residuals directly; how about if we use their absolute values. Whether a number is +5 or −5, it's still an error of 5. But this system is impractical for its own reasons. First, we're going to want to be able to use statistics, calculus, and algebra to be able to draw useful conclusions about this line, what it tells us, and how much we can trust it. The average the squares calculation of MSE has very nice properties for this work, while the absolute value function does not. For example, the absolute value function doesn't have a derivative at zero, which makes it a pain if you're doing calculus.

There's another problem with the average the absolute values approach, too. It doesn't necessarily give you the line that you want. In fact, it may not specify a single line at all. Look at this example. If I asked you to draw the best straight line to match this data, what would you pick? I'm betting it's this. Good choice! Right through the middle! Each point has a residual of either +5 or −5; each one is either 5 above or 5 below the line. So the average distance of the residuals from the line is 5, too.

But look at this. The horizontal line doesn't do nearly as well at capturing this trend of the trend of points, but look at the four residuals. Two of the points are on the line. Residual, 0. The other two points are 10 away from the line, one above, one below. The average distance from this line is the average of 0, 0, 10 and 10, which is also 5. In fact, any line that passes through the goalposts on the left side of the graph and also passes through the goalposts on the right will have an average residual of five. But most of us would say that these lines are not all equally good at characterizing the scatterplot.

On the other hand, the MSE approach to combining residuals, square, them, then average, gets rid of the pesky negative residuals, a square is always nonnegative. It behaves very nicely with calculus and statistics. And the line it prescribes is a close match to what our native intuition says is the right line to draw. For example, it goes right down the middle, splitting both sets of goalposts. Oh, for the sake of completeness, I'll mention that for technical reasons, MSE is often computed by adding the squared residuals and then dividing by the number of points minus two, rather than the number of points itself, as you would for a normal average. This definition is more useful for advanced statistical calculations, and it doesn't change what line minimizes the MSE.

So, we've defined our best straight line to fit to the data, the least squares regression line, and finding and using this line is called linear regression. More precisely, it's simple linear regression. The simple means that we only have one input in our model. Here, that's the duration of the last eruption. If you know some calculus, you can use the definition of the regression line, the line that minimizes MSE, to work out the equation of the regression line, but the work is time consuming and tedious. Fortunately, any statistical software package or any decent spreadsheet, like Excel or OpenOffice's Calc, can find it for you. In those spreadsheets, the easiest way is to right-click a point on your scatterplot, click on "Add Trendline," and for a couple of more clicks, it'll tell you the equation of the line.

For the eruption data, the equation of the line is about $y = 0.21x + 34.5$; x is our eruption duration, and y is the subsequent rest. The equation says that if you want to know how long of a rest to expect on average after an eruption, start with 34.5 minutes, and then add an extra 0.21 minutes for every additional second of eruption. So, if the last eruption lasts 4.5 minutes, 270 seconds, you'd plug 270 in for x, crank out y, and you'd conclude that you'd expect to wait about 90.8 minutes for the next one. You can also get this right off of the scatter plot, like this.

Any software package will also give you another useful number, the r^2 value. Technically, the r^2 value is called the coefficient of determination, since it tells you how much the line determines or explains the data. Most people just call it r^2.

For our Old Faithful data, the spreadsheet reports that the r^2 value as about 0.87. Roughly, that means that 87% of the variation in the heights of the dots can be explained in terms of the line. So, given that we don't really know beans about how geysers work, our predictions are actually quite good! We've got a lot of variation in rest times, from 45 minutes up to 110 minutes. But the up-and-down bouncing around the line is much less. Most dots are above or below the line by no more than 10 minutes, and many are even closer than this. If you take the variation of jiggles above and below the line, it's only 13% as large as the total variation seen in the rest period data. So the linear model is accounting for the remaining 87% of the variation, the model explains 87% of the variations in rest times in terms of the length of the previous eruption.

Compare that to a graph like this one, based on a random sample of my students. It's looking for a link between height and grade point average. Here the r^2 here is 0.02. You can see that there is a trend in the data; it does tend to drop slightly as you move from left to right, but the vertical spread around this line when compared to that drop is extreme. The line explains only about 2% of the variation of in GPA, small enough to be attributed to random chance. We're not seeing any meaningful relationship between height and grades. Which is not as much of a no-brainer as you might expect. Research studies have shown that height and salary show a positive correlation with an r^2 of somewhere between 0.08 and 0.15, even after correcting for gender, age, and weight. While those values are too small to make height alone a good predictor of salary, a value of r^2 in that range, when based on a large enough sample, can still indicate that there is some real connection.

The highest r^2 value that any line could achieve is 1, meaning that 100% of the variance in the output is predicted by the line. That could only be the case if every data point fell exactly on the line. In real life data, whatever it is, that's pretty unlikely. Our r^2 of 0.87 is nothing to sneeze at. Now, people who do regression should always tell you what the r^2 value for the regression line is, and they usually do. A small r^2, unless the sample is huge, means that there isn't much evidence of a linear relationship between the variables in the data. It's worth noting, though, that there may be a very strong relationship between the variables, just not linear or a straight line one.

Here's an example. You throw a ball up in the air and measure its height every fifth of a second till it lands. You get a graph like this. Obviously, there's a strong relationship between time and height, the fact that allows you to actually catch the ball to begin with, but if you apply linear regression to this data, here's what you get: the red line, which corresponds to the ball hovering at an altitude of 35 feet. That model doesn't explain the actual variation in the ball's height at all, so the r^2 is 0. An extreme example, but it drives home an important point. Linear regression assumes that your data is following a straight line, apart from errors that randomly bump the data point up or down around that line. If the model's not close to true, then linear regression is going to give you nonsense. We'll expect data to follow a straight line when a unit change in the input variable can be expected to cause a uniform change in the output variable. For Old Faithful, we saw that each additional second of eruption added about 12 seconds of rest.

So, okay, if the r^2 is low, we're on shaky ground, and that's one thing everyone learns quite early about linear regression. But linear regression is so easy to do, at least with a statistical calculator or computer, that you'll often see people getting quite cocky with it, and getting themselves into some pretty bad trouble. It's like giving a 12 year old a chainsaw and turning him loose. I don't want you to be in that all-too-common situation, so let's take a look under the hood and see what's going on.

For that, let's look at how the populations of U.S. cities compare to their land area. I went to the U.S. Census Bureau site on the Internet. I chose 10 cities and recorded their populations and their areas. Let's see what we can do with that. Intuition would lead us to think that cities with larger populations would also be larger in extent, and we could hope for a linear fit. That would mean that each additional person adds about the same amount of area to the city. Is that intuition borne out in this sample? Well, here's the scatterplot with the trend line, equation, and r^2 value.

Things are looking pretty good. Our r^2 value is even higher than it was for Old Faithful, with about an 88% of the variation in our cities' areas being explained by their populations. The regression equation says that for each additional thousand people, we'd expect the area to increase by about 3.055 square miles. That's the coefficient of x, which gives the slope of the line. The intercept tells you that, for a city of zero people, x equals zero, we'd expect an area of about 95 square miles.

Huh? Zero people in a city of 95 square miles? That's a bit extravagant! Yeah, well, but we really shouldn't be too surprised. We know that the line doesn't fit the data perfectly. So, it's a little off on the intercept, too. The important point is that we've got a reasonably reliable relationship between city population and city area; r^2 was 0.88, after all, so we have an equation that we can use to estimate the area of any U.S. city, at least that's what the conventional wisdom is of people who dabble with regression lines. Like, unfortunately, some business people.

How about if the graphs of the lines were sales volume as a function of advertising, like this? Or sales as a function of price, like this? Knowing the relationship between those variables could be gold for your business. So, given a set of bivariate data, one of the first things that MBA students usually do with it is to perform a linear regression. The spreadsheet is only too happy to do the calculation of the straight line and the r^2. And so for very little effort, they end up with a nice, crisp equation, backed by a computer that they can take to the next planning meeting. What's wrong with that? Well, nothing. Maybe. But, linear regressions aren't always as trustworthy as they seem. Let's see why.

We were assuming that the connection between city area and population can be thought of this way: There's an underlying linear relationship, an expected city size for each city population. Then each real city is bumped off of this line, above or below it, by some random amount, the amount depending on things other than population. Our job is to collect data in order to find the equation of the line that the cities are jiggling around. That's pretty much all that you know when you set out in linear regression.

Here's what you don't know. Let's say that the truth is that the line is really: the area of the city is 3 times the population in thousands of people, but that cities can jiggle above or below that line by up to, say 500 square miles. Let's see what happens if I gathered data on only two cities, rather than 10. Now, using such a small data set would be a very poor way to make predictions, of course, but I'm going to use this extreme example to make it easy for us to analyze the situation. So, maybe the first city has a population of 1 million people, 1000 1000 people. That means the true regression line says that the area is about 3×1000, or 3000 square miles. But that's just the straight line itself, the average. The up-down jiggle for this city might be as big as a 500 square miles. So, your million-person city takes up, in reality, somewhere between 2500 and 3500 square miles. Whatever it is, write it down.

Now, go to a second city. Let's say that this one has a population that's twice as big, 2 million people, 2000 1000. Then the correct straight-line relationship between population and city size says that the area should be about 2000×3, or 6000 square miles. But again, real cities don't all fall right on the line. We said they could be off by as much as 500 square miles, so the population that you measure for city 2 is somewhere between 5500 and 6500 square miles. Let's look at this on a scatter plot.

If we lived in a perfect world, where there was no jiggle, the two cities would be the two blue dots on the plot; they'd be right on the line. Then you'd just draw the straight line between the dots, and voila! You'd have the true trend line, shown in black. But, due to the random variation in city size, that's not how it works. Your million-person city could be anywhere on the left-hand red bar, and your 2-million-person city could be anywhere on the right-hand red bar. So when you draw the straight line between the two cities that you found, you could get a lot of different lines.

Maybe by chance you picked two particularly large cities for their populations. You'd get this purple line, above the true one. Or maybe you picked two cities that were very small for their populations, getting a regression line below the true one, like this blue line. But it gets worse. What if one of your cities had a large size for its population, while the other had a small one? You could end up with one of these lines. The rust-colored one says that a 1000 people need 4 square miles, while the hot pink one says that 1000 people only need 2 square miles. So even though you could draw a straight line between your two cities and get an r^2 of 1, perfect straight-line fit, the line that you find might be a long way from the true line that you want, the one that gives the true, underlying relationship between the population and city size.

Once you have this picture in your mind, you can see that things are even worse if your two cities had more similar populations, like this. The error bars for each city are as tall as before, but those bars are separated by a smaller horizontal distance, so your two cities could lead to guesses for the line's slope that have nothing to do with reality. The hot pink line would lead us to think that cities with larger populations take less room. And still, the r^2 would be 1, because both points lie exactly on your line.

There's one more thing that I want to point out here. Note that, not only might the line that you find differ significantly from the true line, but the further to the right or the left you get from the middle of your data, the larger the gap between the true line and your line can be. This is true for linear regression in general and echoes the intuitive idea that the further you are from the observed data, the less you can trust your prediction.

Now, some of you may feel like I've stacked the deck here, and I have. I just showed you what would happen if you based your regression on only two points of data. It's a general principle of statistics that you get better answers from more data, and that principle applies to regression, too. But if so, how much data is enough? How much can we trust our answers? Well, any software that can find the regression equation for you can probably also give you some insights into the answer to that question. In Excel, it can be done by using the program's regression report generator, part of its data analysis add-in. Put in your x and y values, and it generates an extensive report. Let's look at just one part of that information, for our city data. The information in the coefficients column just reiterates the regression equation that we found earlier today, area = 94.96 + 3.06 × population in thousands.

But look at the last columns. Their job is to give you an idea of how much you can trust that equation. For example, we already talked about our regression equation having a constant term of about 95, meaning that a city of 0 people should take up, on average, 95 square miles. The right value, obviously, is 0 square miles. But the numbers in the last two columns in the first row are telling us that we shouldn't take that 95 square miles too seriously. Given the small data set and the variations from linearity that we see in it, it's safer to say that the constant term is somewhere between about −1500 and +1700. And note that zero, which we know to be the correct value, lies in that range. Actually, the software isn't even guaranteeing that the real intercept lies in that range. But it's making these confidence interval predictions based on some often-reasonable assumptions about how the residuals are distributed. It's giving a range that is 95% likely to contain the real intercept. And in this case, it does contain it.

We have the same kind of information for the population coefficient, which is the slope of the line. Based on our 10 selected cities, the line's slope was 3.06, but that's just the best line for those 10 cities. The best line based on data for all cities would almost certainly be different from this. How different? The 95% confidence limits are about 2.14 and 3.97. That is, the 10 data points told us that, on average, you get 3.06 more square miles for 1000 people, but that's just a best guess from that data. We're 95% confident that the actual average for all cities is somewhere between 2.14 square miles and 3.97 square miles per 1000. That's a lot more variation than the picture might have led you to believe.

And if you look at all 363 of the largest metropolitan areas in the U.S., the slope of the true best straight line is actually quite a bit outside of this range, something like only 0.7 square miles per 1000 people. So, did we just happen to be in the 5% of cases where we randomly fall outside the 95% confidence interval? No, for two reasons.

The first is that I didn't pick my cities randomly. I wanted them to fit comfortably on a graph; I wanted to make a point, so I focused on cities that weren't too large and weren't too small. I took a convenience sample. Bad idea, because all of our analysis in regression assumes that the sample is drawn randomly from the population. In fact, even if we'd used a moderately more extensive data set, drawn randomly, we'd have seen that cities don't actually fit into a straight line that well at all. Here's our original 10-city data in blue, expanded to a random list of 30 U.S. cities. By the way, you might notice that our r^2 has dropped now to less than 20%.

So, let's see how we'd put what we've learned to work on a new problem. The value of a house depends on a lot of factors, but one of them is probably the square footage of the house. I'm going to try to model this in a relationship for U.S. homes, based on a random sample of 1,728 homes. First, the scatterplot, along with its trend line, equation, and r^2. OK, we definitely see a relation with a positive correlation, and the r^2 value is about 50%, which is, frankly, higher than I would have expected. That means we can account for about 50% of the variation in house price from square footage alone. The fact that we don't see any pattern of the residuals as we move from left to right is good too; it suggests that a linear fit to the data doesn't sound like a bad model. There are a lot more houses on the left side of the graph, but that's not a problem.

Importantly, we can also see that the residuals aren't spreading out as we move from one side to another. They suggest more of a cigar shape than a trumpet. That means that we should be able to trust our confidence intervals. Trumpet-like distributions of residuals mean that the data is heteroscedastic, and that messes with the confidence intervals that we compute. The regression line itself says that every square foot of space in a house is worth, on average, $113, but we want to look at the regression report to see how much we can trust that. Here it is.

Again, the highlighted numbers on the left are giving us the equation of the straight line of our graph: price in thousands = 13.44 + 0.11 × square footage. The confidence interval on the right shows that the 13.44 is a little dodgy; we're 95% confident the actual value is between 3.6 and 23.23. The more important coefficient, though, is the slope, and here we're seeing that we're 95% confident that an additional square foot increases the average house price by between $108 and $118. A nice, narrow range.

The uncertainties in slope and intercept translate into uncertainties in what the correct line would predict. As we saw before, any inaccuracy in our line gets magnified as we move further away from the center of the data. The calculations for this are a bit messy, but if your data set is large and you don't go too far from the majority of your sample, the divergence isn't going to be too much. You can see by the red lines that for our data, the 95% interval for what the right line predicts is in a pretty narrow range. For a 3000-square-foot home, the margin of error is about ±$7300. Not bad.

But that's the accuracy on what the true line would say for a 3000-square-foot home, and the true line is only reporting the average price for homes of that size. We know that the actual value of the homes bounce up and down around this average. So suppose I want to be 95% confident about the price of a specific home, given only its square footage? There's a complicated formula for this prediction interval, but if your data set is large, there's a rule of thumb that will give you quite a good working approximation. Look back at your regression report and find one number; it's usually called either the standard error or standard error of the regression. In our current example, the software says its value is about $69,000. Take that number and double it, to get about $138,000. About 95% of the time, the value of a randomly selected home is going to be within $138,000 of what our regression line said. Like this. That's a pretty wide band, but facts are facts, and we see from the graph that there's a lot of variation around the line.

So, there's something else to watch out for if a colleague starts waving regression models around. If you're talking about what happens on average, the regression line is what you want. If you're talking about an individual case, you want the prediction interval.

Nonlinear Trends and Multiple Regression
Lecture 3

There are two important limitations to simple linear regression, both of which will be addressed in this lecture. First, linear regression is fussy about the kind of relation that connects the two variables. It has to be linear, with the output values bumped up and down from that straight-line relation by random amounts. For many practical problems, the scatterplot of input versus output looks nothing like a straight line. The second problem is that simple linear regression ties together one input with the output. In many situations, the values of multiple input variables are relevant to the value of the output. As you will learn, multiple linear regression allows for multiple inputs. Once these tools are in place, you can apply them to nonlinear dependencies on multiple inputs.

Exponential Growth and Decay

- **Exponential growth** is going to show up any time that the rate at which something is growing is proportional to the amount of that something present. For example, in finance, if you have twice as much money in the bank at the beginning of the year, you earn twice as much interest during that year. **Exponential decay** shows up when the rate at which something is shrinking is proportional to the amount of that something present. For example, in advertising, if there are only half as many customers left to reach, your ads are only reaching half as many new customers.

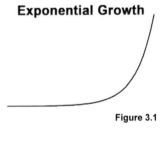

Exponential Growth

Figure 3.1

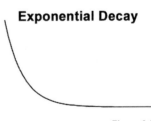

Exponential Decay

Figure 3.2

- For exponential growth, the time taken for the quantity to double is a constant. For example, Moore's law, which states that the number of transistors on a microchip doubles every two years, describes exponential growth. For exponential decay, the amount of time required for something to be cut in half is constant. For example, half-life for radioactivity is exponential decay.

- Anything undergoing exponential growth or decay can be expressed mathematically as $y = c^{ax + b}$, where y is the output (the quantity that's growing or shrinking); x is the input (in many models, that's time); and a, b, and c are constants. You can pick a value for c; anything bigger than 1 is a good workable choice.

- So many things follow the kind of hockey-stick curve that we see in exponential growth or decay that we really want to be able to predict them. Unfortunately, at the moment, our only prediction technique is restricted to things that graph as straight lines: **linear expressions**. In algebra, $y = ax + b$.

- Anytime you do algebra and want to solve for a variable, you always have to use inverse functions—functions that undo what you're trying to get rid of. You can undo an exponentiation by using its inverse: **logarithm** (log). If you take the log base c of both sides, $\log_c y = \log_c (c^{ax + b})$, which simplifies to $\log_c y = ax + b$. This results in a linear expression on the right side of the equation, but y is no longer on the left—instead it's the log of y.

- If y is a number that we know and c is a number that we know, then the $\log_c y$ is just a number, too—one we can find with a spreadsheet or calculator using a bunch of values for x and y. Whereas x versus y will graph as an exponential, x versus $\log y$ will graph as a straight line. And that means that if you start with x and y values that are close to an exponential relationship, then x and $\log y$ will have close to a linear relationship—and that means that we can use simple linear regression to explore that relationship.

- This works for any reasonable c that you pick—anything bigger than 1 will work, for example. Most people use a base that is a number called e: 2.71828.... Using this base makes a lot of more advanced work a lot easier.

- No matter what base we use, we're going to need a calculator or spreadsheet to find powers and logarithms, and calculators and spreadsheets have keys for e. Most calculators have an e^x key, along with a key for the log base e, which is also called the natural logarithm (ln). The $\log_e x$, the natural log of x, or the ln x all mean the same thing. And ln and e to a power are inverses—they undo one another.

Power Laws

- Exponential growth and decay are a family of nonlinear relationships that can be analyzed with linear regression by a simple transformation of the output variable—by taking its logarithm. But there's another family of relationships that are perhaps even more common that will yield to an extended application of this same idea.

- Suppose that we took the log of both the input and output variables. We'd be able to apply linear regression to the result if ln x and ln y actually do have a linear relationship. That is, if $\ln y = a \ln x + b$, where a and b are constants. Then, using laws of exponents and the fact that e to the x undoes ln, we can recover the original relation between x and y, as follows.

$$\ln y = a \ln x + b$$
$$e^{\ln y} = e^{a \ln x + b} = e^{a \ln x} e^b$$
$$y = e^b (e^{\ln x})^a = e^b x^a$$

- Therefore, the relationship between y and x is $y = e^b x^a$, and e^b is just a positive constant, so we're saying that y is proportional to some fixed power of x. A relationship where one variable is directly proportional to a power of another is called a **power law**, and such relationships are remarkably common in such fields as sociology, neuroscience, linguistics, physics, computer science, geophysics, economics, and biology. You can discover whether a power law is a decent description of your data by taking the logarithm of both variables and plotting the results.

- So many relationships seem to follow a rough power relation that research is being done as to why these kinds of connections should appear so often. But whenever they do, a log-log plot can tip you off to it, and linear regression can let you find the equation that fits.

Multiple Regression

- What about allowing more than one input? With a linear relationship, each additional input variable adds one dimension of space to the picture, so the "best straight line through the data" picture needs to change, but the idea of linear regression will remain the same. The mathematics of this plays the same game that we used for simple linear regression.

- Actually doing the math for this becomes quite tedious. The good news is that, again, statistical software or spreadsheets can do the work for you easily. If you're using a spreadsheet, Excel's report has historically been more complete and easier to read than OpenOffice Calc's, but both can do the job. And statistical software like R—which is free online—can do an even more thorough job.

- It's important to note that the **coefficient** of a variable in a model is intended to capture the effect of that variable if all other inputs are held fixed. That's why, when two variables measure almost the same thing, it's often a good idea not to include both in your model. Which one gets credit for the effect can be an issue. This is a special case of the problem of **multicollinearity**.

- Another variant of linear regression is called **polynomial** regression. Suppose that you have bivariate data that suggests a nonlinear relationship from the scatterplot and that your "take the log" transformations can't tame into a straight line. Multiple regression gives you a way of fitting a polynomial to the data. There is a lot going on in multiple regression, and there is some pretty sophisticated math that supports it.

Important Terms

coefficient: The number multiplied by a variable is its coefficient.

e: A natural constant, approximately 2.71828. Like the more familiar π, e appears frequently in many branches of mathematics.

exponential growth/decay: Mathematically, a relationship of the form $y = ab^x$ for appropriate constants a and b. Such relations hold when the rate of change of a quantity is proportional to its current value.

linear expression: An algebraic expression consisting of the sum or difference of a collection of terms, each of which is either simply a number or a number times a variable. Linear expressions graph as "flat" objects—straight lines, planes, or higher-dimensional analogs called hyperplanes.

logarithm: The inverse function to an exponential. If $y = a^x$ for some positive constant a, then $x = \log_a y$. The most common choice for a is the natural constant e. $\log_e x$ is also written $\ln x$.

multicollinearity: The problem in multiple regression arising when two or more input variables are highly correlated, leading to unreliable estimation of the model coefficients.

polynomial: A mathematical expression that consists of the sum of one or more terms, each of which consists of a constant times a series of variables raised to powers. The power of each variable in each term must be a nonnegative integer. Thus, $3x^2 + 2xy + z - 2$ is a polynomial.

power law: A relationship between variables x and y of the form $y = ax^b$ for appropriate constants a and b.

Suggested Reading

Hyndman and Athanasopoulos, *Forecasting*.

Miller and Hayden, *Statistical Analysis with the General Linear Model*.

Questions and Comments

1. The lecture mentioned that one could use linear regression to fit a polynomial to a set of data. Here, we look at it in a bit more detail. Given a table of values for the input x and the output y, add new input variables whose values are x^2, x^3, and so on. Stop when you reach the degree of polynomial that you wish to use. Now conduct multiple regression in the normal way with these variables. The table used in the regression might begin as follows.

x	x^2	x^3	x^4	y
3	9	27	81	17
-2	4	-8	16	40
...	...	...	...	...

Figure 3.3

The same technique can be used to look for interaction effects between two different input variables. In addition to input variables x_1 and x_2, for example, we could include the interaction term $x_1 x_2$. For example, including either mustard or Jell-O in a dish might each be fine individually but might create quite an unpleasant reaction together!

2. In most of its incarnations, regression is pretty specific about what the "random errors" in a model are supposed to look like. You could imagine how they're supposed to work in this way. Suppose that you have a bucket containing a huge number of poker chips, each with a number on it. The numbers are centered on zero, balanced out between positive and negative, and there are a lot more chips with values close to

zero than there are with values of large magnitude. When you need the error for a particular input point, reach into the bucket for a chip, read its number, and then add that number to the calculated linear output. Then, throw the poker chip back in the bucket.

More technically, the errors are supposed to be normally distributed with a mean of zero, a constant standard deviation, and are supposed to be uncorrelated to one another as well as being uncorrelated to the input values—but the error bucket gets the key idea across.

Nonlinear Trends and Multiple Regression
Lecture 3—Transcript

Last time we introduced simple linear regression, a way to quantify the relationship between two numeric variables. But there are two important limitations to simple linear regression, both of which we are going to address today.

One is that linear regression is fussy about the kind of relationships that connects the two variables; it's got to be linear, with output values bumped up and down from that straight line relationship by random amounts. For lots of practical problems, the scatterplot of input versus output looks nothing like a straight line. We'll see that this can often be handled by a technique called transformation of variables.

The second problem is that simple linear regression ties together one input with the output. In many situations, the values of multiple inputs are relevant to the value of the output, a motel room's cost depends on location, number of beds, season, and so on. So we'll turn to the topic of multiple linear regression, which allows multiple inputs. Once we have those tools in place, we can see what we can do with nonlinear dependencies on multiple inputs, a pretty impressive challenge.

But to begin, let's start with a single input variable, in relation with the output that's decidedly nonlinear. Graphs like these.

The one on the left is what the graph of your bank balance might look like over time, under compound interest; or the number of people suffering from a new plague; or the number of watchers of a new video on You Tube. The one on the right is what a graph might look like of the charge left in your car battery in the hours after you accidentally left your lights on; or the amount of a drug still in your system over time; or the new people that your current ad campaign is reaching, day for day.

What ties these things together is the idea of exponential growth and decay. Exponential growth is going to show up any time that the rate at which something is growing is proportional to the amount of that something that's present. If you have twice as much money in the bank at the beginning of the year, you earn twice as much interest during that year. Exponential decay shows up when the rate at which something is shrinking is proportional to the amount present. If there are only half as many customers left for you to reach, your ads are only reaching half as many new customers.

It turns out that this characterization is equivalent to saying this; for exponential growth, the time taken for the quantity to double is a constant. So, for example, Moore's Law describes exponential growth. It says that the number of transistors on a microchip doubles every two years. Of course, there's nothing magic about doubling. I could have equivalently said that the time required for the number of transistors to quadruple is four years, or to increase eight fold is six years. You can describe the process in terms of any base increase that you want. For exponential decay, the amount of time required for something to be cut in half is constant. Remember half-life from radioactivity? That's exponential decay. Carbon-11 has a half-life of about 20 minutes. No matter how much you start with, in 20 minutes, half of it will be gone. But both of these ways about thinking about exponential growth and decay are equivalent to the one that's going to be especially important to us today. Namely, anything undergoing exponential growth or decay can be expressed mathematically as, $y = c^{ax+b}$.

Here, y is the output, the quantity that's growing or shrinking; x is the input, in a lot of models, that's time; and a, b, and c are constants. You can pick a value for c, anything bigger than 1 is a good workable choice. For example, look at $y = 2^{x+2}$. When $x = 0$, $y = 2^2$, or 4. When $x = 1$, $y = 2^3$, or 8. When $x = 2$, $y = 16$. Every time x goes up by 1, y goes up by a factor of 2. And that's exponential growth. And you can see why it's called exponential; the variable stuff is up there in the exponent.

So many things follow this kind of hockey-stick curve that you see in exponential growth and decay; we really want to be able to predict them. Unfortunately, at the moment, our only prediction technique is restricted to things that graph as straight lines—linear expressions. In algebra, $y = ax + b$. Ho! But wait! Write that down! And what was exponential? Argh. So close! That linear $ax + b$ is stuck up there in the exponent, darn it! Ahhhhh, but we can fix that.

Anytime you do algebra and want to solve for a variable, you always have to use inverse functions, functions that undo the thing you're trying to get rid of. If you have $3x = 12$, you can divide both sides by 3 to get $x = 4$. Dividing by three is the inverse of multiplying by three; it undoes the multiplication. If $x^2 = 100$, and you want to get rid of the square, you can undo it by taking the square root of both sides to get $x = 10$.

In a similar way, you can undo an exponentiation, such as 2^x. Its inverse goes by the odd name of logarithm, or log, for short. If you're working with a number c to some power, you undo the c to the whatever by taking its logarithm, base c. So if $y = 2^x$, you can take the log base 2 of both sides, and get $y = 2^x$, and therefore, $\log_2 y = x$. Okay? So how does this help us? Well, go back to the exponential equation: $y = c^{ax+b}$. Take the log base c of both sides. But log base c undoes c to the something, and so we've got $ax + b$ out of the exponent! Yes!

You might be less excited. Yes, I've got a nice linear expression on the right, but now I don't have y on the left anymore. I've got the log of y. Well, true, but if y is a number that I know, and c is a number that I know, then the $\log_c$ of y is just a number too, one that I can find easily on a spreadsheet or a calculator. So, if you give me a bunch of values of x and y that are following an exponential distribution, I can add a new column to your table, one that contains values of $\log y$; x versus y will graph as an exponential. But x versus $\log y$ will graph as a straight line. And that means that if you start with x and y values that are close to an exponential relationship, then x and $\log y$ will have close to a linear relationship, and that means we can use simple linear regression to explore that relationship!

It turns out that this works for any reasonable c that you pick, anything bigger than 1 will work, for example. But most people who do this kind of work use a base that, at first blush, seems stupid. It's not 2 or 10 or some other sensible-seeming number. It's a bit bigger than 2.71828; it's a number called e. You don't have to use e if you don't want to, but that base makes a lot of more advanced work, like work involving calculus, a lot easier. So, I'm going to follow that convention, and it really won't cause you any extra trouble. No matter what base we use, you're going to need a calculator or a spreadsheet to find powers and logarithms, and calculators and spreadsheets know all about e. You calculator probably has an e^x key. It probably also has a key for the log base e, which is also called the natural logarithm, and usually just written "ln." We usually pronounce it "lin," so, the $\log_e x$, or natural log of x, or the $\ln(x)$, they all mean the same thing. The only thing you need to remember is that ln and e to the something are inverses. They undo one another. Okay, let's see how all of this theory works in practice.

Let's look at the up-and-coming technology, photovoltaic cells. In 2009, the European Union declared a target of having 20% of its energy production come from renewable resources by 2020. Let's take a look at the historical data from around that time and see if we can predict what happens next. Here's a graph of the data from 2005 to 2010. It looks rather exponential. But if our data fell on an exponential curve, the relationship would be $y = e^{ax+b}$, where y is photovoltaic capacity, x is the year, and a and b are constants to be determined to fit the data. That's equivalent to $\ln y = ax + b$.

It's the same thing we wrote a minute ago, but since I've committed myself to using base e, the log on the left-hand side is written as ln. I'm going let my spreadsheet compute the ln of all of the megawatt capacities that we had in our last table. Here's what the results look like. I'm plotting the columns in the salmon color this time. That is, on the vertical axis, I'm no longer plotting the actual capacity, but the ln of the capacity. I made another change, too. To make the numbers we'll be dealing with a bit more friendly, I've changed my horizontal axis to record both the year and how much time has passed since 2005. So year one is 2006, year two is 2007, and so on. I'm going to use these smaller numbers as my x values in the work to come.

And look what happened to my graph! Just as hoped, the hockey stick is gone, replaced by points that lie quite close to a straight line. This transformation of variables, working with ln y, rather than y itself, has moved the original problem to a place where linear regression applies. So let's do it!

Well, the r^2 value is quite good, but the data set is small, as well. Let's look more closely, examining the regression report. The 95% confidence intervals are fairly narrow, in fact, the 95% confidence interval for the true straight line looks like this. And you can take this graph back to the original one by undoing the ln with e^x inverse. Here's the 95% confidence interval for prediction.

So for any year, x, calculate $0.5271x + 7.6068$, then raise e to that power. That gives y, the prediction of the photovoltaic capacity in that year. Take 2011, for example; 2011 is 6 years after our base year of 2005, so we need to use $x = 6$. Our exponent expression works out to be about 10.77. So, punch $e^{10.77}$ into a calculator or spreadsheet, and you get about 47,500, the red dot on our graph. The actual European photovoltaic capacity in 2011 was a little higher, the purple cross, about 53,000 megawatts.

Conversely, according to some sources in 2012 and 2013, total capacity was below what we would have predicted, only 70,000 megawatts in 2012, only 82,000 in 2013. The 2012 prediction is still quite close to the trend line, and 2013 was just outside the 95% confidence prediction band. By the way, if you're using a spreadsheet, you can almost certainly get it to do this transformation of variables work for you. In Excel or Calc, for example, you can just ask for the exponential trend line directly on your original graph. It would give you this. Note the same r^2 value that we just found before. The equation looks different from ours, but mathematically, it's equivalent, using laws of exponents.

If you suspect that something may be showing exponential growth, you can do a quick exploratory data analysis just by taking ln of the output variable and graphing it against the input variable. Doing so will turn exponential growth into a linear relation, and it happens a lot. Here's the U.S. GDP, shown on a log graph. GDP is measured in billions of dollars, remarkable agreement with exponential growth. So we've found a whole family of nonlinear relationships, exponential growth and decay, that can be analyzed with linear regression by a simple transformation of the output variable, taking its logarithm. But there's another family of relationships, perhaps even more common, that will yield to an extended application of this same idea.

Suppose we took the log of both the input and output variables. We'd be able to apply linear regression to the result if $\ln x$ and $\ln y$ actually do have a linear relationship. That is, if $\ln y = a \ln x + b$, where a and b are constants. Then, using laws of exponents and the fact that e^x undoes ln, we can recover the original relation between x and y, like this. Raise e to the power of both sides; then, simplify. This is nothing more than laws of exponents, but when the smoke clears, we end up with the relationship between y and x themselves. Namely, $y = e^b x^a$; e^b is just a positive constant, so we're saying that y is proportional to some fixed power of x.

A relationship where one variable is directly proportional to a power of another is called a power law, and such relationships are remarkably common in sociology, neuroscience, linguistics, physics, computer science, geophysics, economics, even biology. The physical sciences are loaded with power laws. Universal gravitation: gravitational force varies as the distance to the −2. Electromagnetic attraction shows the same variation. Projectile motion: The distance an object drops near Earth varies with time as its power of +2. Tidal force: The force of the tide varies inversely as the distance to the power of 3. The luminosity of a star varies as the fourth power of its temperature, and you can look to see whether a power law is a decent description of your data just by taking the logarithm of both variables and plotting the results.

Here's a graph of our solar system on a log-log plot, the log of planetary distance from the Sun on the horizontal axis, and log of their orbital period on the vertical; $r^2 = 1$, an essentially perfect fit. And see that slope of 1.5 on the regression line? That means that orbital period is proportional to the 1.5 power of distance from the Sun. So, Saturn is 9.53 times as far from the Sun as Earth, and so we'd predict a year for Saturn to be $9.53^{1.5}$ years long. That's about 29.4 years. The actual figure: 29.5. We've just discovered Kepler's Third Law of Planetary Motion without breaking a sweat.

There are even good fits to things that involve those most unpredictable of creatures, human beings. Here's one that blows my mind—Zipf's Law. Rank the words in English by frequency of use. Number 1 is "the," which makes up about 5.6% of the words that we write. "Of" comes in second, at 3.4%, followed by "and," and so on. Believe it or not, by the time that you get to 177 words, you've covered about half of all of the words that we write. But look at what happens when you plot the ln of the frequency of these words on the vertical axis, and the ln of their rank on this list on the horizontal.

The first several words on the list are off a bit, but the overall, the fit is incredible. This is the graph of the frequencies of the 3000 most common words in English. The slope of the line as given by the regression equation is almost exactly −1, meaning that a word's frequency is inversely proportional to its rank in the list. So, the 200th most common word, "find," is used about twice as often as the 400th most common word, "mean," and so on. Our language actually works this way. Even more amazing, it seems to apply to most other languages, too, and many other kinds of rankings as well, whether income rankings, the sizes of corporations, or the populations of cities.

Recently, researchers have been looking into how power laws reflect the dynamics of cities. Quite good power relationships can be uncovered linking crime rate, energy consumption, even the number of gas stations to a city's area. Similar relationships hold when describing statistics for various businesses as well. So, many relationships seem to follow a rough power relation, so many, that research is being done as to why these kinds of connections should appear so often. But whenever they do, a log-log plot can tip you off to it, and linear regression can let you find the equation that fits. Okay.

What about allowing more than one input? Let's turn to the other topic for today, and in doing that, go back to linear relationships. Each additional variable adds one dimension of space to our picture, so our best-straight-line-through-the-data picture needs to change a bit. Let's imagine that we have two inputs, x and y. An easy way to imagine this is to imagine a table top. Each point on the table can be identified with an x and y coordinate, the inputs.

Now imagine a flat plane of glass hovering over top of the table at some angle in mid-air. That glass records a linear relationship between the x and y inputs, with the height as the output. A plane, like our flat plane of glass, is the two-dimensional analog of a straight line. With even more inputs, the resulting graph would need more than three dimensions, and the plane becomes an unbending hyperplane, but the idea remains the same. Let's stick with what we can see.

OK, now paint some dots on the pane of glass. Then take each of those dots and bump them vertically a bit, raising them or lowering them off of the plane to hover in midair. These vertical adjustments are the random fluctuations that real data takes from the perfectly linear relationship. Finally, take away the pane of glass, leaving only the dots in space. The result might look like this. The challenge for multiple regression is, given only the dots, come as close as you can to figuring out where the pane of glass was. Find the best linear fit to the data. The mathematics of this plays the same game that we used in the simple linear regression. When you take a guess as to where the glass should be, each floating data point is going to have a residual, a distance that it lies above or below that guessed plane. The best guess for the plane is the one that makes the mean squared error of those residuals as small as possible.

Actually doing the math for this gets quite tedious. The good news is that, again, statistical software or spreadsheets can do that work for you easily. If you're using a spreadsheet, Excel's report has historically been more complete and easier to read than OpenOffice Calc's, but both can do the job. And statistical software, like R, which is free online, can do an even more thorough job.

To see how this works, let's look at some data from a study of households in 1966, in which the head of household had a low annual income, under $15,000. The data set is old, but it demonstrates a number of important points so well that I think we can overlook that. The question was how the average hours worked during the year was affected by variations in hourly wages. To make my charts clear, I'm going to look at a set of 36 observations. For each observation, we have the head-of-household annual income and the hourly wage. But, we also know the head of household's age, earnings of the spouse and of other family members, the number of dependents, and the family's assets, bank accounts and so on.

We can start out with simple linear regression, looking at the connection between wage and hours worked annually. It looks like this, a positive correlation, with wage explaining about a third of the variation in hours worked. Each additional $1.00 wage adds, on average, just under 80 hours to the average annual work time, although the regression report indicates quite a bit of uncertainty on that figure; the 95% confidence interval runs from $39 to $117.

But, let's take a look at what happens if we create a model in which all of what we know about a household is used to predict the hours worked by the head of house. Now we're looking for a best linear fit between all of our input variables and hours worked. Doing this in a spreadsheet is hardly harder than doing a simple linear regression. In Excel, we use the regression tool in the data analysis tool kit; specify where the data for your input variables are, where the data for your output is; and it will generate a report. Regardless of what statistical software you use, you're going to get similar information. Here's part of the standard report. The value of r^2 is being reported, and the interpretation that you're used to, it is the same from linear regression. The fraction of the variation in the output, the work hours, is explained by the linear relation we're proposing with the inputs. It's 80% or so, which is quite respectable.

When working with multiple regression, though, the second figure, the adjusted r^2, is actually more important. Here's why. Suppose you make a prediction based on a bunch of variables that you have available. Now, suppose I ask you to do the same thing again, but this time, I provide you with one additional input variable, one additional piece of information on each individual in the sample. What would happen to the quality of your prediction?

Well, it might go up. If the new information was an important piece of the puzzle, knowing it would let you make a better prediction. On the other hand, your best prediction couldn't possibly get worse. If the new information was completely useless, you could just ignore it and get as good a prediction as you got before. On the other hand, suppose that the new information had some coincidental connection to the output for the values in your data set. Maybe it turns out that, for your sample, people with phone numbers that ended in 9 happened to work a few more hours. Then including the phone number's last digit among your inputs could make your model look a little better. In other words, even if you throw in the kitchen sink as a variable, the only effect it can have on r^2 is to make it go up. And it's this that the adjusted r^2 adjusts for. The more variables you throw in to the inputs without substantively improving the r^2, the smaller the adjusted r^2 gets. So getting a high adjusted r^2 is a target in multiple regression.

Well, we just made our regression by using all of the input variables that we had. Was this a good idea? To answer that, we need to look more carefully at the regression report. Here's some more of it. You've seen this part before in simple regression. The coefficients column tells you the equation of the regression. Here, it's hours worked = 2299 − 4.46 age, + 0.02 spouse, and so on.

But now I want to look at a new column, the one labeled P-value. I'm not going to go into a formal definition of p-value, but the larger the p-value for an input, the less evidence there is that the variable belongs in the regression. A common threshold is 0.05. We can see that four of the variables exceed that threshold, with the largest p-value belonging to spouse, the annual income of the spouse. That is, with the other inputs that we have, spouse doesn't seem to add enough to be kept. Either spousal income doesn't affect hours worked significantly, or other input variables have the contribution covered. So, we're going to do our regression by stepwise elimination, that is, delete the entry with the highest p-value and rerun the regression without it to see where we stand. Okay. We take out spouse and rerun. We get this.

The report also says that the adjusted r^2 is now 0.769, up a tiny bit from our old 0.764. Dependents has the highest p-value, well over our 0.05 threshold. Remove it from the model and see what happens. The adjusted r^2 held at 0.769. Our age is over our 0.05 threshold for p-value. Get rid of it and rerun. Before we do that, though, I'd like you to notice the coefficient of wage. It's about −54. I'll come back to this point in a few minutes. For now, let's rerun.

OK, adjusted r^2 has dropped to 0.747, a little bit of a drop. Wage now has a p-value above our threshold, indicating that it is not statistically significant, to use the technical term. We can delete it from our model and get this. Again, the adjusted r^2 drops by a little bit. It's now 0.731. All of the p-values are now extremely tiny, so we stop.

This last model says that to predict how many hours the head of household will work, begin with 2120.5, subtract off a third of an hour for each dollar that a non-spouse member of the household earns, and then add 0.019 hours for each dollar that the household has in assets. We can look at the graph showing how close to accurate the predictions of this model are.

Here, I've made a scatter plot with predicted hours on the horizontal axis and actual hours on the vertical. A perfect prediction would have the actual and predicted hours identical, so all the dots would be on the diagonal line. Our worst error was about 70 hours, or about 3.5%. If we use the model tied for the highest adjusted r^2 and 4 inputs, we get a slightly better fit, but not much. We can choose the simplicity of the model with two inputs, or a little more accuracy in the prediction with four inputs, at least as far as the sample data goes.

But there remains the question of what all of this means. Our original graph, simple linear regression of hours worked on wage, showed a positive connection, with the line suggesting that each additional $1.00 in hourly wage adds about 78 hours in work. The four-input model had the wage being significant with a negative coefficient, which means that higher wages result in less hours worked. And in the final model, wage doesn't appear at all. It has no direct impact on hours worked at all! So, which one is it?

Well, that's a good question. The first graph was right; knowing nothing else, higher wages correlate to more hours. That doesn't necessarily mean that higher wages cause people to work more hours. For all we know, people who work more hours are more likely to be paid well, but they do go together. Then why did the second model say that increased wages led to fewer hours? Well, you have to remember what the other variables are in the model. For example, assets shows a high correlation with wages, an r^2 of about 0.63. Again, causality is not specified by regression, but it seems reasonable to me that people who make more per hour can save more.

In any case, if I know what your assets are, I know a fair amount about your wage, so even if you are working more hours because of wages, I might be able to explain it in terms of your assets, which are strongly tied to your wages. In a model including assets, wages might be completely unnecessary, as was the case in the last model, or if assets gets a larger coefficient, wages might actually show a negative modifier, as it did in our four-input model.

The coefficient of a variable in a model is intended to capture the effect of that variable if all other inputs are held fixed. You can start to see why. When two variables measure almost the same thing, it's often a good idea not to include both in your model. Which one gets credit can be an issue. This is a special case of the problem of multicollinearity.

OK, we've got time left for one more clever and useful variant of linear regression. Suppose you've got bivariate data that suggests a nonlinear relationship from your scatterplot, and that your take-the-log transformations can't tame it into a straight line. It turns out that multiple regression gives you a way of fitting a polynomial to the data. This is sometimes called polynomial regression. Let's take a look at an example.

We've tried a number of different prices for our product, and the revenues from the resulting sales were recorded for a number of months. It's pretty clear this isn't a linear relationship, but it might fit on a parabola, which is a function that has not only an x in its formula, but also an x^2; $y = ax^2 + bx + c$, if we can find the right values of a, b, and c. Well, we can trick multiple regression into doing exactly that. We tell it that we don't just have one input variable, x, but we tell it we have two, x and x^2. Yes, we know that if you know x, you know x^2, but multiple regression doesn't. It just treats x^2 as a new variable that just happens to have the name of x^2. So if our first data point is $x = 15$, $y = 126{,}000$, we tell multiple regression that the inputs are $x = 15$, $x^2 = 225$ (15×15), and $y = 126{,}000$. We feed the multiple regression algorithm with this data, and it comes back with the coefficients. In this example, the equation is approximately $10{,}000x - 100x^2$, with a very impressive r^2 of over 0.97. And based on this, we could estimate that we should be charging around \$50 if we want to maximize revenues.

You might be tempted to include higher powers of x as inputs, too, x^3, x^4, and so on, but be careful; don't use more than you need. Remember the kitchen-sink effect that we talked about with adjusted r^2? Here, it can show up with a vengeance, in the guise of overfitting the data. Look at this example, where x

is hours worked and y is units produced. We can fit a linear model to this and get a pretty good fit. The line says we'd expect an additional 16.5 units per hour of work, on average, and that r^2 is a very respectable 0.95.

Now let's do it again, but not only with x, the number of hours worked, as the input, but also the powers of x up through x^5. Here's the result. The wiggles in a fifth degree polynomial make it possible to fit the six data points perfectly; r^2 now equals 1. But does anyone actually believe that this snake is a better explanation of the data than the straight line? For example, it suggests that beyond 5.8 hours of work or so, total units produced plummets, while with less than 1 hour of work, it skyrockets. With 0 hours of work, the model predicts 416 units produced!

There's a lot going on in multiple regression, and some pretty sophisticated math that supports it. The topic of linear models can and does fill a graduate level course in statistics, but with what we've covered today, you have the tools to do a great deal of powerful and potentially useful analysis and prediction.

Time Series Forecasting
Lecture 4

The topic of this lecture is forecasting—predicting what's going to happen, based on what we know. In many circumstances, we're looking at historical data gathered over time, with one observation for each point in time. Our goal is to use this data to figure out what's going to happen next, as well as we can. Data of this type is called time series data, and to have any hope of making progress with predicting time series data, we have to assume that what has gone on in the past is a decent model for what will happen in the future.

Time Series Analysis

- Let's look at some historical data on U.S. housing starts—a month-by-month record of how many new homes had their construction start in each month. Housing starts are generally considered to be a leading indicator of the economy as a whole.

- For a **time series**, we can visualize the data by making a line graph. The horizontal axis is time, and we connect the dots, where each dot represents the U.S. housing starts for that month. The basic strategy is to decompose the time series into a collection of different components. Each component will capture one aspect of the historical behavior of the series—one part of the pattern.

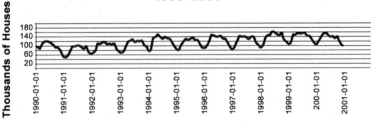

U.S. Housing Starts by Month, 1990–2000

Figure 4.1

- The variation in the data series—the up-and-down bouncing—is far from random. Each January, new housing starts tank, then climb rapidly in the spring months, reaching a peak in summer. Given the weather patterns in North America, this makes sense, and we'd have every reason to expect this kind of variation to continue into the future.

- We've just identified the first component of our time series decomposition: the **seasonal component**. Seasonal components are patterns that repeat over and over, always with a fixed duration, just like the four seasons. But the period of repetition doesn't have to be a year; it can be any regular variation of fixed duration.

- Getting a handle on seasonality is important in two ways. First, if you're hoping to make accurate forecasts of what's going to happen at some point in the future, then you'd better include seasonal variation in that forecast. Second, when trying to make sense of the past, we don't want seasonal fluctuations to conceal other more-persistent trends. This is certainly the case with housing starts and why the government reports "seasonally adjusted" measures of growth.

- The other obvious pattern in the data, once seasonality is accounted for, is that there appears to be a steady increase in housing starts. In fact, we can apply simple linear regression to this line to see how well a linear trend fits the data. In this example, x is measured in months, with $x = 1$ being January 1990, $x = 13$ being January 1991, and so on.

- With r^2 being only 0.36, about 36% in the variation in housing starts can be laid at the doorstep of the steady passage of time. That leaves 64% unaccounted for. But this is what we expect. The data has a very strong annual seasonal component, and the trend line is going to completely ignore seasonal effects. In the sense of tracking the center of the data, the regression line actually seems to be doing rather well.

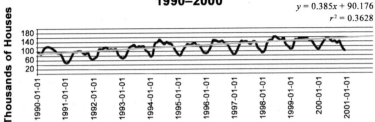

U.S. Housing Starts by Month, 1990–2000

$y = 0.385x + 90.176$
$r^2 = 0.3628$

Thousands of Houses

Figure 4.2

- For this example, the regression line would be the second component of the time series, the trend component. Not all data demonstrates a linear trend, and in general, trend components can actually be quite complicated.

- There's a third component that arises with some time series called the **cyclic component**, and it tracks cyclic variation. While cyclic variation can be thought of as including regular seasonality as one of its subtypes, it's clearer to say that cyclic variation refers to longer-term fluctuations that lack the regularity of seasonal variation. Business cycles are a good example: growth, recession, recovery—but of variable onset, intensity, and duration.

- Our data for housing starts doesn't show any cyclic variation. In fact, many short- and medium-range techniques for forecasting two years or fewer don't include a cyclic component. So, we're currently modeling our housing starts as a seasonal component overlaid on a linear trend.

- Just like in regression, a time series almost never perfectly matches the real-world data. Whatever variation is left unexplained is identified as the error component, which essentially consists of residuals. The component captures all of the variation between what the model predicts and what actually happens.

- When we do a good job with our forecasting, there shouldn't be any significant pattern to the error component. Patterns in the error component mean that the errors contain more information—information that we could have squeezed out of them and included in the other components of the forecast.

- Everything described so far applies to almost every time series forecasting technique, but there are a large number of such techniques, and the variety exists for a reason. Different time series display different characteristics, and the processes that generate them may dictate restrictions on the kind of model that we use.

- Not every model includes all of the possible kinds of components. Some data shows virtually no seasonality, and some shows virtually no trend. But suppose that you have data that includes both. How do you combine them?

- Two common approaches are additive and multiplicative models. In an additive model, you say that the observed value is the sum of the trend component, the seasonal component, and the error component. This is good when the seasonal component stays pretty constant over time. In a multiplicative model, you multiply these pieces together instead of adding them—a better choice when the seasonal component's magnitude varies with the trend.

Measures of Forecast Quality
- People who do forecasting like to be a bit more quantitative when assessing the performance of a forecast. One common measure is called **mean absolute deviation (MAD)**. Start by finding how much each forecast value differed from the actual historical value—the error for each point in time. If it's negative, take its absolute value to make it positive. Finally, average all of these absolute errors.

- In addition to the MAD, people also often report the **mean absolute percentage error (MAPE)**. To do this, find the percent that each forecast was wrong, take the absolute value of each of these in case they're negative, and then average all of these percentages. MAD and MAPE are both particularly popular in the manufacturing sector.

- But the most common way to characterize the fit of a forecast is **mean squared error (MSE)**. That is, take the error for each observation, square each of these errors, and then average all of the squared errors together.

- MSE has much nicer statistical and calculus properties than MAD or MAPE does. It falls short, though, when you try to interpret it simply. You can't compare these different measures of forecast quality to one another, but you can compare two different forecasts using the same measure of quality. The one with the lower MAD—or MAPE, or MSE—should be the better forecast.

- Often, all of the different measures agree on the winner, but not always. MSE tends to care about consistency in a forecast. If you're exactly right a lot of the time but occasionally make howling errors, MSE gives you a big **penalty** for the mistakes. That is, its squaring step turns a very bad error into a very, very bad error. MAD and MAPE are more forgiving of the occasional howler. In many applications, people would prefer many small errors to an occasional large one, which is one of the non-computational reasons that MSE is often the preferred measure.

Characterizing Trends in Data

- One of the dangers you face when you're working with historical data alone is that it lacks a structural model of what factors are influencing the evolution of the data outputs. Time series forecasting is based on the idea that the past is a good model for the future. When something fundamental about the situation changes, if you can't anticipate it, you can be left with forecasts that are really quite dreadful.

- Extending a graph over a much longer number of years also shows how the business cycle (or other cyclic variation) can turn out to be important, even for short-term forecasts. One reason the business cycle is often ignored is that economists find it so difficult to specify precisely. The better we understand the past, the better chance we have of predicting the future. And, sometimes, such as in economic analysis, we're also trying to make sense of the past for its own sake.

- A common way of characterizing a trend in a group of data is by using a **simple moving average**. Simply put, we peg the trend at some point in time by averaging together some number of observations near that point in time. If you're trying to forecast the future with monthly data, you might average together the 12 most recent months to get the forecast of what happens next. If you're trying to make sense of historical data, you might choose a set of observations centered on your current month to average.

- But because each simple moving average forecast is just the average of the preceding 12 months, we don't get a forecast until 12 months have gone by. More importantly, we can only use the technique to forecast one month in advance. If we want to go farther than that—and we probably do—we need more advanced techniques.

- And a plethora of them exist, each suited to different kinds of time series. Some are simple. **Weighted moving average** takes the simple moving average but gives each observation a different weight. These weights tell us the relative importance of the values used in computing the forecast. If what happened one time period ago has twice as much influence on the present as what happened two time periods ago, then their weights would reflect this.

- Of course, you have to find the weights that make the forecast fit the historical data as well as possible. That is, you'd like to find the values of the weights that minimize some measurement of error, such as the MSE of the forecast.

- A close relative of the weighted moving average is called simple **exponential smoothing**. You can think of it as a weighted moving average in which weights grow smaller and smaller in a geometric fashion as we move back in time. Exponential smoothing is an extremely simple forecasting technique, but it's the basis for a lot of more sophisticated and complicated approaches.

Important Terms

cyclic component: The component of a time series forecast that attempts to capture cyclic variation. This differs from seasonal variation by showing nonconstant duration or intensity or unpredictable onset.

exponential smoothing: A time series forecasting technique that forms the basis of many more-complicated models. Exponential smoothing can be thought of as a variant of the weighted moving average.

mean absolute deviation (MAD): A measure of forecast accuracy, MAD is the average amount by which the forecast differs from the actual value.

mean absolute percentage error (MAPE): A measure of forecast accuracy, MAPE is the average percentage by which the forecast differs from the actual value.

mean squared error (MSE): A measure of forecast accuracy that is similar to variance in its calculation, MSE is the average of the squares of all of the residuals for a forecast.

penalty: A modifier to the objective of a problem that reflects that the proposed solution is deficient in completely satisfying a constraint. Used in genetic algorithms and soft constraints.

seasonal component: The component of a time series forecast that captures the seasonality of the data—that is, its regular, periodic variation. Some sources use the term for such variation only if the period length is at least one year.

simple moving average: A forecast for a period in a time series made by averaging together the values for a specified number of nearby time periods. For predictive models, this will mean the n observations immediately preceding the current time period.

time series: A data set consisting of one value of the output variable for each point in time. The points in time are usually evenly spaced. Alternatively, a forecasting technique used on such data.

weighted moving average: A forecast for a period in a time series made by averaging together the values for a specified number of nearby time periods, with each period's importance reflected by its weight. For predictive models, this will mean a weighted average of the n observations immediately preceding the current time period.

Suggested Reading

Hyndman and Athanasopoulos, *Forecasting*.

Questions and Comments

1. If you're going to apply more complicated time series approaches, you're probably going to want to use software to do it. A few resources that you might find extremely helpful in learning more on your own include 1) the open-source statistical software package R, which will run on a wide variety of computer platforms and is fairly straightforward to learn, powerful, and free; and 2) *Forecasting: Principles and Practice*, a free online text that does an excellent job of explaining more sophisticated time series analysis in more detail, including the R commands necessary to conduct the analysis on a time series.

2. It is possible to look at two different time series over the same interval of time and to explore their relationship in a scatterplot. The point (x_i, y_i) is plotted if at time period i the first series has the value x_i and the second series has the value y_i.

Think about what would happen if x and y really had no relationship to one another but both showed a positive trend over a period of time. The resulting scatterplot could have a high correlation for the two variables, which could mislead the investigator into believing the two variables were in fact linked. For this reason, it's dangerous to create such a scatterplot for time series with trends.

Time Series Forecasting
Lecture 4—Transcript

Forecasting: predicting what's going to happen based on what we already know. This has already been a focus of our last two lectures, and our tool for that exploration has been regression, and the idea that we can model our output's value as a linear combination of our inputs. Not all input-output relations are linear, of course, but we found some clever ways to apply linear regression to many nonlinear relationships, too, using transformation of variables and polynomial regression.

But, in a lot of circumstances, our situation is more specific. We're looking at historical data gathered over time, with one observation for each point in time. Our goal is to use this data to figure out what's going to happen next, as well as we can. Data of this type is called time series data, and to have any hope of making progress when predicting a time series, we have to make a rather obvious assumption, that what's gone on in the past is a decent model for what will happen in the future.

To get started, let's look at some historical data on U.S. housing starts, a month-by-month record of how many new homes had their construction start in each month. Housing starts is generally considered to be a leading indicator of the economy as a whole. We'll start by visualizing the data. For a time series, we do this by making a line graph. The horizontal axis is time, and we connect the dots, where each dot represents the U.S. housing starts for that month. Here's what the data looks like from 1990 through the end of 2000.

This is a wonderful data set for introducing some key ideas in time series analysis. The basic strategy is to decompose the time series into different collections of different components. Each component will capture one aspect of the historical behavior of the series, one part of the pattern. So, take a look at this graph. What patterns catch your eye? If you're like most people, one of the first things that you will notice is the up-and-down bouncing of the data of the series. You can see that this variation is far from random. Each January, new housing starts tank, then climb rapidly in the spring months, reaching a peak in summer. Given the weather patterns in North America, this makes sense, and we'd have every reason to expect this kind of variation to continue into the future.

Well, congratulations. You've just identified the first component of our time series decomposition, the seasonal component. Seasonal components are patterns that repeat over and over, always with a fixed duration, just like the four seasons. But the period of repetition doesn't have to be a year. Sunspot activities vary over an 11-year cycle. The number of babies born in U.S. hospitals shows a strong seasonality of seven days, with about 10% fewer babies being born on weekend days than weekdays. Birth rates drop on holidays, too, for that matter. Kids being born might not know what a weekend or a holiday is, but the doctors who schedule the surgical deliveries certainly do. Some people reserve the word seasonal only for variations with a period of a year or more, but others, like me, apply the term to any regular variation of fixed duration.

Getting a handle on seasonality is important in two ways. First, if you're looking at making an accurate forecast of what's going to happen at some point in the future, such as how much electricity you can expect Chicago to need next January, then you'd better include seasonal variation in that forecast. But the second reason that it's important to understand seasonality is the mirror reverse of this. When trying to make sense of the past, we don't want seasonal fluctuations to conceal other, more persistent trends. This is certainly the case with our housing starts, and why the government reports seasonally adjusted measures of growth. Yeah, we know that this January is going to be worse for housing starts than was the preceding June, but taking into account that they are a January and a June, which one looks better? We'll be looking at some ways to do exactly this in a few minutes.

But, for now, let's go back to our time series graph and see what else we can find. The other obvious pattern in the data, once seasonality is accounted for, is that there appears to be a steady increase in housing starts. In fact, we can apply simple linear regression to this line to see how well the linear trend fits the data, like this. Here, x is measured in months, with $x = 1$ being January 1990, $x = 13$ is January 1991, and so on.

Your first impression might be that this is a terribly unimpressive performance by linear regression. With r^2 being only about 0.36, about 36% of the variation in housing starts can be laid at the doorstep of the steady passage of time. That leaves 64% unaccounted for. But this is the sort of thing we'd expect, actually. The data has a very strong seasonal component,; the trend line is going to completely ignore seasonal effects. In the sense of tracking the center of the data, the regression line, actually, seems to be doing rather well. For our current example, the regression line would be the second component of our time series, the trend component. We've already seen that not all data demonstrates a linear trend, and in general, trend components can actually be quite complicated. We'll take a look at some less well-behaved examples in a few minutes.

There's a third component that arises with some time series called the cyclic component. As the name suggests, it tracks cyclic variation. While cyclic variation can be thought of as including regular seasonality as one of its subtypes, it's clearer to say that cyclic variation refers to longer-term fluctuations that lack the regularity of seasonal variation. Business cycles are a good example, growth, recovery, recession, recovery, but of variable onset, intensity and duration.

Here's the U.S. unemployment rate since 1948, and it's hard to miss the repeated peak-and-trough nature of the graph. But, while the general character of the cycle is known, there are highs and lows over longer stretches of time, their intensity and duration generally vary from cycle to cycle. Our data for housing starts doesn't show any cyclic variation. In fact, many short-term and medium-range techniques for forecasting of two years or less don't include a cyclic component.

So, we're currently modeling our housing starts as a seasonal component overlaid on and linear trend. Anything else? Well, yes. Just like in regression, a time series almost never perfectly matches the real-world data. Whatever variation is left unexplained is identified as the error component. The error component essentially consists of all the residuals that we talked about in the regression. The component captures all of the variation between what the model predicts and what actually happens. When you do a good job with your forecasting, there shouldn't be any significant pattern to the error component. Patterns in the error component mean that the errors contain more information, information we could have squeezed out of them and included in the other components of the forecast.

Everything I've said so far applies to almost every time series forecasting technique, but there are a large number of such techniques, and the variety exists for a reason. Different time series display different characteristics, and the processes that generate them may dictate the restrictions on the kind of model that we use. Let's look at an example. I'm going to show you some time series data for the earnings on an initial investment of 100 bucks, and I want you to guess what happens next.

In this one, it looks like earnings get to about 13, then level out. Best guess for the future? Keeping level. How about this one? After some original hard times, we see a long trend of growth, although we've had some backsliding from time to time. Evidently something good happened around day 85, triggering an earnings spurt. Best guess? Continued growth. OK, one more. What do you think? Well, my thought is, stay away from this one. While we've got bumps and wiggles, we see a downward trend that is, if anything, getting worse. That was pretty easy, wasn't it? Except that, in every case but the first, we've got the trend quite wrong. All three of these data sets were generated in exactly the same way. Like this.

Flip a coin. If it comes up heads, earnings increase by $1.00 for that day; if it comes up tails, earnings decrease by $1.00. The symmetry of the process means that, over any period of time, your chances of gaining a certain amount of money and losing that same amount of money are exactly the same. That is, on average, you break even. This kind of process is called a simple random walk. If I understand that the process driving my observations is such a random walk, there won't be a trend, and so my forecast model should not include a trend component. A simple random walk like this is an example of a stationary process, and the mean and variance of such a process don't change over time.

In fact, the single, best forecast for what happens next in all three of these pictures is what's called the naïve forecast. That simply means that our forecast for future times repeats whatever the earnings were today. It's like forecasting the weather in the future as more of the same. Now, in real life, one almost no one proposes the naïve forecast seriously, but its simplicity makes it a good benchmark for stationary time series data; and with difficult data, it can be surprisingly difficult to beat it! Some people think that the short-term fluctuations in the Dow Jones Industrial Average are essentially a random walk, so you can understand why predicting the stock market is such a challenge.

This random walk example brings back into focus another idea from simple linear regression, confidence intervals and prediction intervals. If the future ups and downs of my earnings are really nothing more than coin flips, it is absolutely true that the average value of the earnings at all future points in time, across all possible futures, would be whatever they are today. But of course, we're not going to see all possible futures; we're only going to see one. And a terribly important question is, how far from our predicted value can the observed value reasonably be expected to go? This is the question of prediction intervals. For our last random walk example, the naïve forecast, with the 80% prediction interval, would look like this.

The horizontal line shows the naïve forecast, and the actual value for each point in the future has an 80% probability of falling in that gray zone. As you can see, other than the current value of the earning, $16.00 below where we started, the history of the random walk is irrelevant to the prediction. Still, we can say that it's less than 10% likely that we'll get back above −3 or so by day 300. The shaded area covers 80% of the possible futures, so there are 10% that fall above the gray zone and 10% that fall below it. And that's really about the best we can do with a random walk.

Okay, so, not every model includes all of the possible components. Some data shows virtually no seasonality, some, virtually no trend. But suppose you have data that includes both. How do you combine them? Two common approaches are additive and multiplicative models. And the names are well chosen. In an additive model, you say that the observed value is the sum of the trend component, the seasonal component, and the error component. This is good when the seasonal component stays pretty constant over time.

If you multiply these pieces together instead of adding them, a better choice when the seasonal component's magnitude varies with the trend, you get a multiplicative model.

For example, imagine that we are tracking the revenues generated by businesses in a beachside city in Virginia that's been growing over time. Well, we'd expect revenues to be growing too, although we'd also expect revenue to show a seasonal variation. During the warm months, the city is going to be raking in money from tourists, as well as locals. In the winter, it has to get by with local trade. So as the city grows, what's going to happen to the size of the seasonal variation? Well, that depends. If the growth of the city is almost all due to growth in year-round population, say, due to a computer firm moving into the area, then the dollar variation in revenue due to tourism won't vary much as the years pass.

In this case, we'd want to use an additive model. The seasonal component adds about the same amount of business each summer, year after year. On the other hand, if the growth were spread across sectors of the city's economy, all over, we'd want a multiplicative model. Here, a 10% increase in overall revenues for the city would also show a 10% increase in the size of the seasonal fluctuations. A multiplier of 1 is the base, and a multiplier of 1.3 for July would mean that revenues in July are 30% higher than this base level.

Let's return to our housing starts data and apply this new information. The size of the seasonal variation is just about constant, even though the average number of starts per month has increased by about 50% over the 10 years. So we'd want to use an additive model. Our forecast at any point in time is going to be the sum of the trend component and a seasonal component. Any deviation between the predicted and actual values will be captured by the error component.

Here is maybe the simplest way to make this forecast, taking advantage of the fact that our data seem to be displaying a linear trend. First, we use linear regression to fit a straight line to our data. We already did this a moment ago. Now, subtract the linear regression prediction from the observed data values for each point in time. Numerically, this is easy to do, since we have an equation of the regression line and the historical data values. Graphically, we're just removing the trend and focusing on how the data wiggles up and down around the trend line. The result, the de-trended data, looks like this.

OK, now we'll capture the seasonal variation. The simplest way to do this is to come up with one seasonal index for every month of our forecast. For example, we can see on the seasonal graph that January is consistently low, about −30, so the season index for January is going to be about −30. To be more precise, we're going to average the values for all the January observations on this de-trended graph and use the results as the seasonal adjustment, or seasonal index, for January. It comes out to be −29.2, meaning that the average January is about 29,200 houses below the regression line that we just found. May ends up being the busiest month, with a seasonal index of 17.9. There are, on average, 17,900 more housing starts in May than you'd guess just by looking at the regression equation.

When you put all of the seasonal indices together, you've created the seasonal component of our forecast. Here it is, superimposed on our last graph, the de-trended data series. The blue wiggle shows the actual, historic data moving up and down around the trend line. The red wiggle is our seasonal component, repeating the same pattern over and over. The dip that you see on the red line at each January is the seasonal index of −29.2, because the average of all of the historical, blue-line Januarys was −29.2. OK? Because it repeats over and over, this seasonal forecast can easily be extended into the future. The seasonal component for every January is the same, for every February is the same, and so on. This is the simplest kind of seasonal forecast to do. And as you can see, it does a pretty good job.

But not a perfect one. We can see the discrepancy by looking at the vertical separation between the blue line and the red line, blue minus red; that'll be the error component, the stuff that our model doesn't explain. Here it is, in a line graph of its own. And what we're seeing here is pretty much what we're supposed to see. The first couple of months have noticeably high errors; housing starts were [considerably] higher than we would have expected, but beyond that point, there seems to be no pattern to our errors. They hover around zero, zigging and zagging in what appears to be a random walk. And happily, they tend to be small. This suggests that we've gotten out of our data about as much as we can.

So, let's see how well the forecast actually did when compared to the real data. I'm going to construct our forecast with the additive model we discussed, a linear trend, modified by adding a seasonal component. Here's the forecast, superimposed on the actual data. And you can see how well it did in the 90s, and you can also see what it predicts for the next five years. Based on how well our forecast has done so far, I'm feeling pretty good about the reliability of our prediction.

Not surprisingly, people who do forecasting, like this, for a living, want to be a bit more quantitative when assessing the performance of a forecast. One common measure is called mean absolute deviation, or MAD Start with finding how much each forecast value differed from the actual historical value, the error for each point in time. If it's negative, take its absolute value to make it positive, so an error of +5 or −5 still counts as an absolute error of 5. Finally, average all these absolute errors. For our housing starts data, the MAD turns out to be about 7.3, or 7300 houses. That is, sometimes our prediction was too high, sometimes it's too low, but on average, we missed the mark by 7300 houses during the 11 years for which we have data. Coming to it cold, learning that the MAD is 7300 houses, might not be as informative as you'd like. Is that a lot or a little?

Because of this, people also often report the MAPE, or mean absolute percentage error. To do this, find the percent that each forecast was wrong, take the absolute value of these, in case they're negative, then average all of these percents. For our data, it comes out to be 6.8%, not bad. Again, sometimes we're too high, sometimes we're too low, but on average, we missed the actual housing starts by 6.8%. MAD and MAPE are both particular popular in the manufacturing sector. But the most common way to characterize the fit of a forecast should sound quite familiar to you from our discussion of regression. It's MSE, mean squared error. That is, take the error for each observation. Square each of those errors. Then average all the squared errors together. That's MSE.

We talked about the advantages of MSE when we discussed regression. It has much nicer statistical and calculus properties than do MAD or MAPE. It falls short, though, when you try to interpret it simply. I gave you a nice, intuitive interpretation of what an MAD of 7.3 and an MAPE of 6.8 mean. It's much harder to explain what MSE of our housing forecast actually means. It's 88.7 million square houses. Yes, I said square houses. The error was in houses, and we squared that before we averaged. Obviously, you can't compare these different measures of forecast quality to one another; that's comparing apples and oranges, or apples squared. What you can do is to compare two different forecasts using the same measure of quality. The one with the lower MAD, or MAPE, or MSE should be the better forecast.

Will all of the different measures always agree on the winner? They often do, but sadly, no. MSE tends to care about the consistency of a forecast. If you're spot on a lot of the time, but occasionally make howling errors, MSE gives you a big penalty for the mistakes. That is, its squaring step turns a very bad error into a very, very, bad, BAD error. MAD and MAPE are more forgiving with the occasional howler.

In many applications, people would prefer a lot of small errors to the occasional large one, which is one of the non-computational reasons why MSE is often the preferred measure. I'd rather have my budget for my business off by $5000 per month, every month, than have it be exactly correct in January through November and $60,000 off in December. MAD says that both of these forecasts are equally good; MSE says that that end of year surprise is 12 times worse than the small, consistent errors.

But no matter which measure you use, if you look more closely at how we assessed the quality of the forecast, you might notice that I've played a little bit of a shell game. The calculations don't talk about what's going to happen after the historic data I've got. They talk about what's already happened. In a sense, I was forecasting the past. I'm deciding how good my model is at how well it fit the historic data, when I used that data to make the forecast to begin with. Isn't that kind of like forecasting yesterday's weather?

Well, yes, and no. I applied my forecasting techniques to model the 132 months of historical data that I had available. That's what I used to assess its quality, to compute MSE and so on. The model itself had a linear trend component and a seasonal component. To specify this model, I had to come up with value of 14 quantities, the slope of the regression line, the intercept of that line, and 12 seasonal indices for the 12 months. Actually, since the seasonal indices will add to 0 in a problem like this, I only needed to specify 11 of them. So, 13 numbers total, what are often called model parameters, had to be found.

At the end of our discussion of polynomial regression, we saw the danger of over-fitting the data, which really means using a model with too many parameters to explain a set with too little data. But here our model has 13 parameters, and we're using it to explain a data set more than 10 times that size, 132 months. So the fact that we got a very good forecast over that entire range gives us some confidence in its quality.

Still, the proof, really, is in how well the forecast does on new data. What happens after 2000? Well, we know our forecast out through the end of 2005. How well did it do in predicting the actual events of the early 21st century? Take a look. The model characterized the patterns of the past, extended them into the future, and the result was, I think, a very nice forecast indeed. Unfortunately, not every time series is so well-behaved. In fact, even this series has some surprises for us. Let's look at housing starts over a larger time horizon, from 1959 to 2013. Here's that bigger picture.

The part in green is the part we've already been discussing, and you can see that once you move past 2005, the forecast that we made has a lot of problems—the huge decline in housing starts that began in 2006 and persisted for the next two years, for starters. This is one of the dangers you face when you're working with historical data alone. It lacks a structural model of what the factors are that are influencing the evolution of the data outputs. Time series forecast is based on the idea that the past is a good model for the future. When something fundamental about that situation changes, if you can't anticipate it, you can be left with forecasts that are really quite dreadful.

Extending our graph over a much longer number of years shows how the business cycle, or other cyclic variation, can turn out to be important, even for short-term forecasts. One reason the business cycle is often ignored is that economists find it so hard to specify precisely. But just by inspecting the extended graph, we can see that housing starts peaked and turned downward three times, in 1973, 1978, and 1986. And we can see that the level of those peaks together define a fairly narrow horizontal zone, where we might have begun to anticipate a possible cyclic downturn. Maybe lower, as in the 1986 level, or maybe higher, at the 1973 level. Clearly, the better we understand the past, the better chance we have of predicting the future. And sometimes, such as in economic analysis, we're also trying to make sense of the past for its own sake. But how do we get a trend out of something like this?

A common way of characterizing the trend is by using a simple moving average. Simply put, we peg the trend at some point in time by averaging together some number of observations near that point in time. If you're trying to forecast the future with monthly data, you might average together the 12 most recent months to get the forecast of what happens next. If you're trying to make sense of historical data, you might choose a set of observations centered on your current month to average. Here's an example for our housing data. It's the simple moving average forecast with $n = 12$, meaning that the preceding 12 observations are used to predict the new value. The actual data is in blue, the forecast is in red. And the forecast doesn't do a very good job of capturing the details of the actual events.

Its most obvious fault is that it ignores seasonality. And it should! Remember how we got this forecast; we predicted starts in any given month as being the average of the starts over the preceding 12 months. By averaging over the preceding years, we wiped out any seasonality effects from our forecast. And really, that was the idea. If we're interested in how the market for housing starts was doing over this span of years, we don't want to focus on the seasonal variation that we know is going to be present. That's why the government always uses deseasonalized data when viewing housing starts as a barometer for the economy. They also control for the number of so-called trading days in each month.

If we're trying to do a trend-seasonal decomposition of data, the red line gives us the trend line that we could use for our trend component, and on top of this component, we could put a seasonal variation, computed as we did in our original example. There's one more problem, though. Since each simple moving average forecast is just the average of the preceding 12 months, we don't get a forecast until 12 months have gone by. More importantly, we can only use the technique to forecast one month in advance. If we want to go farther than that, and we probably do, we need more advanced techniques.

And a plethora of them exist, each suited to different kinds of series. Some are simple. Weighted moving average takes the simple moving average that we just did, but gives each observation a different weight. These weights tell us the relative importance of the values that are used in computing the forecast. If what happened one time-period ago has twice as much influence as what happened two time-periods ago, then the weights would reflect this. W1 would be twice as big as W2. In a 12-month weighted moving average, there'd be weights W1 through W12. W1 isn't the weight for January; it's the weight for whatever happened one month ago. So W12 tells us the relative importance of what happened a year ago.

With data that's strongly seasonal, like ours, W12 would probably be the largest weight. That's because what will happen this month is probably most strongly tied to what happened at the same month last year. If the data isn't seasonal, then what happens this month might be most strongly tied to whatever happened last month. In that case, W1 would be the largest, and the weights would grow smaller as we move backward in time from the time for which we're making a forecast. Of course, you'd have to find the weights that make the forecast fit the historical data as well as possible. That is, you'd like to find the values of the weights that minimize some measurement of error, like the MSE in the forecast.

This is an example of an optimization problem, since we're trying to find the best weights for our model. The middle section of this course is going to be dedicated to setting up and solving optimization problems like this. By the time we get there, you'll be finding the best weights is going to be a piece of cake. It's worth noting, though, it's something that we'll see again and again in this course; the synergy among the different analytic techniques. All our tools fit together. Time series forecasting, for example, might use linear regression for the trend component or might need nonlinear optimization for finding the weights of the weighted moving average.

Here's a close relative of the weighted moving average, called simple exponential smoothing. You can think of it as a weighted moving average in which the weights grow smaller and smaller in a geometric fashion as we move back in time. For example, consecutive weights might differ by a factor of two; W1 is $\frac{1}{2}$, W2 is $\frac{1}{4}$, W3 is $\frac{1}{8}$, and so on. The shrinkage rate is generally conveyed by giving the forecast's smoothing constant, which is called alpha (α). If each weight is 90% of the weight before it, the value of alpha is $1 - 90\%$, or 0.1. Exponential smoothing is an extremely simple forecasting technique, but it's the basis for a lot more sophisticated and complicated approaches.

It can be surprising that time series work so well. In fact, even when the data does not show a noticeable seasonal or trend component, time series can still be useful, in two regards. First, you can look for near-term correlations between the residuals of a forecast and the identical series of residuals displaced one or more periods in time, what are called the lagged residuals. Such correlations are called autocorrelations and are quite common in many time series. On the one hand, autocorrelation has an important impact on confidence and prediction intervals obtained from the forecast. It'll generally lead to intervals that are narrower than they should be. That means that autocorrelation in the data can give us excessive confidence in our forecast. Sometimes this problem can be addressed by differencing, creating a new data set that tracks not the value of the quantity over time, but its change from one time period to the next. Moreover, a series may show, for example, that the residual in the forecast at one time period tends to be strongly linked to the residual from three time periods before. Such information can be used, with more complicated techniques, to make a better forecast.

Second, time series techniques, like the moving average and exponential smoothing techniques, can remove high-frequency variations in the data and allow longer-term drifts to become evident. A business cycle, for example, such as we found in the data for housing starts, might be noticed first by looking at a smoothed curve.

Sometimes we're lucky; sometimes we know enough about the mechanism and process that we're able to build a structural model of it, a mathematical representation of how it interacts with the outside world and aspects of itself. In such cases, tools such as optimization and simulation can give us insights in to what we might expect from that process, including behavior we've never yet observed. But often, we're not that lucky. Sometimes, all we have to go on is the historical data and very little insight into the mechanisms that actually drive the process underlying it. In these cases, time series forecasting can often fill the gap. It's especially good for forecasting short-term economic demand for a product or service, such as airline seats, electricity, or water usage. So long as we can trust the assumption that the past is a good indicator of the future, time series forecasting can be a powerful tool for quantifying what that future may hold.

Data Mining—Exploration and Prediction
Lecture 5

S tatistical techniques like regression and time series forecasting were strongholds of mathematical prediction for much of the 20th century. But they have been supplemented in recent decades by additional techniques from the exciting and fast-growing area known as data mining, which focuses on large data sets. As you will learn in this lecture, the job of data mining is to find what useful patterns and connections large data sets contain—connections and patterns that might otherwise be missed. Data mining is especially important in the 21st century because of two advances: computational power and a veritable explosion in the quantity of collected data.

Data Mining

- One of the key differences between many classical statistical techniques and data mining is the quantity of data available. Throughout much of history, data was scarce and hard to come by. And that meant that all of the data that was available had to be used in the analysis—both to create the model and to test its accuracy.

- In order for that analysis to get very far, some assumptions had to be made, including that the errors are normally distributed (in technical terms, the data is supposed to be **homoscedastic**) and that the errors are independent from one another. The violation of these assumptions leads to questions of how much to trust the predictions.

- In data mining, we often have an embarrassment of riches when it comes to data—if anything, we have too much. The first thing that means is that we can use part of the data to come up with our model and a completely different set of data to test the reliability of that model. This also means that we aren't bound by a lot of assumptions, such as the distribution of errors.

- In other techniques, such as regression, we can spend a lot of time transforming the data so that we don't have **heteroscedasticity**, autocorrelation, or error terms that are distributed, for example, like lottery payoffs—many small negative values sprinkled with occasional big positive values. Fixing these problems can be quite a headache. With data mining, the headache largely goes away.

- Practitioners sometime debate where exactly the margins of data mining lie, but there are three objectives that are central to it: data visualization and exploration, **classification** and **prediction**, and association among cases. Time series analysis is sometimes considered a part of data mining, too.

Data Visualization

- When we're exploring data in data mining, we're probably going to start by looking at one variable at a time. If a variable is categorical, bar charts are a good choice. In the following chart, we can see at a glance that heart disease and cancer—"malignant neoplasms"—are overwhelmingly the most common causes of death in the United States, and seeing this in a chart conveys this fact much more quickly and memorably than a table of numbers does.

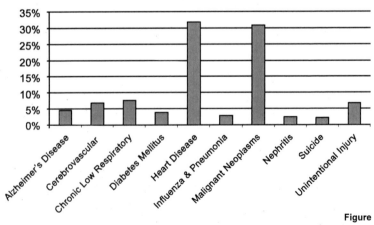

Causes of Death in the United States

Figure 5.1

- Increasingly, software supports the ability to look at interesting aspects of the data in greater detail. For example, from the previous graph, we can break things down by age, as follows. Excel's pivot table and pivot chart can give you some of this functionality.

Causes of Death in the United States

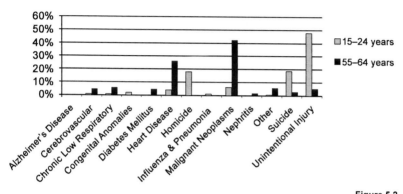

Figure 5.2

- For a single numeric variable, the most common choice is the histogram, which is like a bar chart in which each bar represents a range of values. For example, the following is a histogram of the distribution for how many Facebook friends a group of students have.

Number of Facebook Friends

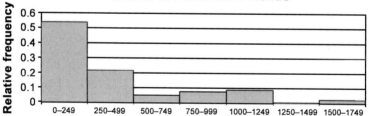

Figure 5.3

- For relationships between two numeric variables, we generally use scatterplots. This initial exploration often uncovers connections between the variables—connections we can exploit in further analysis.

- For relationships among three variables, we can use a three-dimensional plot. This is less useful when the software renders such data as a static plot on a two-dimensional screen, or piece of paper. But there are still options. We can code additional information with color or size. Many weather maps do this. Such representations are often called **heat maps**, even when the color is coding for something quite different than temperature or when the thing being colored is not a map.

- We can also use the time dimension, combining multiple images into a movie. Swedish academician Hans Rosling created an astounding video that you can find on the BBC Four website in which he tracks the wealth and longevity of people in 200 countries over 200 years of history. He uses a scatterplot with an axis for lifespan, a logarithmic axis for wealth, color to indicate continent, size to indicate population, and time to indicate the passage of years. The film lasts only four minutes and is stunningly effective at communicating all of that data.

- Processing visual information in two or three dimensions—recognizing structures and patterns—is something that human beings are uncannily good at, which is why charts and graphs exist to begin with. But as the number of variables increases, obvious ways of representing data graphically fail. Each variable—each category of information—needs its own dimension in space, and for visualization, three dimensions of physical space is our limit. So, what can we do?

- To begin, we can conduct a kind of dimensional reduction by using mathematical procedures. It's like using a student's GPA or SAT score in place of every detail of his or her academic history. The trick is to find a way to summarize multiple pieces of data in a way that loses as little information as possible.

- The computations required may not be simple, but because data mining by definition deals with large data sets, no one would perform the calculations by hand anyway. Statistical software such as SAS or SPSS, or the free package known as R, are perfectly capable of this kind of dimensional reduction; in fact, it can even be done in Excel with an add-in package such as XLMiner.

Measuring Variation and Normalizing Data

- We measure variation by using the most common statistical measure of dispersion: the **variance**. The calculation for the variance bears a great similarity to the calculation of MSE—that "average the squared errors" calculation for regression. For variance, the only difference is that the "error" is taken to be the distance of each data point from the mean of the data. If you take the square root of the variance, you get the **standard deviation**, something that comes up in most statistical discussions.

- A technique called **principal components analysis (PCA)** creates new variables that partition the variation between two variables in a better way, with the first variable capturing as much of the variation as possible. PCA begins by finding the line along which that variation—that variance—is maximized. The math can be done by a computer, which gives us a line that would normally be called the z_1 axis when doing PCA. Impressively, this line captures over 97% of the total variance in the two variables. To capture the remaining 2.6%, PCA would construct a second axis, perpendicular to the first, called the z_2 axis.

- In complicated examples, with more than a handful of variables, PCA finds the direction in multidimensional space that captures the largest possible fraction of the variation in the data and calls that the z_1 axis. It then looks for a direction perpendicular to that one that captures the maximal amount of the remaining variation. That's the z_2 axis, and with more than two dimensions in the original problem, there are more perpendicular directions to choose among. PCA continues like this. It next finds a direction for a z_3 axis that is perpendicular to both z_1 and z_2 and captures as much of the remaining variation as possible—and so on.

- We often find that after only a handful of principal components are found, the vast majority of the variation is accounted for. If that's the case, the original variables, however many there were, can be discarded and replaced with this handful, losing almost nothing about the original variation in the data and greatly simplifying the future work.

- But there are some disadvantages to doing this, too. For example, PCA won't work well for data whose underlying pattern is not linear. But that can be overcome by modeling a nonlinear pattern using local patterns that are linear, which is analogous to multiple regression.

- **Normalizing** data means standardizing it. We do this by first subtracting the mean of a variable from all values of that variable, and then dividing each of those differences by the standard deviation of that variable. No matter what data you start with, after you've normalized it, the mean of the data is zero, and its standard deviation is one. In that sense, then, each variable is now on equal footing. Generally, if the two variables are measured in units that aren't comparable or on dramatically different scales, normalization is a good idea.

- Another factor that can be useful in reducing the dimension of the data is domain knowledge. Someone familiar with the field of investigation can often narrow our focus before we even start with the math. Experts can tell us what quantities are almost certainly relevant—or irrelevant—to the task at hand. They can also let us know when one of our variables is useless as a predictor because we won't be able to gather it in time to make use of it. Domain knowledge can also allow us to inspect our final answer and see if it makes sense.

Classification and Prediction

- Perhaps the most common kind of task for data mining is using the data about an individual to predict some unknown characteristic about them. An "individual" here could be a person, or a river, or a bottle of wine, or anything else. The thing we're trying to predict might be a categorical variable, such as what brand of car the person owns, or whether he or she would accept a particular offer to refinance their home, or whether he or she committed fraud on their federal income tax. When that's the case, the prediction is often called classification—sorting individuals into the classes in which we think they belong.

- On the other hand, we may be trying to predict a continuous variable, such as a person's life expectancy, or the probability that a firm will fail within the next five years. Some people use the term "prediction" only for such continuous variable cases, while some use prediction for sorting individuals into classes as well.

- **Classification trees** are a great data mining technique that we can picture as repeatedly subdividing rectangles into smaller and smaller rectangles until all of the points in any given rectangle fall into the same category of the output. If you're working with continuous variables rather than classes, they're usually called **regression trees**, but the idea is the same. This technique can be used on many different kinds of problems, it's easy to explain to nontechnical people, the results are directly usable by the nonexpert, and it doesn't care whether the data is normalized.

Important Terms

classification: Using the information available about an individual to predict to which of a number of categories that individual belongs.

classification tree: A data mining classification technique. The algorithm repeatedly splits a subset of the individuals into two groups in a way that reduces total "impurity."

heat map: A data visualization technique in which colors or shades are used to convey information about a variable in a chart, graph, or map.

heteroscedastic: A collection of random variables is heteroscedastic if their standard deviations are not all equal.

homoscedastic: A collection of random variables is homoscedastic if they all have the same standard deviation. Linear regression generally assumes that the error terms in the forecast are homoscedastic.

normalizing: Also called standardizing. Linearly rescaling a variable to make its mean 0 and its standard deviation 1. This is done by taking each of the variable's values, subtracting the mean of the variable, and then dividing the result by the variable's standard deviation.

prediction: In data mining, using the information available about an individual to estimate the value that it takes on some continuous output variable. Some people use the term to include classification, as well.

principal components analysis (PCA): A technique for reducing the number of variables in a data set by identifying a collection of linear combinations of those variables that capture most of the variation of a larger set of the original variables. These new variables can then be used in place of the larger set.

regression tree: A classification tree with a continuous output.

standard deviation: A measure of dispersion, it is the square root of the variance.

variance: A commonly used statistical measure of the dispersion, or spread, of data. For a population, the variance is computed by deviation of each observation from the population mean, squaring those differences, and averaging the squares. For a sample, the same calculation is performed, but the result is multiplied by $n/(n-1)$, where n is the sample size.

Suggested Reading

Berry and Linoff, *Data Mining Techniques.*

Dunham, *Data Mining.*

Shmeuli, Patel, and Bruce, *Data Mining for Business Intelligence.*

Questions and Comments

1. The spam filter in the video lecture had 57 input variables to determine the output variable. Most of these record either the proportion of certain words in the email ("money" = 0.01 would mean that "money" made up 1% of the words) or the proportion of certain characters in the email (such as ! = 0.01 meaning that exclamation points made up 1% of the characters in the email). Notably different is the variable TOTCAPS, the total number of capital letters in the email. When PCA is done with the variables not being normalized, the first principal component is dominated by TOTCAPS and captures 92.7% of the variance. When the variables are normalized, the first component captures only 11.6% of the variation and is not dominated by anything. Can you explain why this kind of result is to be expected?

   Answer:

   The value of the TOTCAPS variable, when unnormalized, shows much more spread than the percentage variables, which run only between 0 and 1. As a result, the vast majority of the variation is in the TOTCAPS variable. When variables are normalized, each variable is on equal footing, with a variance of 1. Now the variation in each variable contributes equally to total variation.

2. A second use of the classification tree technique that we developed is to help with the reduction-of-variables problem. The variables appearing early in our classification tree will probably be important in further analysis of the problem, while those that appear quite late might be safely ignored.

Data Mining—Exploration and Prediction
Lecture 5—Transcript

Statistical techniques, like regression and time series forecasting, were the strongholds of mathematical prediction for much of the 20th century. But they have been supplemented in recent decades by additional techniques from the exciting and fast-growing area known as data mining.

Data mining focuses on large data sets, and its job is to find what useful patterns and connections they contain, connections and patterns that might otherwise be missed. Some of the techniques used in data mining have been around for many years, including regression and time series analysis, but others are quite new. But data mining is especially important in the 21st century because of two advances, computational power, and a veritable explosion of the quantity of collected data.

When I was a kid, stores had to close from time to time to take inventory, to physically examine the goods on their shelves and in their warehouses so that they knew what they had. Those days are largely gone. Now, when any item is purchased, a bar-code scanner records its sale, and all of that purchase information is stored in a database, along with a lot of other information, like what else you bought, and what it cost, and whether you used coupons, and, assuming that you have a discount card, what other items you've bought in the past.

It adds up to a lot of data. According to *The Economist*, Walmart adds about a million transactions every hour to its 2.5 petabyte database. A petabyte is a million gigabytes, or a thousand million megabytes. To put such a large number into context, that's about 167 times as big as all the books in the Library of Congress put together. And that's just one company. And retailers like Walmart aren't the only ones. In today's world, an amazing amount of information is collected every day by many types of organizations, from satellite photos, internet traffic, GPS devices, telephone conversations, medical tests, and on and on. The trouble, of course, is how to make sense of it all. And that's where data mining comes in.

One of the key differences between a lot of classical statistical techniques and data mining is the quantity of the data available. Throughout much of history, data was scarce and hard to come by. And that meant that all of the data that was available had to be used in the analysis, and used in a double duty, both to create the model and to test its accuracy. In order for that analysis to get very far, some assumptions had to be made.

For example, the confidence intervals that we've built with our regression analysis assumed that our errors were normally distributed. That means that if, for example, you are applying linear regression to predict a man's weight based on his height, that the there's a straight-line relationship between height and average weight for that height, but that if you measured the weight of a lot of men of the same height, their variation from that average height, the ups and down random bumps that we talked about, would be distributed as a bell-shaped curve—a normal distribution. And the spread of that curve isn't supposed to depend upon what height you focus on, in technical terms, the data is supposed to be homoscedastic. The work also assumes that one observation being above or below the line by a certain amount doesn't imply anything about whether another observation will be above or below the line. That is, the errors are assumed to be independent from one another.

Some divergence from these assumptions don't cause too much trouble, but if they're seriously violated, you can't trust your confidence intervals, which means you don't know how much you can trust your predictions. In data mining, we often have an embarrassment of riches when it comes to data; if anything, we have too much. The first thing that means is that we can use part of the data to come up with our model, and a completely different set of data to test the reliability of that model. This also means we aren't bound by a lot of assumptions, such as the distribution of errors.

In other techniques, such as regression, we can actually spend a lot of time transforming the data so that we don't have heteroscedasticity, and we don't have autocorrelation, and don't have error terms that are distributed, for example, like lottery payoffs, lots and lots of small negative values sprinkled with occasional big positive values. Distributions like that are said to be highly skewed, values on one side of the average clumped together, while on the other side, they're spread out—a long way from the assumption of normally distributed errors. Fixing these problems can be quite a headache. With data mining, the headache largely goes away.

Practitioners sometime debate exactly what the margins of data mining are, but there are three objectives central to it—the first, data visualization and exploration; the second, classification and prediction; and the third, association among cases. Time series analysis, which we've already discussed, is sometimes considered a part of data mining, too. Let's start with visualization.

When we're exploring data in data mining, we're probably going to start by looking at one variable at a time. If a variable is categorical, bar charts are a good choice. We can see at a glance, here, that heart disease and cancer, malignant neoplasms, are overwhelmingly the most common causes of death in the U.S., and seeing that in a chart conveys this information much more quickly and memorably than does just a table of numbers. And increasingly, software supports the ability to drill down, to look at interesting aspects of the data in greater detail. Like this. From our last graph, we can break things down by age. In my age group, the greatest threats to health may be heart disease and the big C, but we see from the green bars that, for teens and young adults, the list is quite different, unintentional injury, suicide, and homicide.

Excel's pivot tables and pivot chart can give you some of this functionality in this area. For a single numeric variable, the most common choice is the histogram, which is like a bar chart in that each bar represents a range of values. Here you can see a histogram of the distribution for how many Facebook friends the students in my university classes have. The relationships between two numeric variables, as we've seen, we generally use scatter plots. This initial exploration often uncovers connections between the variables, connections that we can exploit in a further analysis.

We can push it one more dimension to look for relationships among three variables in a three-dimensional scatter plot. That's less useful when the software renders such an image as a static plot in two-dimensional screens, like on a screen or piece of paper. But there are still options. We can code additional information with color or size. A lot of weather maps do this, where temperature at each point in the country is given a color corresponding to it; the hotter the color, the hotter the location. Such representations are often called heat maps, even when the color coding is something quite different from temperature, like ozone concentration, or when the thing being colored isn't a map at all. For example, there are heat maps of the stock market, arranged by market sectors, sized by company value, and colored by change in stock price.

We can push the envelope still farther by using the time dimension, combining multiple images into a movie. Swedish academician Hans Rosling created an astounding video that you can find on the BBC Four website, where he tracks the wealth and longevity of people in 200 countries over 200 years of history. He uses a scatter plot with an axis for lifespan, a logarithmic axis for wealth, color to indicate continent, size to indicate population, and time to indicate the passage of time. The film lasts only four minutes and is stunningly effective at communicating all of that data.

Processing visual information in two or three dimensions, recognizing structures and patterns there, is something that humans are uncannily good at, which is why charts and graphs exist to begin with. It's also why, whenever possible, I'm going to use pictures to get across new ideas in this course.

But, as the number of variables increases, obvious ways of representing data graphically fail. Each variable, each category of information, needs its own dimension of space, and for visualization, three dimensions of physical space is our limit. So what can we do? Well, for starters, we can conduct a kind of dimensional reduction by using mathematical procedures. It's like using a student's GPA or SAT score in place of every detail of his or her academic history. The trick is to find a way to summarize multiple pieces of data in a way that loses as little information as possible. The computations required may not be simple, but since data mining by definition deals with large data sets, no one would perform the calculations by hand anyway. Statistical software, such as SAS or SPSS, or the free package known as R, are perfectly capable of this kind of dimensional reduction; in fact, it can even be done in Excel with an add-in package, such as XLMiner.

Because the tools always do it for you, I don't want to focus on the computational details of the techniques we'll discuss. Instead, my intent is to give you an understanding of when the techniques are useful and some intuitions about how they work, and, some examples of the tools in action. So, let's take a look at an example.

Suppose I'm engaged in a project to predict whether a person owns a computer tablet. I have a large dataset with information about thousands of people, with dozens of pieces of information about each person. Suppose that among them are the person's age and the number of Facebook friends that they have. For demonstration purposes, let me focus on only these two variables, and to only 100 people. The restrictions won't alter the idea of what's going on.

Since the data set is fairly small, I'd prefer to display it as a dot plot than a histogram, so that I can see every person and yet still perceive the pattern. As you can see, there's considerable variation in the ages in our data, and also in their number of Facebook friends. But if we make a scatterplot of these two variables, we see that they actually have quite a strong correlation with one another, with an r^2 of about 0.76.

So while either of these variables might be quite important in predicting whether someone owns a computer tablet, a lot of their variation is actually co-variation. Young users tend to have more Facebook friends than older ones. That means that useful information contained in the one variable is, to a significant degree, repeated in the other. It's for situations like this that a technique called principal components analysis, or PCA, exists. The idea is to create new variables that partition the variation in a better way, with the first variable capturing as much of that variation as possible.

I've chosen this simple, two-variable example because it allows your natural visual processing to be brought to bear. Looking at the picture, it's clear that most of the variation is occurring along a diagonal line from upper left to lower right. How do we measure variation? By using the most common statistical measure of dispersion, the variance. The calculation of the variance bears a great deal of similarity to our calculation of MSE, remember that average the squared errors calculation from regression? For

106

variance, the only difference is that the error is taken to be the distance of each data point from the mean of the data. If you take the square root of the variance, you get the standard deviation, something that comes up in most statistical discussions.

So, PCA begins by finding the line along which that variation, that variance, is maximized. If you're a glutton for punishment, and are good with matrix algebra, you could do this yourself by finding the eigenvector of the covariance matrix of the data that has the largest eigenvalue. Yeah. Personally, I'd let the computer do the math, which gives us this line, which would normally be called the z_1 axis when we're doing PCA.

Impressively, this line captures over 97% of the total variance in the two variables. That's great. In an important sense, we're already done our work. But if you want to capture that remaining 2.6%, visible here as the spread of the points around the z_1 axis, PCA would construct a second axis, perpendicular to the first, called the z_2 axis. I've shown it here in green. Converting data from the original age and Facebook friends values to the z_1 and z_2 values is just basic algebra. The z_1 value for a point can be computed as a specific linear combination of age and Facebook friends, and the same is true for z_2. And part of the PCA analysis is finding and reporting these combinations.

For our example, to find z_1, you take 94.8% of a person's Facebook friends value and subtract 31.8% of his or her age, and then subtract 66.1. For z_2, it's 94.8% of the person's age, plus 31.8% of his or her Facebook friends count, then subtract 60.4. By the way, you may have noticed the similarity of the coefficients in the two expressions. It isn't coincidence, but it's a consequence of the two axes being perpendicular, so their slopes are negative reciprocals. If we were combining more than two original variables, we generally wouldn't see such an obvious pattern.

OK, let's try this out. For example, the red point in our scatterplot represents a person who's 70 years old and has 51 Facebook friends. We'll use our two magic equations to find z_1 and z_2 for this person. Like this: age and Facebook friends in, z_1 and z_2 out. You can see what the new coordinates mean, as is shown on the graph; 40 units down on the z_1 axis, and 22 units over on the z_2. Well, swell.

But what did we really gain here? We replaced two variables that we understood with two new variables that are hybrids of the original two, and who knows, intuitively, what $z_1 = -40$ even means? This is progress? Well, if we keep both variables, no. But given that over 97% of the variation in the two original variables is captured by the one new variable, z_1, it becomes a possible approximation just to jettison z_2 and use z_1 in place of the original age/Facebook friends pair.

And remember, it's in more complicated examples, with more than a handful of variables, that this becomes particularly useful. In those cases, PCA finds the direction in multidimensional space that captures the largest possible fraction of the variation in the data and calls that the z_1 axis, then looks for a direction perpendicular to that one that captures the maximal amount of the remaining variation. That's the z_2 axis, and with more than two dimensions in the original problem, there are more perpendicular directions to choose among. PCA continues like this. Next it finds a direction for a z_3 that's perpendicular to both z_1 and z_2 and captures as much of the remaining variation as possible, and so on.

We often find that after only a handful of principal components are found, the vast majority of variation is accounted for. If that's the case, the original variables, however many there were, can be discarded and replaced with this handful, losing almost nothing about the original variation in the data and greatly simplifying future work.

But there are some disadvantages too. Principal components analysis won't work well for data whose underlying pattern is not linear. But that can be overcome by modeling a nonlinear pattern using local patterns that are linear, analogous to what we learned in multiple regression. A more fundamental disadvantage is difficulty in interpretation. We understand what an age of 70 is. But what is a z_1 score of -40? Often we can make some progress on this by looking at the size and sign of the coefficients in the equations for the various z_i.

For example, suppose we are looking at airline passengers and we find that z_1 has large positive coefficient for on time and short wait on check-in and a large negative weight on baggage lost and bumped. These are all service/convenience/time issues; z_2 may have large, positive weight on free upgrades and complimentary drink and a large negative weight on ticket price. Then z_2 could be summarized in terms of cost/value issues.

One other problem with PCA, at least as I've presented it, has to deal with the units of measurement of my variables. In my age/Facebook friends example, if I had measured people's ages in months, rather than years, I would have gotten very different principal components. Now all of the age numbers are 12 times as large as they were, and so the variation in ages has grown correspondingly. The z_1 component is now much more concerned with capturing that increased variation. Whether this is a good thing or not depends on the particular problem, and brings up a question that needs to be considered before applying most data mining techniques. To normalize or not to normalize?

Normalizing data means standardizing it. We do this by first subtracting the mean of a variable from all values of that variable, and then dividing each of those differences by the standard deviation of that variable. No matter what data you start with, after you've normalized it, the mean of the data is zero, and the standard deviation is one. In that sense, then, each variable is now on equal footing. A person of age 51 with 120 Facebook friends is equally remarkable for being one standard deviation above the average age of 36.3 and being one standard deviation above the average number of Facebook friends, 81.9. He or she would have a normalized score of 1 for each of the two variables.

Generally, if the two variables are measured in units that aren't comparable or on dramatically different scales, normalization is a good idea. Take hourly wage and outstanding mortgage value. Both are in dollars, yes, but a $10 change in the wage is huge, while a $10 change in mortgage is insignificant. Normalize the data. With age and Facebook friends, I don't even have the same units, so normalization is probably what should have been appropriate for my example. I didn't normalize because, with only two variables, the resulting PCA is always rather boring. The two principal components are essentially just the sum and the difference of the two normalized variables.

There's another, very human factor that can be useful in reducing the dimension of the data, domain knowledge. Someone familiar with the field of investigation can often narrow our focus before we even start with the math. They can tell us what quantities are almost certainly relevant, or irrelevant, to the task at hand. They can also let us know when one of our variables is useless as a predictor, because we won't be able to gather it in time to make use of it. Power outages may be a wonderful predictor of home generator sales, but by the time that you know about the outage, it's too late to stock more generators. Domain knowledge can also allow us to inspect the final answer and see if it makes sense.

Okay, so far we've discussed visualization, exploration, and dimensional reduction of data. But in a way, that just gets us ready for the real work. Perhaps the most common kind of task for data mining is using data about an individual to predict some unknown characteristics about them. An individual here could be a person, or a river, or a bottle of wine, or anything else. The thing I'm trying to predict might be a categorical variable, such as brand of car a person owns, or whether they would accept a particular offer to refinance their home, or whether they committed fraud on their federal income tax. When that's the case, the prediction is often called classification, sorting individuals into the classes where we think they belong. On the other hand, we may be trying to predict a continuous variable, like a person's life expectancy, or the probability that a firm will fail within the next five years. Some people use the term prediction only for continuous variables, while some use prediction for sorting individuals into classes as well.

So, let's start with a classification example, and one of my favorite techniques from data mining, classification trees. If you're working with continuous variables, rather than classes, they're usually called regression trees, but the idea is the same. I like the technique because it's a great off-the-shelf approach. It can be used on a lot of different kinds of problems, it's easy to explain to non-technical people, the results are directly usable by the non-expert, and it doesn't care if the data is normalized or not. If you want to use categorical variables that have more than two categories, like, kind of pet, you have to create dummy variables to replace it—dog, yes or no; cat, yes or no; hamster, yes or no; and so on. But that's not really a problem.

For my example, I'm going to use a data set obtained from the UCI Machine Learning Repository at the University of California at Irvine website. It lists 58 characteristics of over 4600 email messages. Included in the list of characteristics is whether or not the email was spam. We're going to use classification trees to see how well we can do in building our own little spam filter. To get an idea of how classification trees work, I'm going to focus our attention on only two of the variables in the problem and look at only a small number of the 4600 cases. My reason is the same as with PCA, to let you use your visual intuition to understand the process. Mathematically, the procedure in higher dimensions and larger sets is a perfect analog of what we're going to be doing here.

So, the variables I'm going to focus on are x, the percentage of characters in the email that are exclamation points, and caps, the average length of a string of capital letters in the email. Note how domain knowledge is already at work here. I've all seen enough emails to suspect that these might be useful flags for spam. Let's start with a scatter plot; exclamation point percentage on the horizontal axis, and average cap lengths on the vertical. The red dots are spam; the blues are non-spam.

You undoubtedly see some patterns. Spam tends to be to the right and top of this picture, non-spam tend to be on the bottom and left. But for classification trees, we're going to pay attention to only one variable at a time. So let's start by focusing on x, the percent of exclamation points in the email. I want to draw a vertical line that, so much as possible, separates the spam and the non-spam. To my eye, it belongs around here.

So, how did I do? Well, on the left side of the line, there are 24 points, and 23 of them are real emails. So by classifying those points as emails, I'd have a 23 out of 24, or 96%, success rating. On the other side, though, I don't do as well. There are 26 points on the right of my line, and only 19 of them are spam. So by classifying the right side of my graph as spam, my success rate is only 73%, overall success rate, 84%.

You might think that overall success rate is the right way to proceed here, but not so fast. There are actually several problems with that. First off, how about if what you're looking for is either very common or very rare, like say, an unusual cancer. Since so few people have it, there would be very few red dots in my scatterplot. The lowest overall error rate might just be to classify everyone as not having cancer. If only 1 person in 1000 has the disease, my error rate for doing so is 0.1%. Of course, if what I'm trying to do is to detect cancer for treatment, this is worthless.

Instead of overall success rate, we often talk about sensitivity and specificity. If what we're looking for is people with the cancer, the sensitivity is the probability that we correctly identify someone with cancer as having it; 100% sensitivity would mean that we found everybody who has cancer, though maybe we worried a bunch of people with false scares, too. The specificity is the probability that we correctly identify people free of cancer so that we don't give a healthy person cancer therapy. So if we classed everyone as being free of cancer, we'd have 100% specificity, we got all of the healthy people right, but 0% sensitivity, we got all of the cancer victims wrong, hardly a satisfactory screening procedure. The best classification procedure usually involves a tradeoff between sensitivity and specificity.

The cancer example brings up a second point, too. It's not just the fraction of the population having the characteristic that you're looking for, it's also how important it is that you find them. Not finding a person who has cancer might be a death sentence for that person. On the other hand, identifying a person who's free, as possibly having it, will lead to more tests, not pleasant, but much less severe a consequence than death. So when deciding our best classification rule, we're going to need to take into account the relative costs of making mistakes in different categories. We'll come back to this idea later. For the moment, let's just treat spam and not-spam as being equally important, so a screw-up in either direction is equally bad.

We've just divided the original rectangle into two by drawing a line. The idea behind classification trees is to keep doing this in a way so that each line drawn results in new rectangles that maximally increase the purity of the partitioning that we've done so far. There are a couple of common measures for this purity, one called the entropy of the rectangle, and one called the Gini index. The Gini index is a little easier to understand, so that's what I'll use.

To find the Gini index of a rectangle, take the fraction of points in that rectangle that are classed as each of the possible output classes, square these proportions, add them up, and subtract the total from one. Let's see how that works. In our original picture, before we drew our line, there were 20 spam points, 40%, and 30 non-spam points, 60% of the points. To find the Gini index, take the 40% and 60%. Square them to get 0.16 and 0.36. Add those up to get 0.52. And the last step: Subtract it from 1 to get 0.48. So the original Gini index is 0.48. Smaller is better. A perfectly pure rectangle with only one color dot would have a Gini index of 0.

OK, so after I drew my vertical line, what's the Gini index of each of the two subrectangles I made? The one on the left is 96% blue and 4% red, for a Gini index of $1 - 0.96^2 - 0.04^2 = 0.08$, nice. The right-hand rectangle is 73% red and 27% blue, for a Gini index that works out to 0.39. The rule for combining these is to weight them by the number of observations in each rectangle; 48% were in the left rectangle, and 52% were in the right rectangle, so the Gini index for the two together is $0.48 (0.08) + 0.52 (0.39) = 0.24$.

Cutting to the chase, by drawing a vertical line where we did, we dropped the Gini index from 0.48 to 0.24, a reduction of 0.24. It turns out that no different horizontal or vertical line would have reduced the Gini index by more, and it's for that reason that the classification tree's algorithm would draw the first line that I did, by eye.

The algorithm just keeps on repeating this process, checking each possible horizontal or vertical line in each new rectangle, and choosing the dissection that causes the greatest drop in the Gini index. The best split for the left-hand rectangle would be to draw a horizontal line splitting the two top points off from the rest, but this line drops the overall Gini index by a paltry 0.008. How about the right rectangle? Well, it's not hard to believe that for it, the best choice is a horizontal line a bit above 3, like this.

The upper rectangle has 10 points, all red, for a Gini index of 0. The lower rectangle is a fairly mixed bag, but when you find its Gini index and compute the weighted average for the pair, it turns out that splitting the right-hand rectangle this way drops the Gini index by 0.047, much bigger than the 0.008 we saw for the left-hand rectangle. So this horizontal line is our next dissection. Next comes another vertical line for the lower rectangle, and we can keep doing this in this way until we have classified everything, like this.

We've done this work with only two variables, but the procedure is identical in higher dimensions. Instead of two dimensional rectangles, we get n-dimensional ones, but you don't need to be able to see them to apply the procedure. At each step, you're looking for a cutoff value of one variable, applied to one rectangle, that cuts that rectangle in two. For each possible division, compute the Gini index of the partition, and choose the one that gives the smallest Gini index. You're going to find the rectangle, the variable, the division point that drops the Gini index as much as possible, and then, repeat the whole process again. When you're done, every box will have dots of only one color in it. Note, too, that there's nothing in the procedure that says you have to classify into only two classes, like spam and non-spam. We could have dots of more than two colors and classify into more than two categories, and the technique still works fine.

All of these lines that we drew can easily be turned into a series of binary decisions, and those decisions can be arranged in a tree that captures the decision rules we've created, a classification tree. For the work we just finished, here's the tree. Let's zoom in for a closer look and see how we'd use it to classify an email that was 0.07% exclamation points, and in which the average length of a string of caps was four. Here's the first question; it's about x, the percentage of exclamation points. If it's less than 0.0865, go left. More than 0.0865? Go right. At 0.07% exclamation points, I'm left of the cutoff, so I go left. The next node is about caps. If the average caps length of a string is less than 3.6, go left; otherwise go right. Mine was four, so I go right. And we just keep going. The next node is also about caps, and since 4 is less than the 5.44, the threshold, so I go left. My message is classified as spam.

This is all very nice, but it's not enough. What we just built was a classifier that's perfectly accurate in separating spam from non-spam, but only for the set of 50 emails that we started with. Look at the last node that we worked with and you'll see that something looks fishy. It says that if my emails had contained longer strings of capitals, it would have been classed as good! Our tree says this because our original data included one good email with a string of 7 caps and no exclamation points, and our tree generalized from that.

That rule happened to work for our limited emails that we used to create the tree, but we want to build a classifier that will do well with new data, not just be flawless on the data that it's already seen. The answer is going to be more data, not just more training data that we can use in creating the tree, but more new data that we can use to assess how well our classifier will do in the field. This approach applies not only to classification trees, but to most data mining applications. We'll be looking into this in our next lecture, along with the other important task of data mining, association. For example, how does Amazon decide what offers to make you? How does Pandora Internet radio create custom stations to reflect each listener's musical tastes? Data mining has raised the business of anticipating customer preferences to a high art. Next time, we'll look into how it's done. See you then.

Data Mining for Affinity and Clustering
Lecture 6

W   ith the mathematics of algorithms and the power of computers,
         data mining confers upon us an ability that could seem almost
         magical: the ability to find patterns in a wilderness of data that
by all rights should defy our conception. Specifically, data mining has raised
the business of anticipating customer preferences to a high art. For example,
how does Amazon decide what offers to make to you? In this lecture, you
will learn about some important tasks of data mining, including association
and clustering.

Training Data
- Data mining involves using mathematical algorithms to extend our native ideas of pattern recognition. We can then use computers to implement those algorithms on large, multidimensional sets of data, letting us find relationships and patterns that the data contains.

- With classification and prediction, we first show the algorithm a large set of training data. This data contains both the input and output values for a large collection of individuals. It's like showing the algorithm a bunch of questions, along with their answers. The algorithm then generates a way of predicting the output value of new individuals, based on their input values.

- Any such activity can be called prediction, although if the output value is a category—a class—it's usually called classification. All such procedures are called supervised learning, which merely means that you have to supply both the input and output values for each individual in the training data.

- When we get our data—and in data mining, we often have a lot of it—we don't use all of it for training. Instead, we partition it randomly into two or three groups. The first (and almost always the largest) is used for training. But the second set is called the **validation set**, and it's not used until after the data mining algorithm has been applied to the training data to generate its rules.

- The performance of the model on the validation data is a fair test of the model's quality, because that data had nothing to do with the model's creation. This gives us a way to benchmark the results of different data mining algorithms and to evaluate which one looks best—the one that does best on the validation data.

- However, selecting the model this way still doesn't tell you how good you can expect it to be. The model you pick is the one that did best on the validation set, but would it do as well on other, new data? You picked the winning model because it was the best on the validation set, but maybe it just got lucky with that particular set.

- That's where the third set comes in: the **testing set**. If we have used the validation data to pick a best model, we've more or less poisoned its usefulness for assessing how the model does with new data. So, the final step is to evaluate how well the selected model does on brand new data—the testing data. Given a decently sized testing set, we should be able to use our model's performance on it as a fairly reliable measure of its overall performance.

- Applying this to classification trees, we begin by following the procedure that continues by adding one node after another to the classification tree—in other words, by subdividing rectangles again and again—until every point is correctly classified. Eventually, we reach the point where the original algorithm is no longer training to the real relationships between inputs and outputs; instead, it's training to fit the noise in the **training set**.

- One way to handle this issue is called **oversampling**, and it's used in particular when one of the classes of interest is especially uncommon in the data. The training set might include so few examples of the rare class of interest that a straightforward application of data mining techniques would fail. The "best" rule for the technique might be to classify everyone as belonging to the largest class and no one to the class of interest.

- Oversampling includes a number of observations in the rare class disproportionate to its frequency in the **population**. You can do this by either taking a random sample (with replacement, if necessary) of individuals in the rare category and adding them to the training data or by duplicating each rare class entry in the training set a fixed number of times.

- In prediction problems, the quantity to be predicted is a continuous variable, rather than a category, like in classification problems. We use the same kind of techniques, but we use different measures of error. Most software that can handle the one kind of problem can handle the other as well.

Association Rules

- There are other interesting pattern recognition problems besides classification and prediction. Data mining also deals with problems of **affinity analysis**, or "what goes with what." The recommender systems at online stores are an example. They use **association rules**. The system recognizes that you were looking at A and B, and then points out to you that people who liked them tend to be interested in C and D, too.

- In retail, association rules are sometimes called market basket analysis. If you know that people who buy these two products also tend to buy that one, you can use that information—for product placement, for promotional sales, and so on.

- The strategy behind a basic association rule generator is actually pretty simple. In terms of retail sales, you have, for each customer in your database, a list of all of the items that the customer bought from you. You're going to be looking for rules of the following form: "If a customer buys all of these, then I guess that he or she wants these, too." The game is probabilistic in nature, of course. You won't be right every time, but you want your rules to give you useful guesses as to what might interest your customer.

- What do we need for a useful rule? First, there's no point in considering an item—or group of items—that appear very rarely, at least not unless the combination is very important. So, you want to consider the support of a collection of items, which is merely the number of your customers—or fraction of your customers—who bought everything in the collection. If you try doing this in every possible way for a store with more than a small number of items, you're going to run into trouble. A common work-around is to use the Apriori method.

- Association rules software will scan the data to create frequent sets of items and then generate association rules from them and compute the **confidence** and lift index for each. Those with high values are good candidates for rules that the company may want to note and use. There is definitely a place for human beings in this process; there are some rules that can be either discarded or merged with others.

Clustering
- Association rules are looking at information in a very specific way: that one set of attributes gives you reason to anticipate another set of attributes as well. But other techniques for looking for "things that go together" are based on the idea of clustering the data—of creating categories and then sensibly sorting the individual data points into them. Sometimes this is a first step to further analysis that will be based on these clusters.

- The following two questions, of course, arise: what the categories should be and how we decide into which category an individual should be sorted. And the key idea to answering both of these questions is the same: distance.

- The most straightforward way to define the distance between two individuals is Euclidean distance, the straight-line distance separating the points in an n-dimensional scatterplot. That's difficult to visualize in more than three dimensions, but it's no more difficult to compute with a souped-up version of the Pythagorean theorem from geometry.

- To find the distance between two people, given that we have information about six characteristics, A through F, we find out how much they differ on each of the six traits, square those differences, add them up, and take the square root.

- Although this is the most obvious way to define the distance between two observations, there are some problems with it. There are other ways to define distance, such as the **statistical distance**, which both standardizes and accounts for **correlation**.

- Before we cluster data points together, we have to answer a question: What do we mean by the distance between two clusters? The definition we choose depends on what we know about the data.

- Suppose that we're trying to track pollution levels. If a river is contaminated, we would expect to find pollution close to the river. You don't have to be close to all points on the river—just to some point on the river. Your distance from the river, then, would be measured by the distance to the closest point on the river. That's one definition for the distance between two clusters. It's called the minimum distance, or the **single linkage distance**. If you use it, you can get clusters that tend to be ropey.

- A second possibility goes to the opposite extreme. You could define the distance between two clusters to be the largest distance between any two points in the clusters. So, the distance from California to Washington is the distance from the southern tip of California to the Canadian border in Washington. This method usually gives small clusters to begin with, and clusters tend to be spherical.

- There are other methods, too. We can define the distance between two clusters as the **average distance**, computed for each pair of points with one in each cluster. With large clusters, that's a lot of calculation. Or we can find the middle of each cluster, its so-called **centroid**, by finding the average value of each variable for points in that cluster and then define the distance as the distance between the centroids. Other, more complicated methods seek to minimize the amount of information lost by creating new clusters from old ones.

- How many clusters do you want? There is no universal answer to this. Sometimes your domain knowledge will inform you of how many clusters should be in the data, but more often it comes down to examining the groupings resulting from different levels of clustering. Ideally, the clusters pass three checks. First, they are interpretable in some sensible way. Second, the clusters should be stable to minor changes in the data used to create them. Finally, you'd like the variation between clusters to be significantly greater than the variation within a cluster.

Important Terms

affinity analysis: Umbrella term for the data mining techniques that seek to determine "what goes with what."

Apriori method: A method for generating frequently occurring subsets of items for an association rules application.

association rules: A data mining task of creating rules of the following form: "If an individual has *this* collection of traits, predict *that* individual also has that set of traits."

centroid: In data mining, the centroid of a cluster is the point for which each variable is the average of that variable among points in that cluster.

confidence (data mining): For an association rule, the probability that the consequent is true given that the antecedent is true.

distance, average: In cluster analysis, the definition of the distance between two clusters as being the average of the distances between each point in one cluster and each point in the other.

distance, single linkage: In cluster analysis, the definition of the distance between two clusters as being the minimum of the distance between a point in one cluster and a point in the other.

distance, statistical: Also called Mahalanobis distance. A distance measure that takes into account the variance and covariance of the various variables used to define it.

oversampling: In data mining, the choice of representing a class in a data set at a relative frequency higher than its occurrence in the population. Oversampling is appropriate if the class of interest forms only a small proportion of the entire population.

population: The set of all individuals of interest for purposes of the current study, or the set of all relevant variable values associated with that group.

training, validation, and testing sets: In data mining, where data is plentiful, model quality is usually assessed by using part of the data as a training set, to create one or more models, and then assessing their quality with the validation set. If more than one model is being assessed, the best is chosen based on validation results, and the final choice is then evaluated on a third set of data, the testing set.

Suggested Reading

Berry and Linoff, *Data Mining Techniques*.

Dunham, *Data Mining*.

Shmeuli, Patel, and Bruce, *Data Mining for Business Intelligence*.

1. Why is an average distance measure more appropriate than a single linkage measure for the grade example in the video lecture?

Answer:

Because being part of a student group of a given performance profile (a cluster) should mean that the student is similar to what is typical for that group, not merely similar to one other student in that group.

2. The grade example was done without normalizing the data. Do you think that normalizing the data would have been a good idea? Do you think that it would have made a big difference in the resulting clusters?

Answer:

Both variables are on a scale of 0 to 100, and we have no idea of their relative importance in this problem. Quiz averages appear to be slightly more dispersed than test averages, so normalizing would be equivalent to squeezing the picture slightly from top to bottom. It does not appear that this would significantly affect the clusters.

Data Mining for Affinity and Clustering
Lecture 6—Transcript

The last lecture was an introduction to data mining. We looked at how to use mathematical algorithms to extend our native ideas of pattern recognition. We can then use computers to implement those algorithms on large, multidimensional sets of data, letting us find relationships and patterns that the data contains.

So far we've focused on classification and prediction. In a nutshell, these algorithms proceed like this. First I show the algorithm a large set of training data. This data contains both input and output values for a large collection of individuals. It's like showing the algorithm a bunch of questions, along with their answers. The algorithm then generates a way of predicting the output to individuals, based on their input values. Any such activity can be called prediction, although if the output [value is] a category, it's usually called classification. All of these procedures are called supervised learning, which merely means that you have to supply both the input and the output values for each individual on the training data.

The techniques that we explored last time included the construction of a classification tree, which we could picture as repeatedly subdividing rectangles into smaller and smaller rectangles until all of the given points in any given rectangle fall into the same category of output. When we applied this to our 50-point dataset on spam emails, we ended up with this picture. Given the picture, the rule that it implies for new data is the obvious one. Any new data point falls into one of these rectangles. Assign the new point to be in the same class as the others in that rectangle.

Like this. If you're in a blue-dot square in which all the emails are good, your email is classed as good. If you're in a red-dot square in which all of the emails are spam, your email is classed as spam.

Classification trees aren't the only way of making such an assignment, of course. For example, there's the nearest neighbor classifier, which is pretty much what it sounds like. First, normalize all of the variables so that they have the same mean and standard deviation; this puts them on the same even ground. Next, given a new data point, find its nearest neighbor in the training set. Assign the new point to the category of that nearest neighbor.

For our spam training data, the result would look like this. If your email ends up in a blue region, it's classed as good. In a red region, it's classed as spam. The picture is similar to the classification tree of a moment ago, but far from identical. This simple nearest-neighbor rule can do a surprisingly good job when given lots of data, as can a slightly more complicated version of it, called k-nearest neighbors. If we decided to use a five-nearest neighbors rule, for example, then, given a new point, we'd find the five nearest points to it in the training data. We'd then check the output categories for those five neighbors and let majority rule in assigning a category to the new point.

There are other classification algorithms as well, such as discriminant analysis. In our graphical interpretation, discriminant analysis finds a flat surface, in our two-dimensional example, a line, that tries to separate the one category from the other. For our data, it would look like this. Each has its own advantages and disadvantages. A quick glance at our three examples, though, might make you think that discriminant analysis has to be the loser of this competition. Yes, a single line gives a simple rule, but discriminant analysis misclassified some points. The other two methods got everything right.

And that point actually deserves quite a bit more attention, because the regions that we see in these three examples were created by using the training data. The first two techniques kept grinding away until they had gotten rid of all of the misclassifications. For those techniques, every region in the picture consists of dots of a single color.

Well, what's wrong with that? Well, it would be pretty naïve to think that these super classifiers are going to do as well on new data as they did on the training data. For example, let's take a look at the nearest neighbor picture again. See that large blue patch on the right-hand side? If an email falls anywhere in that range, it's classified as being good, non-spam. And that's all because of the single blue dot on the right-hand side of the field. Obviously, it's possible for a point in that area to represent a good email; the blue dot that's shown is just such a point. But is it sensible to expect that that blue patch is really dominated by good emails? One cause of the problem, an obvious one, is that I restricted us to only 50 emails for the training data. Nearest-neighbor approaches, in particular, like lots of data, and this chart shows you why. But it's not the only problem.

Look at the border between red and blue in the lower left-hand corner. Do you really think that the best demarcation of spam from non-spam follows such a tortured contour? No. The problem is one that we discussed before when we used polynomials to fit a set of data points in regression, the problem of over-fitting the data.

You can think of it this way; some of the observed relations between inputs and outputs in the training data are due to genuine relations between the two. But there are random fluctuations in observations, too, noise, if you like. And if we go too far in our classification algorithm, we can end up fitting the noise as well as the true relation. And that's bad, because noise is random, and new observations are unlikely to follow the quirky pattern that let us correctly classify training points based on its noise.

So how do we avoid this? The key idea is validation. When we get our data, and remember, in data mining we often have a lot of it, we don't use all of it for training. Instead, we partition it randomly into two or three groups. The first and almost always the largest is used for training, in the way that we've already described. But the second set is called the validation set, and it's not used until after the data mining algorithm has been applied to the training data to generate the rules. The performance of the model on the validation data is a fair test of the model's quality, because that data had nothing to do with the model's creation. This gives us a way to benchmark the different data mining algorithms and to evaluate which one looks best, the one that does best on the validation data. Except, except that if you think about it, selecting the model this way still doesn't tell you how good you can expect it to be. Yes, the model that you pick is the one that did best on the validation set, but would it do as well on other, new data? Face it, you picked the winning model because it was the best on the validation set; maybe it just got lucky with that set.

And here's where the third set comes in, the testing set. If we've used the validation data to pick the best model, we've more or less poisoned it for usefulness in assessing how that model does on new data. So the final step is to evaluate how well the selected model does on brand-new data, the testing data. Given a decently sized testing set, we should be able to use our model's performance on it as a fairly reliable measure of its overall performance.

So, how does all of this apply to a technique like classification trees? Well, we begin by following the procedure that we discussed last time, a procedure that continues by adding one node after another to the classification tree, in other words, by subdividing rectangles again and again until every point is correctly classified. But, we keep track of the order in which we add the nodes to our tree. To see how this works, let's take the training wheels off of our spam filter. My actual data file had 57 input variables and over 4600 records. I used half of these for the training set, 30% more for validation, and the last 20% for screening.

The validation set was 41% spam, so if you know nothing about an email, classifying it as "good email" has a 41% error rate. But the first division ordered by the classification tree algorithm was to split email based on the density of exclamation points, dividing at a line of at about 0.03%. If we assign non-spam to emails below this threshold and spam to emails above the threshold, we misclassify 22% of the validation set. The next split in the tree lowers the Gini index but leaves the error rate at 22%; the one after that lowers it to 16%; and we keep going. There are some special rules that can alter the order in which we add nodes to our growing tree, but they're not important here. What is important is the result. Not surprisingly, as we add additional splits, we fine tune our algorithm and get less and errors on the validation set. Up to a point.

After 66 nodes, the misclassification rate on the validation set is down to 9.05%. But at that point, the error rate levels off, and adding more nodes leaves it at 9.05%. Then, adding the 70th node actually increases the misclassification rate on the validation set. Beyond that point, additional nodes just make it worse. We've gotten to the point where the original algorithm was no longer training to the real relationships between inputs and outputs; it was training to fit the noise that was in the training set.

Because we are deciding where to stop based on the validation set, we actually could be guilty of over fitting it; with 66 nodes, we may have gone farther than new data would warrant. To account for this, we usually prune the tree by more than the best fit to the validation data. The amount of pruning is computed based on a statistical formula for how much we can trust the validation set error rate, and for our data, this ends up pruning the tree quite extensively, leaving a 30-node tree that has a 9.56% error rate on the validation data.

It turns out that to use this pruned tree, you only have to know 10 things about your email, not 57. The character densities of exclamation point, hash tag, and dollar sign, the total number of capital letters, the longest string of caps, and average length of a caps string, and what fraction of the words in the email were "edu," "hp," "money," and your first name. You never have to answer more than eight questions about your email to get a classification. And I need to hardly point out that with eight questions, that could take you a bit of time, but it's an eye blink for a computer, as is collecting the data required by scanning your email to begin with. Predictably, the filter doesn't do as well on test data as it did on the validation data, but it's not bad. The overall error rate was just over 11%. Our filter messed up on 6.4% of the good emails and almost 19% of the spam. Filters built with more care could do better; of course, spam writers learn what the filters are looking for and the spam war continues.

But let's stick with our filter, which is messing up on three times as many spam emails as it does on good emails. That lopsided balance results in the minimum overall error rate of 11%, but is it necessarily what we want? Well, in general, no. We've been assuming, so far, that the errors of both kinds are equally costly. And yet that isn't necessarily true. Which would you consider a more serious issue, finding a spam email in your inbox, or never receiving a real email because it was rerouted into your junk folder? I'm betting that losing real mail is much worse. You'd be willing to see a bit more spam to significantly reduce the chance of losing a real message. Can data mining handle this preference?

It certainly can. When the pruned tree looks at the training data, it gets some things wrong. Suppose, for example, that after asking all of its questions, the pruned tree lumps your email in with 25 training emails, 25 emails for which the answers to the questions were the same as they were for your new email. I'll call that your email's bin. Maybe you check those 25 emails from the training set and find that 14 of the 25 were spam. Then the bin is 64% spam, so it's identified as a spam bin. Majority rules. So your new email is classed as spam, too. The algorithm could go on, in fact, to estimate that there's a 64% chance that your email is spam.

But, you can change the bin's assignment rule, in light of the fact that misclassifying a real email is more serious than misclassifying spam. Suppose, for example, that you decided that having a real email classed as spam was twice as serious as having a spam classed as real email. Then you could set a threshold for your bin to be classed as spam at $^2/_3$, not 50%. If a bin isn't at least $^2/_3$ spam on the training data, it's classed as good mail. So, by the new rule, your email would now be classed as good, since the bin was only 64% spam. Most classification and prediction techniques have an equivalent ability to set a slider for the threshold. This answers one of the questions we discussed last time, about screening for cancer. We want to set our threshold so that the bin that has any sizeable number of cancer victims in it is classified as potential cancer rather than as clear.

There's a second way to handle this issue, and it can be used alone or in conjunction with adjusting the classification threshold. It's called oversampling, and it's used, in particular, when the classes of interest are especially uncommon in the data, like the rare cancer example we were discussing, or for that matter, like picking terrorists out of a crowd of airline passengers. The training set might include so few examples of the rare cases of interest that a straightforward application of data mining techniques would fail. The best rule for the technique might be to classify everyone as belonging to the largest class and no one to the class of interest.

Oversampling does exactly what it sounds like it does; it includes a number of observations that are in the rare class disproportionate to its actual frequency in the population. You can do this either by taking random samples, with replacement, if necessary of individuals within the rare category and adding them to the training data, or by duplicating each rare entry in the training set a fixed number of times. In our cancer example, suppose missing a cancer victim is 20 times more costly than treating a healthy person as potentially having cancer. Then in the training set, each cancer victim could be replicated 20 times. The model could then be validated on a sample reflecting real-life proportions. Alternatively, it could be validated with an oversampled validation set, and then the results could be rebalanced to reflect the actual population proportions.

These examples have been focusing on classification problems, rather than prediction ones, mainly for clarity of presentation. The prediction problems are basically the same, but the quantity to be predicted is a continuous variable, like income, rather than a category, like spam. We use the same kind of techniques but different measures of error.

In classification trees, the purity of a rectangle depended on what fraction of its contents belonged to each output class. In the continuous case, the regression trees might be based on the variance of the output values in the rectangle. In classification trees, the class is assigned to the bin based on the vote of the training data in that bin. In regression trees, the prediction is the mean of the training data of the bin, or some similar measure of central tendency. Happily, most of the software that can handle one kind of problem can handle the other as well.

But there are other interesting pattern-recognition problems besides classification and prediction. Data mining also deals with problems of affinity analysis, or put simply, what goes with what. The recommender systems at online stores are an example. They use association rules. The system recognizes that you were looking at A and B, then points out that people who liked them tend to be interested in C and D, too. In retail, association rules are sometimes called market basket analysis. If you know that people who buy these two products also tend to buy that one, you can use that information for product placement, for promotional sales, and so on.

The strategy behind a basic association rule generator is actually pretty simple. Let's discuss it in terms of retail sales. You have, for each customer in your data base, a list of all of the items that they've bought from you. You're going to be looking for rules of the form, "If they buy all of these, then guess that they want this, too." The game is probabilistic in nature, of course. You won't be right every time. But you want your rules to give you useful guesses as to what might interest your customer.

Well, what do we need for a useful rule? First, there's no point in considering an item, or group of items, that appears very rarely, at least not unless the combination is very important, like one that leads to you selling a pro football team. So, you want to consider the support of a collection of items, which is merely the number of items, or fraction of your customers, who bought everything in that collection. Believe it or not, if you're doing this in every possible way for a store with more than a small number of items, you're going to run into trouble.

For example, if there are 100 items in your catalog, there are 2^{100-1} different possible collections of items that could be purchased. If you consider a trillion such possible orders every second, and you got an early start, namely, when the universe was created, you'd be a bit over $^1/_{10}$ of the way now through the job. A common work around is to use the Apriori algorithm. Decide what your support threshold is going to be, like 2% of your customers. We'll call any item or collection of items that were purchased in at least 2% of your transactions, frequent. So, look through your database and identify each frequent item. Now, any subset of two items that appears in at least 2% of your transactions will consist of two frequent items. Check each pair, and keep the frequent ones, the ones that appear in at least 2% of the orders. Now, a frequent triple of items will have a single item paired with a frequent pair. Generate all such triples and identify the frequent ones. You keep going until you can make no more frequent sets.

But knowing, for example, that A and B are frequently bought together isn't enough to make a good prediction rule. Imagine that 40% of people who shop in a store buy bread, and 35% buy butter. Then you can show that 14% of people will buy both, even if purchasing one has absolutely nothing to do with purchasing the other. No. If we're going to propose a rule like bread implies butter, we want more than bread and butter showing up together. We'll want to know how much that bread bumps the probability of butter showing up.

To find this, we first compute the confidence of the inference. That is, if you know that they did buy bread, what's the probability that they also bought butter? Then we divide this value by the chance of buying butter at all, to get the lift ratio of the bread-implies-butter rule. A value close to 1 means the rule is not pinpointing a very strong association. A large value means that if a customer is meeting the "if" conditions, the antecedent, then the customer is much more likely than a typical customer to also meet the then part, the consequent.

Let's check out an example. Suppose that our grocery store database has the records for 1000 transactions, 400 bought bread, 350 bought butter, and 340 bought both; 340 of the 400 bread buyers also bought butter, that's 85%. That's the confidence of the bread-implies-butter rule, 85%. But without knowing that the customer bought bread, the chance of butter is $^{350}/_{1,000}$, or 35%. The lift ratio is the ratio of these two, $^{85\%}/_{35\%} = 2.42$. That means that finding out that a person bought bread increases their chance of buying butter by 142%. That's a good rule. This example had the antecedent and the consequent, both being single items, but they can just as easily be sets of items. Association rules software will scan the data and create frequent sets of items, then generate association rules for them and compute the confidence and the lift indices for each. Those with high values are good candidates for rules that the company may want to note and use.

There's definitely a place for human beings in this process, since the procedure I've just described can result in some rules that can be either discarded or merged with others. For example, "people buying lemons also tend to buy citrus." Undeniably true, but useless as a prediction. Association rules are looking at information in a very specific way—that this set of attributes gives you reason to anticipate that set of attributes as well. But, there are other techniques for looking at things going together that are based on the idea of clustering the data, of creating categories, then, sensibly sorting the individual data points into them. Sometimes this is a first step into further analysis that will be based on these clusters. Two questions, of course, arise. What should the categories be, and how should we decide which category an individual should be sorted into? And the key idea to answering both of these questions is the same, distance.

We've already seen how distance played a role in our k-nearest neighbors approach to classifying data. And the issues that we raised there are relevant here as well. The most straightforward way to define distance between two individuals is Euclidean distance, the straight-line distance separating the points in an n-dimensional scatter plot. Yeah, that's hard to visualize in more than three dimensions, but it's no harder to compute with a souped-up version of the Pythagorean Theorem that you learned in geometry. To find the distance from me to you, given that we have information about six characteristics, A through F, we find out how much we differ on each of the six traits, square those differences, add them up, and take the square root. So if my scores on A through F are 3, 4, 5, 6, 7, 8, and your scores for the five traits are 5 on all six traits, then the differences in the six scores are $-2, -1, 0, 1, 2, 3$, and we square all of these, and add them, then take the square root to get a distance of 4.36 or so.

Although this is the most obvious way to define the distance between two observations, there are some problems with it. One is the issue that we discussed before, scale. Imagine that we are measuring distance from Washington, DC, but for some reason, we are measuring north-south separation in miles and east-west separation in degrees of longitude. A change of 40 in the north-south direction would take us 40 miles north, a bit beyond Baltimore. A change in 40 in the west direction, and we'd be 40 degrees west, somewhere in Nevada. It's often the case that there is no possible or sensible common unit of measure for each variable, and to level the playing field, we can normalize our variables in the way we discussed for the PCA and nearest neighbor algorithms.

Normalizing may not be enough, if some of the variables are strongly correlated. As an extreme example, if I recorded city locations by their miles east-west of DC, their miles north-south of DC, and their longitude, well, the first and last variables are perfectly correlated. The result of the computations will be a distance measure that considers east-west separation as being twice as important as north-south separation. Nobody wants that.

There are some nice ways of addressing the problem by using different definitions of distance, such as the statistical distance that both standardizes and accounts for correlation. Pandora Internet radio uses this kind of distance when picking songs to play for you. Their database contains the results of the Music Genome Project. The project has hundreds of humans analyze 400,000 songs by 20,000 contemporary artists. They tracked hundreds of different characteristics, from the gender of the lead singer to whether the piece includes accordion. Each of these characteristics, the variables, is scored on a scale of 0 to 5. When you tell Pandora that you like a song, it looks for another song near to the one you liked by its distance measure. It plays that. If you say you like it too, it adds it to your cluster of liked songs and looks for another near that cluster, and so on. Not surprisingly, it gets a little more complicated than that, like how to handle songs that you say you dislike, but that's the essential idea.

To explore how this works, I'm going to work through an example small enough that we can rely on your visual intuition to help us along. Here's a two-dimensional example of a small data set, and I'll use our usual meaning of distance in this example. But it will give you a good intuitive feel for how our clustering algorithms work with any data set. So, here's my data set from the students in one of my classes, showing each student's final exam scores and their average on quizzes that were given throughout the semester. Your eye is probably going to pick out some clusters, but we want to see how we can do this mathematically so that it can be done with much larger data set with many more variables. Multiple techniques to do this do exist, including k-means clustering, where we specify the number of clusters that we want in advance.

The approach I'm going to show you today, though, is called agglomerative hierarchical clustering, and it starts with each point being its own cluster. The next step is rather obvious, really. Check the distances between every pair of points, and put the two that are the closest together into a cluster. Since I'm using the normal Euclidean distance as my distance measure, you can do this by eye, like this. And, we continue this way, looking at the things that are closest together. But, before we can do that, we have to answer a question. What do we mean by the distance between two clusters?

Well, it's up to us, and that definition that we choose depends on what we know about the data. Suppose we're trying to track pollution levels. If a river is contaminated, we'd expect to find pollution close to the river. You don't have to be close to all points on the river, just close to some point on the river. Your distance from the river, then, would be measured as the distance of the closest point on the river. That's one definition of distance between two clusters; it's called the minimum distance, or the single linkage distance. If you use it, you get clusters that tend to be ropey.

A second possibility goes to the opposite extreme. You could define the distance between two clusters as the largest distance between any two points in the clusters. So the distance from California to Washington State is the distance from the southern tip of California to the Canadian border of Washington State. This method usually gives small clusters to begin with, and the clusters tend to be spherical.

There are other methods, too. We can define the distance between two clusters as the average distance, computed for each pair of points with one in each cluster. With large clusters, that's a lot of calculation. Or we can find the middle of each cluster, the so-called centroid. And by finding the average value of the variables in the points in the cluster, and then define the distance as the distance between the two centroids, and other more complicated methods seek to minimize the amount of information lost by creating new clusters from old ones.

For demonstration, let's stick with the minimum distance definition, as it's perhaps the easiest to see by eye. Here's the evolution of the clusters of our students. When you watch this progression, from each cluster being a single point until, eventually, all the points being in a single cluster, it prompts a rather natural question. When do you stop? How many clusters do you want? Well, there is no universal answer to this. Sometimes your domain knowledge will inform you of how many clusters there should be in the data, but more often, it comes down to examining the groupings resulting from different levels of clustering. Ideally, the clusters pass three checks. First, they are interpretable in some sensible way. The Nielsen PRIZM segmentation of American consumers divides them into 66 demographic groups, such as Young Digerati and Pools and Patios, and each group can be described roughly in a sentence or two. Second, clusters should be stable to minor changes in the data used to create them. One way to test this is to partition the data before starting into training and validation sets. Begin by using all of this data to define clusters. Set this result aside. Now build the clusters based on the training data alone; then assign each point in the validation data to the nearest of these training-based clusters. Compare the result to the clusters you got from starting with all the data. You don't want to see dramatic differences between the two approaches.

Finally, you'd like the variation between clusters to be significantly greater than the variation within a cluster. You can examine this last point by a dendrogram, a chart that's essentially the script for the movie that we just saw. As you move from bottom to top on the dendrogram, you see what clusters are merged next to form a new cluster. The vertical dimension is recording the distance between the clusters, so merging toward the top indicates that clusters are far apart. I'm stopping at the red line, before joining clusters at a distance of 18 or more from one another. That gives a reasonably small number of clusters that are fairly far apart, this one.

And the groups are sensible. The rectangle holds the high achievers, the triangle the low achievers, the ellipse the students who can't be bothered to study for the daily quizzes but do well on tests, and some outliers. Actually, it probably makes sense to group the small rectangle below the ellipse with it, although the distance rule for nearest point between clusters would not do so next. If I had used the average distance measure, which is actually more sensible in this application, the result would be identical, except the ellipse and the small rectangle would be merged, a better clustering, given the nature of the data.

For only two variables and a small data set, the creation of clusters could have been done by hand, although it's time consuming. For real-world scale data sets with tens or hundreds of variables, you need the mathematics of our algorithm and the power of a computer to implement them. But with those algorithms and that power, data mining confers upon us an ability that could seem almost magical, the ability to find patterns in a wilderness of data that by all rights should defy our conception.

Optimization—Goals, Decisions, and Constraints
Lecture 7

W hen given a troublesome situation, the goal is to identify its key factors. What is it we're trying to accomplish? What controls do we exert over the situation? What rules are we compelled to obey? And how do all of these pieces fit together? As you will learn in this lecture, a good model is going to strip away all of the extraneous distractions and focus on the heart of the problem. And once we have the model, we'll be in a position to apply some accessible and powerful mathematical machinery— machinery that will provide us with the best-possible answer to the problem.

Optimization Problems

- All optimization problems have the same purpose: to find the best-possible answer to the problem at hand. This idea of obtaining the best-possible answer leads to three questions that we ask for every optimization problem: What are you trying to accomplish? What do you have the power to decide? What rules do you have to obey along the way?

- This set of questions can be surprisingly useful in getting a handle on problems—both professional and personal—without any mathematics. So, it's amazing how rarely people take the time to explicitly identify their options, their restrictions, and what they're actually trying to accomplish.

- How do we find our best option? We can approach our optimization problem with a mathematical model. We begin by framing the problem—by identifying its decisions, **constraints**, and **objective**. In order to apply mathematics to the result, we're going to need to express all of the components of the model in terms of measurable quantities. A **measurable quantity** is basically anything that you could sensibly assign a numerical value to in a specific situation.

- Although there are exceptions to every rule, you can think of a measurable quantity as basically anything that you could specify with a phrase starting with "number of ...." Examples include the number of employees we hire or the number of dollars we spend on equipment.

Objective

- The first part of the model is the objective, or goal, represented by a measurable quantity. For any particular proposed solution, the objective will be a number. We'll judge the quality of the solution by the size of this number. While some problems might have the goal of keeping a value within a certain range, most can be expressed with the goal of either maximizing something or minimizing something.

- What if you have more than one goal? Perhaps you're primarily concerned with minimizing costs of operations, but you also want to make sure that customers are satisfied. If you have more than one goal, you're in a bit of a jam, and there are only a few of ways out. One way out is to combine the goals in some way—for example, by taking a weighted average of them with appropriate weights. In this case, your multiple goals become one goal: minimize or maximize this weighted average.

- A second option is to prioritize your goals—for example, that your first goal is to reduce costs to no more than this amount and your second goal is to maintain a customer satisfaction level that's as high as it can be.

- A last option is to look at trade-offs, focusing on the question of how much of one thing you're willing to give up for how much of another. But if you have more than one goal, the goals are usually going to conflict with one another, so you have to make a decision about how to resolve those competing objectives.

Decisions

- The next step is to identify the decisions we make. Again, these are going to be represented by measurable quantities. There is generally one **decision variable** for each quantity over which you have direct control. Focusing on the word "direct" here will help you avoid some common formulation errors in optimization problems.

- When you go to the store, for example, you *do* control how much money you spend—but not directly. Your total bill depends on the money you spend on each type of item, and that in turn depends on how many of each type of item you buy. And what does that depend on? Nothing. You get to pick. That's the hallmark of a decision variable: If you repeatedly ask yourself, "what does that depend on?", the answer finally comes back as, "nothing, I get to decide that." That's the point at which you're on the level of a decision variable.

Constraints

- Next, we have to consider the constraints, which are the rules that we are required to obey. Again, if we're going to be solving problems quantitatively, our restrictions have to be expressed in terms of measurable quantities. In almost every case, our constraints are basically going to specify how the size of one measurable quantity compares to the size of another one. Are they equal in size? Or is the first one less than or equal to the second one? Or is the first one greater than or equal to the second one?

- For example, a line worker may have a constraint that the number of units that he or she assembles in a shift must be greater than or equal to the prescribed shift quota. A soccer league may have a constraint that the number of games played by team A during a season is the same as the number of games played by team B. Constraints tell you what you *may not* do or what you *must* do. They're the rules you have to obey.

Modeling Optimization Problems

- Not every problem can be naturally described with a quantitative objective, a set of quantitative constraints, and a set of decision variables representing measurable quantities. For example, if your goal in life is to be happy—or to maximize your overall happiness—how do you measure it? You might have no idea how to effectively measure your happiness numerically, or how you could accurately connect it to the decisions that you make.

- Part of the problem is that you can't effectively model your happiness because you don't understand how the pieces fit together. And understanding how the pieces fit together—creating a structural model—is essential if you want to find an optimal solution, a best set of decisions in the face of the problem at hand.

- Even when we can sensibly model a problem mathematically, we're going to want to look carefully at our answer before implementing it, or we may be in for a surprise. What we modeled might not quite be what we want.

- A mathematical optimization finds the best answer to the problem as posed. To the extent that the question is not well posed, the answer is in doubt and may be nearly useless. But even when a mathematical solution doesn't take into account some human detail, a small adjustment by a human decision maker can often result in an excellent solution.

Types of Programming

- There are many types of programming that we will encounter throughout this course, including linear programming, nonlinear programming, integer programming, mixed integer programming, and goal programming. Keep in mind that the term "programming" in all of these topics doesn't refer to computer programming. Many of the early optimization problems that were worked out involved scheduling and logistics, such as figuring out what factories should be making during World War II. The term "programming" is meant in the sense it has in the phrase "television programming." It means scheduling.

- That being said, it's undeniable that today almost any real-world scheduling or optimization problem is going to be solved by harnessing the power of computers, whether it's a PC or a mainframe. And for many smaller problems, you don't need to be a computer programmer to take advantage of the computer's power. Problems that took months to solve before computers can be done on any desktop computer in a fraction of a second today, once it is set up. And setting it up is a matter of minutes, not hours.

- Spreadsheets were useful for our regression and data mining work, and they'll continue to be useful for optimization. Excel, in the Microsoft Office suite, does a lovely job for most of our optimization work, and most people already have it on their computers. An alternative is Calc, the free alternative to Excel that is part of OpenOffice, downloadable from the web.

- Spreadsheets are useful not just because they're accessible and familiar to many people—they're also transparent. That means that you'll be able to analyze the details of the problem at hand, and you'll be able to see the model as it evolves over time.

Important Terms

constraint: A rule that must be obeyed by any acceptable solution.

decision variable: In a mathematical model, any measurable quantity over which you have direct control. A solution to a model is a specification of the values of all of its decision variables.

measurable quantity: The quantities represented by numbers in a mathematical model. In most cases, a measurable quantity can be defined as "the number of (unit of measure)" of something, such as the number of hours spent on a project.

objective (function): The mathematical expression that represents the goal of an optimization problem. The better the objective function value, the better the solution. Objective function values are usually either maximized or minimized in an optimal solution.

Suggested Reading

Anderson, Sweeney, Williams, Cam, and Cochran, *Quantitative Methods for Business*.

Butchers, Day, Goldie, Miller, Meyer, Ryan, Scott, and Wallace, "Optimized Crew Scheduling at Air New Zealand."

Cox and Goldratt, *The Goal*.

Samuelson, "Election 2012."

Questions and Comments

1. Consider three possible objectives for a firm: minimize cost, maximize profit, and maximize net worth. In general, these objectives do not lead to the same optimal solution. Give an example where minimizing cost does not maximize profit. Now consider a firm that is given a budget by the state and at the end of the year must return any part of that budget that is not spent. Assume that the firm is maximizing its net worth. Why might it buy a $3 screwdriver even if it has to spend $50 to get it?

Answer:

A firm may well be able to minimize its cost by producing nothing at all! Maximize profit and minimize cost are equivalent goals only when revenue is fixed. In the second example, if the firm buys the screwdriver, it has an asset worth $3 as part of its net worth. If it does not buy the screwdriver, the money it would have spent on it is lost. In other words, $3 is better than $0. This is the kind of situation that can result in runaway prices.

Lecture 7: Optimization—Goals, Decisions, and Constraints

2. Grocery store proprietors need to create the floor plans for their stores, and part of their solution is the location of the frozen foods section. Its location is consistent in most stores. Can you identify a goal that would suggest this placement? Do you think this is the only objective of the grocery store's management?

Answer:

The frozen foods section is usually near the end of the standard path that customers take through the store. The goal is to minimize the amount of thawing that occurs before the frozen items bought can be put in the freezer at home. Obviously, this is not the only objective. Profit is certain to be a concern, but market share, expansion, and customer satisfaction might also be objectives.

Optimization—Goals, Decisions, and Constraints
Lecture 7—Transcript

This time, we begin our exploration of the traditional heart of operations research, the once top-secret techniques that helped win World War II. These are practical techniques capable of handling thousands of variables, but the distinction between what we've done so far and what's coming up isn't really the number of variables. Instead, the big difference is in the models, both the type of models and the purpose of those models.

Thinking about models doesn't come automatically. I was surfing the Internet the other day, and I came across a plea for guidance from a troubled soul at an answers website. Someone trying to help replied with the old adage that the first step in solving any problems is admitting that a problem exists. True enough, certainly. But I was rather amused by the response from the troubled soul. She said, Yeah, I got that one. But what's the second step?

Talk to anyone who tackles problems professionally, and I think you'll find there's a strong consensus about what that second step should be. Define the problem. Determine where you stand now, and what it is, precisely, that you're trying to accomplish. But then we get to the third step, and there are a number of choices. In a business environment, for example, it's common to conduct a SWOT analysis, short for strengths, weaknesses, opportunities, threats. But anyone tackling a problem from an analytics perspective is going to make a different choice, consistently. The third step? Adopt a model.

We've done this repeatedly in our lectures so far, from regression models, to time series models, to exploratory models in data mining. And in not one of these cases have we captured every tiny detail of the real-world situation that the model is being used to investigate. In a real sense, we don't want to. A map that tells you everything, isn't a map, it's the reason you wanted a map in the first place. Any real-world situation, viewed in its entirety, is likely to be incredibly messy. Our intent has been to extract from that tangle, the qualities, quantities, and relationships that are at the heart of what's going on, and then to focus our efforts on those.

There are, of course, good models and bad models. And the test of a model is the extent to which its predictions agree with observation. If the agreement between the two is good, the model can be used in the better understanding of the reality and in predicting the consequences of our actions. A good model doesn't necessarily even have to be right in an absolute sense, as long as its results give an accurate enough fit to reality. Newtonian mechanics and classical electromagnetic theory have been shown to be wrong, and have been superseded by relativity and quantum field theory, but the simpler, earlier models are still used in guiding spaceships to other planets and in the designing most electronic equipment.

That's because, when moving from a theoretical to applied perspectives, we want a model for something different than maximal agreement with observation. We want one that's simple enough to be readily applied in a practical context, yet accurate enough to make quite useful forecasts as to how things will play out. The mathematical models that we develop in this course aren't intended to capture every nuance of reality. If we do our job well, they'll meet two criteria—tractable enough to be solved, accurate enough to be useful.

Those requirements still leave open a lot of possibilities, and we're exploring some of the best of them in this course. Up to this point, all of our models have included an element of randomness, but restricted to a particular context. Everything we've done so far has been taken from a random sample from a population and tried to find a connection among the quantities represented in that sample. We addressed the randomness with a statistical analysis of error that told us how much we could trust the result when applied to new data. Later in the course, we're going to let the stochastic genie out of the bottle and look at the impact of randomness on a much broader context, situations where random events play a central role in the problem that you're trying to analyze—machine failures, service times, public opinion, financial speculation, acts of God, or of government. I'm sure you can think of a lot of examples from your own experience.

But for this central section of the course, I want to go in the other direction, because there's a lot in life that's not driven by random fluctuations. We manage to navigate the world because we can often rely on cause-and-effect relationships among the events that play out in it. And when we can quantify the nature of those relationships, we can build a structural model of the situation that we face. There's no randomness here, we're talking deterministic systems, where a particular set of inputs lead to a predictable set of results. And if you have any doubt about the power or usefulness of that kind of modeling, take a look at any of the physical sciences or engineering. Our technology, from a match to a supercomputer, hinges on this kind of predictability.

What we've been doing so far is like science, looking at data and trying to uncover relationships that tie things together. In science, you try to figure out how the pieces are connected, to find rules that some aspect of the universe obeys. But our approach now, by contrast, is going to be more like engineering, in that I'm going to assume that you already know about how the pieces are connected. So our job is to figure out what to do with them. Because you're going to try to accomplish something, and the question is, what actions you should take so that you accomplish that something as well as you possibly can? And that's going to be our focus in this central part of the course—optimization.

All optimization problems have the same purpose, to find the best possible answer to the problem at hand. Think about that phrase for a moment, the best possible answer. Because it's really all that you need to know to get a clearer image of how we model an optimization problem. It's actually easier to start at the end of best possible answer and work our way back toward the beginning. Start with answer. What's an answer?

Well, you're analyzing a situation because you want to know what to do, right? You're going to need to take some kind of action, and the question is, what kind? The answer is the specification of the decisions that you'll make. Those decisions might be simple to describe, such as, don't buy the car. Or they could be more subtle and complex. If the Dow stays above 12,000 and we hold a market share of at least 30%, then approve the new construction project with a budget of 5% of the gross income from last year. Simple or complex, though, your answer is your set of decisions, your course of action in the upcoming situation.

But of course, you can't just do whatever you want. Virtually every real-world decision-making situation comes along with its own limitations on what you can do. It comes with its own set of constraints. And there really isn't much point in spending a lot of time considering answers that are impossible to implement. So, we want to focus on answers that don't break any of the constraints—on possible answers.

Okay, so we've narrowed the field down to only possible answers to the situation. In the end, we're going to have to select one course of action from among these candidates. We want the best possible answer. And that's not as straightforward as you might at first imagine. Look, I've got a set of decisions over here, A, and another set of decisions over here, B. Both are possible to implement. Which one's better? Well, I have to have a way of comparing any two such answers and choosing between them. We'll generally do this by identifying our goal, our objective, and then seeing which of the two solutions better achieves that goal.

So, this idea of best possible answer leads pretty directly to three important factors in analyzing a situation. What are you trying to accomplish? What do you have the power to decide? And what rules do you have to obey along the way? And these are the three questions that we ask for every optimization problem. And, it's been my experience that this set of three simple questions can be surprisingly useful in getting a handle on problems, both professional and personal, without any mathematics at all. So, it's amazing to me how rarely people take the time to explicitly identify their options, their restrictions, and what they're actually trying to accomplish.

One example. In his book, *The Goal*, Eliyahu Goldratt has a protagonist who's a plant manager in a manufacturing firm. The manager is going to a conference on robotics, and he's very proud of the efficiency of his robots in his factory. In order to get these efficiencies, though, the robots had to be used pretty much 24-7, and since there wasn't enough demand for their output from the processing steps further down the line, the robots ended up making parts that nobody really needed. When quizzed about this by his old college professor doing operations research work, the manager admitted that they were getting no more product to market, that they had not reduced

147

employees, and that they had actually built up a considerable and expensive inventory of parts that they had no immediate use for. But he had been proud, because the robots were so efficient.

It kind of reminds me of comedian Rita Rudner's dieting solution. At restaurants, she says, she orders dessert, but only eats half. But if it's really good, she orders two. Well, jokes aside, it's surprisingly easy to get stuck doing something that is sub-optimal. So, how do we find the best option? For purposes of having something concrete to talk about, I'm going to tell you about an aviation company, Air New Zealand. They knew what flights they wanted their jets to fly, but there are some obvious logistical matters to be worked out. You can fly a plane and crew from A to B, always assuming that you have the plane, A. But then after the flight, you've got a crew and plane at B. If you're going to have a repeatable schedule, that plane eventually has to get back to A again, as does the crew.

In the airline biz, the collection of flights and rest periods that make up a circuit are called a tour of duty, and the schedule of a member of the crew that brings them back to their crew base at the end is called a line of work. So how do you figure out all of that? Let's see how we can approach this problem with a mathematical model. We're going to begin by framing the problem, by identifying its decisions, its constraints, and its objectives. In order to apply mathematics to the result, we're going to need to express all of the components of the model in terms of measurable quantities.

For this section of the course, we won't be looking at categorical data. A measurable quantity is basically anything that you could sensibly assign a numerical value to in the specific situation. Although there are exceptions to every rule, you can think of a measurable quantity as pretty much anything that you could specify with a phrase starting of number of this, or number of that, the number of employees we hire, the number of days before an order arrives, the number of dollars we spend on equipment, the number of fatalities in a car accident, the number of pounds I put on over the winter holidays. That kind of data.

So, how do we express the Air New Zealand problem in terms of measurable quantities? Well, hold on a minute. As I've described it, it might be more tractable to approach it as two problems, rather than one. First, find what tours of duty you want to use to minimize the cost of your flights and layovers. Then figure out how you can use the people that you have to crew those flights. It might be better to solve these two problems together. We might find some really clever ways to schedule crew if we have a particularly friendly tour of duty schedule, but it's also a lot harder to solve. So, for the sake of tractability, the airline and the University of Auckland, who were collaborating on this project, treat it as two separate problems. Well, eight, actually. Because the rules concerning pilots are different than for flight crews, and the rules for international flights are different than from purely domestic ones. But to keep our discussion from getting out of control, let's just talk about scheduling flight crews on international flights.

Okay, the first part of the model is the objective, the goal. Remember, we're going to represent the objective by a measurable quantity, like the number of dollars of profit, or the number of people seeing my ad, or the number of minutes a patient spends in an emergency room waiting area. Notice that each of these starts with the phrase "number of," which is how all of our measurable quantities will be expressed. For any particular proposed solution, the objective will be a number. We'll judge the quality of the solution by the size of the number. If Air New Zealand cares about profit, it's pretty clear they'd like to maximize it—the bigger, the better. On the other hand, if their revenues are more or less fixed, they might prefer to analyze the problem in terms of total cost. This would be a minimization problem; the goal is to minimize the expenditure.

While some problems have the goal of keeping a value within a certain range, most can be expressed with the goal of either maximizing something or minimizing something. This actually brings up an important point. How about if you have more than one goal? For example, the Air New Zealand management is primarily concerned about money, specifically, minimizing the costs of operations. The crew, on the other hand, is much more concerned about pleasant or unpleasant schedules for them. The management is calling the shots, but it's obviously good business practice to keep your employees as happy as you can. So Air New Zealand management actually has at least two goals. We'll be looking at how to handle this kind of thing in detail when we discuss goal programming and how to address more than one goal later in the course. But in brief, if you've got more than one goal, you're in a bit of a jam, and there are only a few of ways out.

One way out is to combine the goals in some way, for example, by taking the weighted average of them with appropriate weights. In this case, your multiple goals become one goal: Maximize or minimize this weighted average. A second option is to prioritize your goals, to say, for example, my first goal is to reduce costs to no more than this amount, and my second goal is to maintain a crew satisfaction level that's as good as I can get.

The last option is to look at tradeoffs, focusing on the question of how much of this you're willing to give up for how much of that. But if you do have more than one goal, the goals are usually going to conflict with one another, so you have to make a decision about how to resolve those competing objectives. It's interesting to listen to political speeches with this observation in mind. You'll often hear politicians make statements such as, "We're going to provide the best possible health care at the lowest possible cost." Well, the lowest possible cost is free, and the easiest route to that lowest cost is by offering no health care at all. This obviously is not the best possible health care. What's really being said, in a fuzzy way is, "We're going to provide the best balance of health care quality and cost." But what is that balance? Who decides? These are hard questions, but the phrasing of the original statement too often lets the speaker avoid them entirely.

But, back to Air New Zealand. The objective is going to be, essentially, to minimize cost, although we'll include crew dissatisfaction in the objective, appropriately weighted, as a cost. What's next? Well, the decisions we make. Again, these are going to be represented by measurable quantities, number of this, number of that. There is generally one decision variable for each quantity over which you have direct control.

I put that word, direct, in there quite deliberately and remembering it will help you avoid some of the most common formulation errors in optimization problems. When you go to the store, for example, you do control how much money you spend, but not directly. Your total bill depends on the money you spend on each type of item, and that in turn depends on how many of each type of item you buy. And what does that depend on? Nothing. You get to pick. That's the hallmark of a decision variable. If you repeatedly ask yourself, what does that depend on? the answer finally comes back as nothing, I get to decide that. And then you're at the level of a decision variable.

For the airline, we might have decision variables such as the number of flight attendants that were assigned to each tour, or the like. But here, our job is to decide how to schedule the specific people that we have available as flight attendants, so our variables are both simpler and more subtle than the number assigned to each flight. We already solved the tour of duty problem, so Air New Zealand knows the tours that they're going to fly. It knows the available staff. The decisions we have to make are really at a grass-roots level. Here is a particular flight attendant, Alexis. Here is a particular tour of duty, a six-day loop from Auckland, to L.A., to Sidney, and back to Auckland, with layovers. What we decide is, does Alexis fly this tour, or doesn't she?

It's a really simple variable; it doesn't even sound like a number. But we can use 1 to mean that, yeah, she does fly the tour, she flies it one time, and 0 to mean that she does not. This kind of variable is called a binary, or Boolean, or 0/1 variable, and we'll look at them carefully in a later lecture on integer programming. So, I might have a variable called ALEX07, which is 1, if Alex flies tour number 7 and 0 if she doesn't.

Are you starting to get scared yet? The researchers on the project were, because these 0/1 problems can actually get pretty time consuming to solve when they get big, and our problem is looking like it's going to get big. As an example, suppose that Air New Zealand has 100 tours of duty and 300 people available as flight crew. Then there is a 0/1 variable for every combination of person and tour, meaning our problem has 30,000 variables. And that's just for the flight crew! There might be several thousand more variables for the pilots.

OK, that brings us to the constraints, the rules that we are required to obey. Again, if we're going to be solving problems quantitatively, our restrictions have to be expressed in terms of measurable quantities. In almost every case, our constraints are basically going to specify how the size of this quantity compares to the size of that one. Are they equal in size? Or is the first less than or equal to the second? Or greater than or equal to it? For example, I personally have a constraint that the money I spend has to be less than or equal to the money that I have. The government evidently doesn't have this particular constraint. A line worker might have the constraint that the number of units that he or she assembles in a shift must be greater than or equal to the prescribed shift quota. A soccer league may have the constraint that the number of games played by Team A during the season is the same as the number of games played by Team B. We're going to have a lot more to say about constraints a little later on, but you get the idea. Constraints tell you what you may not do or what you must do. They're the rules you have to obey.

For Air New Zealand, what do we have to do, and what can't we do? Well, our particular problem is only about flight attendants, so, we have one set of rules that says this, a flight attendant can't be doing two different things at the same time. Tours of duty are mutually exclusive, so if she's on tour 7 on Wednesday morning, she can't also be on tour 16 at that time. That's one constraint for each flight attendant for each time window, again, a lot of constraints. But equally important is a second set of constraints saying that the flights must be crewed. If a particular tour of duty needs three flight attendants, then there have to be at least three people assigned to that particular tour. That's going to be one constraint per tour.

If the first program, which was making up the tours to begin with, took into account all of the regulations concerning what tours are and aren't legal, and which ones break labor agreements and so on, then these two kinds of constraints may pretty much be all that we need in our current program. And this is essentially what Air New Zealand ended up with—conceptually, simple; computationally, a little terrifying. But with computer power and mathematical cleverness, it turns out that it's quite possible to solve. The solution, coupled with the solutions to the 7 other related problems that they solved, is estimated to have saved Air New Zealand over $12 million in U.S. dollars a year, and to have provided better attention to the crews' preferences in scheduling.

Now, I'm not saying that every problem can be naturally described in this way, with a quantitative objective, a set of quantitative constraints, and a set of decision variables representing measurable quantities. For example, if your goal in life is to be happy or to maximize your overall happiness, how do you measure it? Well, before this lecture started, I was feeling a little down, about 30 happiness points. But now, my happiness rating has skyrocketed to 73.1! I'll be honest, really, I have no idea how I'd effectively measure my happiness numerically, nor how I could accurately connect it to the decisions that I make. Part of the problem is I can't effectively model my happiness because I don't understand how the pieces fit together. And understanding how the pieces fit together, creating a structural model, is essential if we want to find an optimal solution, a best set of decisions in the face of the problem at hand. Even when we can sensibly model a problem mathematically, we're going to want to look carefully at our answer before implementing it, or we may be in for a surprise. What we modeled might not be quite be what we want.

One of the pioneers of mathematical optimization, George Dantzig, found out the hard way. Dantzig was a mathematical scientist, and important to us, because he developed a procedure, the simplex method, for solving a class of optimization problems called linear programs. Linear programs are going to be extremely important to us in this course, and when we get into the details, I'm going to be letting you in on what was originally top-secret stuff. Much of the early work in the field was classified until 1947. The Allies used linear programming to reduce the cost of World War II operations and increase enemy casualties. Important stuff.

Anyway, in 1947, Dantzig used his simplex method, which in principle could solve any linear program, to solve a linear program that had earlier been formulated by another researcher in linear programming. The decision variables were the number of units of various foods that you would feed a person per day. The constraints were that the minimum daily adult requirements for various vitamins, minerals, calories, and so on.

The goal was to meet these constraints at minimum possible cost. Even though the program was formulated with only 77 foods and only 9 requirements as far as nutrition, computers weren't available at the time, so, even with Dantzig's lovely simplex method, it had to be solved using desktop calculators; 120 man-days of work later, Dantzig's team came up with the answer. It wasn't an exciting diet, but for people going hungry, excitement is not usually the primary concern. It consisted of only five foods, wheat flour, evaporated milk, dried navy beans, cabbage, and spinach. It could feed a man for only $39.69 per year; 39 dollars and 69 cents in 1947 prices. That's still well under $600 per year to feed a person in current dollars. Under $2.00 a day, not bad.

Skip ahead a few years, in the 1950s. Computers are coming into their own. Dantzig, like many of us, had put on some pounds that his doctor thought he'd be better off without. So, thinking of the success of the earlier diet program, Dantzig wrote a linear program to find his optimal diet. The goal was to maximize the feeling of being full, which he estimated by the weight of the serving of food minus its water weight. He imposed nutritional constraints, as well as the constraint that the total calories consumed couldn't be more than 1500 calories a day. He announced to his wife that whatever the computer generated as his optimal diet was what he intended to eat. She was willing to play along and prepare what the computer ordered.

The program was run, and Dantzig informed his wife of the suggested diet. She said it was weird, but doable. Then Dantzig told her that there was one more item on the menu, 500 gallons of vinegar. Evidently, vinegar was considered a weak acid, with virtually no water, so it looked great in the objective function. Dantzig decided vinegar wasn't a food after all and reran the model the next day. This time, the solution was reasonable, except for including 200 beef bouillon cubes. Still, Dantzig gave it a whirl, starting breakfast the next morning with four bouillon cubes dissolved in water. He couldn't drink it; it was pure brine. And just as well, you can imagine what his salt intake would have been. No sane person wants to eat that much salt, so doctors weren't warning people against this ridiculous level of salt intake at the time.

So, new constraint: No more than three bouillon cubes a day. The computer found the optimal solution for the new problem. It included two pounds of bran per day. Imagining the effect of eating two pounds of bran, he limited the amount of bran allowed in the solution, and the computer switched to two pounds of blackstrap molasses. At which point, Mrs. Dantzig said, George, I love you, but enough is enough. She said she'd gotten some good ideas from the solutions that the program had proposed, but that she was going to decide the meals. She did. Dantzig lost 22 pounds.

The moral of the story? A mathematical optimization finds the best answer to the problem as posed. To the extent the question is not well posed, the answer is in doubt, and may be nearly useless. But even when a mathematical solution doesn't take into account some human detail, the suggestions are rarely as surreal as Dantzig's diet. A small adjustment by a human decision maker can often result in an excellent solution.

Before we go on, I think I'd better take a minute and talk about programming, and what it means in this course. For example, I told you that the feed the hungry diet problem, and Dantzig's less successful personal diet, were examples of linear programs. The Air New Zealand problem [we] discussed was integer program, because it had lots of ones and zeroes. There are other kinds of programming as well that we'll be coming across, nonlinear programming, mixed integer programming, goal programming. All of this programming makes it sound like we're going to be working with computers. Well, that's true, but the term programming in all of these topics doesn't refer to computer programming. A lot of early optimization problems were worked out involving scheduling and logistics, like figuring out what factories should be making during the war. The term programming is meant in the sense that it has in the phrase television programming—it just means scheduling.

That said, it's undeniable that today almost any real world scheduling or optimization problem is going to be solved by harnessing the power of computers, whether it's a PC or a mainframe. And for a lot of smaller problems, you don't need to be a computer programmer to take advantage of the computer's power. The feed the hungry problem that took George Dantzig and his team 120 man-days to solve could be done on any desktop computer in a fraction of a second today, once it was set up. And setting it up would be a matter of minutes, not hours.

Spreadsheets were useful in our regression and data mining work, and they'll continue to be useful in optimization. Excel, in the Microsoft Office suite, does a lovely job for most of our optimization work, and most people already have it on their computers. An alternative is Calc, a free alternative to Excel that is part of OpenOffice, downloadable from the web. I use spreadsheets not only because they're accessible and familiar to a lot of people. They're also transparent. That means I'll be able to show you the details of what I'm doing, and you'll be able to see the model as it evolves over time.

So, we've got the big picture. Given a troublesome situation, we're going to identify its key factors. What is it we're trying to accomplish? What controls do we exert over the situation? What rules are we compelled to obey? And how do all of these pieces fit together? A good model is going to strip away all of the extraneous window dressing and focus on the heart of the problem. And once we have the model, we'll be in a position to apply some accessible and powerful mathematical machinery, machinery that will provide us with the best possible answer to the problem.

That's the skill set we'll be building in this central part of the course, and the most important skill is the one that I mentioned in the first lecture, that act of translating the problem from English that we use in everyday descriptions, to the mathematese that will allow us to bring our analytical tools to bear, formulating the optimization model.

And we're going to begin, next lecture, by looking at a set of models that's near the head of the list of mathematical tools widely used by businesses, big and small; they're very beautiful, remarkably applicable, perfectly solvable class of problems called linear programs. See you then.

Linear Programming and Optimal Network Flow
Lecture 8

A linear program is an optimization model, but not all optimization problems are created equal—no more than all hikes are created equal. Strolling along a flat beach that gently slopes into the sea is much easier than scaling a rough and craggy mountain. A major issue in how difficult an optimization problem is to solve comes down to its mathematical geography. And linear programs are the simplest optimization problems to solve because they have the simplest geometry. That's going to make the math easier to write and the model easier to solve, meaning that problems big and small can be solved reasonably quickly. In this lecture, you will learn the basics of linear programming.

Linear Programs: Geometric

- The word "linear" suggests a line—and it's worth keeping that association. Two variables are linearly related if and only if their graph is a straight line. Simple linear regression finds the best linear relation between input and output; it's the straight line that best matches the data on a scatterplot.

- In multiple regression, we can talk about a linear function of more than one variable. If we have two inputs, we can visualize this as a tabletop in which every position on the table is represented by a pair of x and y coordinates. A function of these two variables is represented by a surface hovering over that tabletop.

- Suppose that we have a function like $3x + 2y$. Pick a point on the tabletop—for example, $x = 5$, $y = 10$. To find how high above the tabletop that hovering landscape is over that point, plug $x = 5$, $y = 10$ into the function. So, $3x + 2y$ becomes $3(5) + 2(10)$, or 35. That means that the landscape is hovering 35 units above the tabletop at that point. And the entire landscape is created by doing this for each x-y pair. In general, the graph—that surface—might resemble almost any landscape that you can imagine.

- But if the function is a linear function, the landscape has a very special appearance: It's like a flat pane of glass, tilted at some angle. In particular, a flat plane of glass has this special property: Pick any point on the glass, pick a direction, and then draw on the glass, continuing in that direction. What you draw will be a straight line.

- That's really what we mean by "flat": No matter what line of sight you choose along the surface, you're always looking along a straight line that stays in the surface. That same idea can be used as a definition of flat in higher dimensions—no bends, no jumps, no kinks. Linear functions always graph as flat surfaces, in this sense. That's what linear is geometrically. But it's even easier to characterize algebraically.

Linear Programs: Algebraic

- Suppose that you have a collection of input variables and you want to create an output variable, but you want the graph to be flat, in the sense that was just described. You want a linear relation. You can build such a function in an extremely simple way. Multiply each variable by a number, and then add the results. If you like, add one more number, by itself, at the end, for good measure. The numbers could be positive, negative, zero—whatever you like. The result is always a linear function of the inputs.

- So, $3x + 2y - 17z - 2$ is linear. Or $w + 8$. Or $x/2 + y/3$, because $x/2$ is the same as $1/2$ times x. $2/x$ is not linear; you can't divide by a variable. In addition, xy isn't linear; you can't multiply two variables together. It's just a number times a variable, plus a number times a variable, and so on, maybe with a number by itself added to the end.

Linear Programs: Conceptual

- What is a linear function conceptually? Let's begin with a concrete example with only two variables: an input and an output. Suppose that your significant other sends you to the grocery store with a specific list of things that he or she wants. In addition to that, you can buy as many cans of tomato soup as you want. The input variable is how many cans you buy, and the output is the total grocery bill.

- Each time you add one more can of soup to the cart, you increase the total bill by the same amount: the cost of a can of soup. And if you change your mind, each time you take a can out of the cart, you always decrease your bill by that same amount. These statements are true, regardless of what else is already in the cart—whether we're talking about your first can of soup or your ninth. That's what characterizes a linear relationship. On the other hand, if you got a discount for buying four or more cans, for example, that would make the relationship nonlinear.

- What about if there's more than one input? Suppose that for your grocery run, you have more latitude. You can buy soup and hamburger meat in whatever quantities you like, along with the stuff on your significant other's list. If you change only your soup purchases, you're going to see the linear behavior just discussed.

- On the other hand, if you change only your hamburger purchases, you see the same kind of behavior. Each additional pound increases your bill by the cost of a pound of ground beef, and so on. Again, these numbers don't depend on what else is in the cart; they only depend on the fact that you're changing the quantity of only one thing at a time.

- Whenever you have these kinds of relations, then the output—here, your grocery bill—is a linear function of those inputs.

- Finally, suppose that the grocery list of items from your significant other totals $50. In addition to these items, you can buy as much soup and hamburger meat as you want. Soup costs $1.05 per can and hamburger costs $3.29 per pound. What will the total bill be? Obviously, that depends on how much soup and hamburger meat you buy.

- Let S be the number of cans of soup that you buy and H be the number of pounds of hamburger meat that you buy. Then, the total bill is going to be the following.

$$\underbrace{1.05S + 3.49H + 50}_{\text{number of dollars spent}}$$

- As easy as this is, it's worth looking at in a new way, because it's a little road map for linear programming formulation. First, look at the algebra. It is linear: a number times a variable, plus a number times a variable, plus a number. But here's the important thing: Each of those numbers actually answers a specific question.

  1.05: If I buy one more can of soup, how many dollars does my bill go up?

  3.49: If I buy one more pound of hamburger meat, how many dollars does my bill go up?

  50: If I buy no soup and no hamburger meat, how much is my bill?

- We can generalize this. In *any* linear expression, the numbers answer one of two questions. The constant term—the number with no variable—answers this question: "Suppose I set all of the variables to zero. What should the thing underneath be?" The "thing underneath" is the measurable quantity you're interested in—here, that's the total bill.

- This problem has two variables, S and H, representing cans of soup and pounds of hamburger meat, respectively. If they're both zero, you don't buy soup or hamburger meat, so your total bill is just the cost of everything else in your cart, which is $50—the cost of the original shopping list.

- What about the other numbers, the coefficients? (A coefficient is just a number multiplying a variable.) They answer this question: "If I get one more of this and make no other changes, how much does the thing underneath go up?" So, that 1.05 in front of S means, "One more S, and cost goes up 1.05."

- This is how it works for all linear expressions. The coefficient of a variable tells you how much the "thing underneath" goes up if the variable increases by one. For most variables, this will either mean that you get one more, or use one more, or need one more of something, etc.

An Application

- Suppose that during a 40-hour workweek, a building inspector can inspect either 80 houses or 10 farms. What math represents the number of hours the inspector spends inspecting this week if he or she inspects H houses and F farms?

- The problem implies that each farm takes the same amount of time and each house takes the same amount of time. That makes the expression linear. Then, algebraically, the answer will look like the following.

$$\underbrace{\underline{\quad}H + \underline{\quad}F + \underline{\quad}}_{\text{number of hours used}} = $$

- And it's just a fill-in-the-blanks game. Find the constant term first. If we set both H and F equal to zero, the inspector doesn't inspect anything, so he or she spends no time inspecting. That's how we find the constant term: It's zero.

$$\underbrace{\underline{\quad}H + \underline{\quad}F + \underline{0}}_{\text{number of hours used}} = $$

- Next is the coefficient of H. The question is as follows: If H goes up by one—which means that the inspector inspects one more house—then how much does "the thing underneath"—the hours used—go up? The inspector can do 80 houses in 40 hours, which means that it takes him or her 1/2 an hour per house. So, if H goes up by one, time goes up by 1/2 an hour.

$$\underbrace{0.5\,H + \underbrace{F} + \underbrace{0}}_{\text{number of hours used}}$$

- And, of course, we use the same reasoning for F. One more farm increases time used by how much? Ten farms in 40 hours means 4 hours per farm, so that's the coefficient of F. And we're done. $0.5H + 4F$ gives the time used. If the inspector only works 40 hours in a week, our constraint would be as follows.

$$0.5H + 4F \le 40$$

- Notice that the coefficients showing up in the constraint aren't the numbers that we started with. Those were 80 and 10. We found the needed coefficients by asking the right questions.

- This time constraint we just created is an example of the most common kind of constraint in a **linear program**, called a **limited resource constraint**. It says that you can't use more of something than you have—or, more formally, number of units used ≤ number of units available.

Important Terms

limited resource constraint: The most common kind of linear programming constraint. It says that you can't use more of something than what is available.

linear program: A model in which the objective function and the two sides of each constraint are all linear expressions. Linear programs are especially easy to solve.

Suggested Reading

Narisetty, Richard, Ramcharan, Murphy, Minks, and Fuller, "An Optimization Model for Empty Freight Car Assignment at Union Pacific Railroad."

Stevens and Palocsay, "A Translation Approach to Teaching Linear Program Formulation."

Questions and Comments

1. In the railroad example in the video lecture, instead of shipping empty cars, imagine that we were shipping fruit and that the variables represented the number of tons of fruit shipped from city to city. Now imagine that at a transshipment point, the fruit must be offloaded and reloaded, resulting in 5% of the fruit being ruined. How would you modify the linear program to reflect this loss?

Answer:

The "measurable quantity formulation" would now read

$$\underbrace{\qquad\qquad}_{\substack{\text{\# of tons of fruit}\\\text{leaving Chicago}}} = 95\% \text{ of } \underbrace{\qquad\qquad}_{\substack{\text{\# of tons of fruit}\\\text{entering Chicago}}}$$

and "95% of" becomes "0.95 ×." So, we get

$$\underbrace{CE + CF + CG}_{\substack{\text{\# of tons of fruit}\\\text{leaving Chicago}}} = \underbrace{0.95\left(AC + BC\right)}_{\substack{\text{\# of tons of fruit}\\\text{entering Chicago}}}$$

All "percentage constraints" follow this same pattern.

2. A common mistake in formulation is what is sometimes called a "recipe error," because it results when one attempts to write a recipe as a constraint. For example, suppose that you build chairs, and a chair consists of four legs, one back, and one seat. It is tempting (but altogether wrong) to write $CHAIR = 4LEGS + BACK + SEAT$.

a) Assuming that *CHAIR* is the number of chairs made, *LEGS* is the number of legs used, and so on, show that this constraint is incorrect.

b) Write the correct constraints to connect *CHAIR*, *LEGS*, *BACK*, and *SEAT*. (Hint: you'll need three of them.)

Answers:

a) The easiest way is "validation": plugging in numbers that *should* work and checking to see if they *do* work. For example, 1 chair needs 4 legs, 1 back, and 1 seat, so the values $CHAIR = 1$, $LEGS = 4$, $BACK = 1$, and $SEAT = 1$ *should* work in a correct equation. But plugging these values into the equation above gives the following.

$$1 = 4(4) + 1 + 1 = 18$$

This is obviously untrue. So, the formula is wrong.

b) $BACK = CHAIR$, $SEAT = CHAIR$, $LEGS = 4CHAIR$. The key idea here is that once you know *CHAIR*, you know all four variable values, which means that each of the other three variables can be defined in terms of *CHAIR* alone.

Linear Programming and Optimal Network Flow
Lecture 8—Transcript

Union Pacific Railroad, it's the largest railroad in North America, over 30,000 miles of track, 8500 locomotives. Union Pacific moves more than 100,000 freight cars across the western $^2/_3$ of the U.S. Its annual payroll alone is almost $4 billion.

When you think of the services that Union Pacific provides, you might miss an important one. Delivering empty freight cars to customers who don't have their own and who want to move their goods by rail. The trouble is that the cars that Union Pacific has can be scattered all over the country. Using an empty car in San Francisco to meet a demand in Little Rock is costly in a number of ways: the cost of the transportation, obviously; also the delay in the car arriving where it's needed, which can generate both loss of customer good will and explicit penalties; then there's the congestion of the railways, and so on.

Union Pacific's network is huge, and its situation is dynamic. New orders can come in at any time, and new cars, released from an earlier job, can suddenly become available. Because of this, Union Pacific was looking for a way to determine the assignment of empty freight cars to customers in a real-time environment, using a solution procedure that was accurate enough to be useful and fast enough to keep up. Working with [Purdue University], they did just that, using the exceptional power and speed of a linear program.

A linear program is an optimization model, but not all optimization models are created equal, no more than all hikes are created equal. Strolling along a flat beach that gently slopes into the sea is a lot easier than scaling a rough and craggy mountain. As it turns out, this analogy between optimization problems and geographic topography is a surprisingly good one. A major issue in how hard an optimization problem really is to solve really comes down to its mathematical geography. And linear programs are the simplest optimization problems to solve because they have the simplest geometry.

That's going to make easier to write the math and to solve the model. The relative simplicity of linear programs means that problems that aren't huge can be solved blindingly fast, an important consideration in situations like that faced by Union Pacific, when there isn't a lot of time between when you get a rundown of the current situation and you need a strategy to implement. Even huge problems with a collection of decisions and constraints that stagger the imagination can be addressed in a reasonable amount of time. So, what does a linear geometry look like?

The word linear, of course, suggests a line, and it's worth keeping that association. Two variables that are linearly related will always have their graph being a straight line. This isn't something new for us. Simple linear regression found the best linear relationship between input and output, the straight line that best matched the data on the scatterplot. But we also saw, in multiple regression, that we could talk about a linear function of more than one variable. If we had two inputs, we visualized this as a table top in which every position on the table was represented by a pair of x and y coordinates. A function of these two variables was represented by a surface hovering over that table top.

Let's be a bit more specific about this visualization. Suppose we have a function like $3x + 2y$. Pick a point on the tabletop, maybe $x = 5$, $y = 10$. To find how high above the tabletop that hovering landscape is at that point, plug in $x = 5$, $y = 10$ into the function. So, $3x + 2y$ becomes $3(5) + 2(10)$, or 35. That means that the landscape is hovering 35 units above the tabletop at that point. And the entire landscape is created by doing this for each x-y pair. In general, the graph, that surface, might resemble almost any landscape that you can imagine.

But if the function is a linear function, that landscape is very special; it's like that flat pane of glass, tilted at some angle. And you know what that's like. In particular, a flat pane of glass has this special property; pick any point on the glass, pick a direction, and then draw on the glass, continuing in that direction. What you draw will be a straight line. That's really what we mean by flat; no matter what line of sight you choose along the surface, you're always looking along a straight line that stays in the surface. The same idea can be used as the definition of flat in higher dimensions, no bends, no jumps, no kinks. Linear functions always graph as flat surfaces, in this sense.

Okay, that's what linear is geometrically. But it's even easier to characterize them algebraically. Suppose you have a collection of input variables, and you want to create an output variable, but you want the graph to be flat, in the sense that we just described. You want a linear relation. You can build such a function in an extremely simple way. Multiply each variable by a number, then add the results. And, if you like, add one more number, by itself, at the end, for good measure. The numbers could be positive, negative, zero, whatever you like. The result is always a linear function of the inputs.

So, $3x + 2y - 17z - 2$ is linear; or $w + 8$; or $x/_2 + y/_3$, since $x/_2$ is the same as $\frac{1}{2}$ times x; $2/_x$, not linear. You can't divide by a variable; xy isn't linear either, you can't multiply two variables together. It's just number times variable, plus number times variable, and so on, maybe with a number by itself added at the end. Piece of cake.

Okay, one last way to imagine linearity, and we'll have all the pieces that we need. What's a linear function conceptually? Well again, let's start off easy. A concrete example with only two variables, an input and an output. Your significant other sends you to the grocery store with a specific list of things they want. In addition to that, you can buy as many cans of tomato soup as you want. How nice. The input variable is how many cans of soup you buy, and the output is the total grocery bill.

Well, each time you add one more can of soup to the cart, you increase the total bill by the same amount, the cost of a can of soup. If you change your mind, each time you take a can out of the cart, you always decreases your bill by that same amount. These statements are true, regardless of what else is already in the cart, whether we're talking about first can of soup or your twelfth. And, that's what characterizes a linear relationship. On the other hand, if you got a discount for, say, buying 4 or more cans, that would make the relationship nonlinear.

How about if there's more than one input? Suppose that in your trip to the grocery store, you have a little more latitude. You can buy soup and hamburger meat in whatever quantities you want, along with the stuff that your S.O.'s wants. If you change only your soup purchases, you're going to see the nice, linear behavior we just discussed. On the other hand, if you change only your hamburger purchases, you see the same kind of behavior. Each additional pound increases your bill by the cost of a pound of ground beef, and so on. Again, these numbers don't depend on what else is in the cart; they only depend on the fact that you're changing the quantity of only one thing at a time.

Whenever you have these kinds of relations, then the output, here, our grocery bill, is a linear function of those inputs. Simple? Well, let's put it all together. One more trip to the grocery store, the grocery list of items from your significant other totals $50. In addition to these items, you can buy as much soup and hamburger as you want. Soup costs $1.05 a can, and hamburger costs $3.29 a pound. What's the total bill going to be? Well, obviously, that depends on how much soup and hamburger you buy. So, let's let S be the number of cans of soup that you buy, and let H be the number of pounds of hamburger that you purchase. Then your total bill is going to be this: $1.05S + 3.49 H + 50$. Easy as it is, it's worth looking at this in a new way, because it's a little roadmap for linear programming formulation. First, look at the algebra. It is linear: number times variable, plus number times variable, plus number.

But here's the important thing; each of those numbers actually answers a specific question: 1.05: If I buy one more can of soup, how many dollars does my bill go up? 3.49: If I buy one more pound of hamburger, how many dollars does my bill go up? 50: If I buy no soup and no hamburger, how much is my bill?

We can generalize this. In any linear expression, the numbers answer one of two questions. First, the constant term, the number with no variable, answers this question: Suppose I set all of the variables to 0. What should the thing underneath be?" The "thing underneath" is the measurable quantity I'm interested in, here, that's the total bill. This problem has two variables, S and H, representing cans of soup and pounds of hamburger. If they're both zero, I don't buy soup or hamburger, so the total cost of my bill is just the cost of everything else in your cart, which I told you was $50, the cost of the original shopping list.

How about the other numbers, the coefficients? A coefficient is just a number multiplying a variable. They answer: "If I get one more of this, and make no other changes, how much does the thing underneath go up?" So, the 1.05 in front of S means, "One more S, and cost goes up $1.05. And this is how it works for all linear expressions. The coefficient of the variable tells you how much the thing underneath goes up if the variable increases by 1. For most variables, this will either mean that you get one more, or use one more, or need one more of something, or something like that.

OK, let's try a less obvious one. Suppose that during a 40-hour work week, a building inspector can inspect either 80 houses or 10 farms. What math represents the number of hours he spends inspecting this week if he inspects H houses and F farms? The problem implies that each farm takes the same amount of time, and each house takes the same amount of time. That makes the expression linear. So algebraically, the answer will look like this: And it's just a fill-in-the-blanks game. Constant term first. If we set both H and F equal to zero, he doesn't inspect anything, so he spends no time inspecting. And that's how you find the constant term. It's 0.

Now, coefficient of H, the question is: if H goes up by one, which means, he inspects one more house, how much does the thing underneath, the hours used, go up? Well, he can do 80 houses in 40 hours, which means it takes him ½ hour per house. So if H goes up by 1, time goes up by ½ hour, like this. And, of course, we use the same reasoning for F. One more farm increases time used by how much? Well, 10 farms in 40 hours means 4 hours per farm, so that's the coefficient of F. And we're done; $0.5H + 4F$ gives the time used. If he only works 40 hours in a week, our constraint would be $0.5H + 4F \leq 40$. Notice that the coefficients showing up in the constraint aren't the numbers that I originally gave you this time. Those were 80 and 10. We found the needed coefficients by asking the right questions.

This time constraint we just created is an example of the most common kind of constraint in a linear program, what I call a limited resource constraint. It says you can't use more of something than you have, or more formally number of units used [is] less than or equal to number of units available. So, how would all of this stuff look in a problem like the one that Union Pacific had to handle? Well, to demonstrate, I'm going to present a problem that's both simpler and harder than the Union Pacific one. It'll be simpler in that it will be much, much smaller than the real problem and have a simplified cost structure. But it will be more complicated in that we're going to work out not only where the empty cars should be assigned to what destinations, but how they're going to get there. We'll have not only supply-and-demand constraints, but also transshipment points in between, which makes this a transshipment problem. Ladies and gentlemen, I give you TGC Railroad, The Great Courses Railroad.

As you can see, we have two supply points, in Atlanta and Baltimore, A and B over on the east coast, and, with 300 supply in Atlanta and 200 in Baltimore. As it turns out, all of our demand for empty cars happens to be in the west, Eugene, Oregon (E), Fresno, California (F), and Great Falls, Montana (G). There's a direct line from Baltimore to Fresno, but otherwise the cars are going to have to pass through two transshipment points, Chicago (C), or Dallas (D). There's no rail running from Dallas to Great Falls, so if you're going to Great Falls, it has to be by way of Chicago. To keep it simple, I'll have all of our rail routes running from east to west.

The three demands are for 100 cars, 150 cars, and 200 cars for the three demand points in the west, which add to 450. Since our total supply was 500 cars, we can meet all of the demand and still have 50 cars on the east coast that will stay in their original positions. The numbers on each edge tell us the cost of moving one car along that particular arrow, for example, it costs $70 to move a car from Baltimore to Chicago. Obviously, we'd need to get these figures before we could solve the problem. I made a simple assumption that the cost of shipping from one city to another was dependent only on distance and costs 10 cents a mile.

All right, let's put the tools that we've developed to work. We're going to create a linear program, step-by-step, whose solution will tell us how we should ship our cars. First, we're going to need to find our objective, our constraints, and our decision variables. To start, let's put the issues into a more friendly form. What am I trying to accomplish? (objective) What rules do I have to obey? (constraints) What quantities do I have direct control over? (decision variables)

Well, the goal in an LP, a linear program, is always to maximize or minimize something. What's the goal for our car-shipping problem? We're trying to minimize the cost of getting the job done. We're assuming here that the only costs we need to worry about are the actual transportation costs of moving the cars. OK.

How about the rules that we have to obey? People often get stuck at this point. What must you do? What can't you do? Well, one thing that can help us is to look for words and phrases that carry this have to/cannot kind of feel. Words and phrases like, at least, at most, no more than, no less than, must, cannot, limited, restricted. Rules can also come from words like supply, demand, budget, required, and so on.

The second thing that can help is to remember the three basic kinds of constraints and what they say. I've already told you about one of these, the limited resource constraint, which says you can't use more of something than you have. A second kind is the minimum performance constraint, also called a quota constraint. A minimum performance constraint says that you have to get at least what you need. Being more formal, it says, number of units obtained is greater than or equal to the number of units required. So, limited resource constraints and minimum performance constraints are two out of the three. Let's work with those, and we'll pick up the third class of constraints a bit later.

Before we go hunting for constraints, though, I have one more helpful hint. There are two questions that can help you find constraints. Here's the first one: What, if anything, prevents me from doing nothing at all? I grant you, I might not want to do nothing, but are you allowed to? If you aren't, then whatever stops you from doing nothing is going to be a constraint.

So how about our shipping problem. Can we do nothing? No! We're required to get cars to the western cities, and if we don't do anything, they won't get there. We have to give Eugene, Fresno, and Great Falls at least the cars that they need, and if you listen to that, they have to get at least what they need; you know that we're talking about a minimum performance constraint. Three, actually, since there are three different conditions that need to be satisfied. It doesn't matter if Eugene gets a surplus of cars if Fresno is shorted. So we have one minimum performance constraint for each city, saying that each city must get at least the cars that it demanded. So, good. Asking what stopped us from doing nothing was a worthwhile question. Sometimes, you are allowed to do nothing, but this time, we found three constraints that prevented it.

OK, second question, useful question. What, if anything, prevents me from blowing the doors off of the situation? That's a very vague term, but what I mean is, what if anything prevents me from making quantities in the problem absurdly huge? Again, it might be stupid to do so, but is it permissible? Let's apply this to the shipping problem. Maybe you see a bunch of quantities in this problem already, like, how many cars we ship from place to place, but for now, I'll assume that you don't, that you just know one quantity, shipping cost.

Now, I certainly wouldn't want to make my shipping billions and billions of dollars, but could I? Well, the only way that I can generate cost is by shipping cars, so a mammoth cost requires a mammoth number of cars shipped, and I can't do that. I only have 500 freight cars that I can move, because the east coast cities have limited supplies. And there we are. We can't exceed the supplies of our cities. Note again that the words "limited" and "supply" show up, suggesting that we're on the right track. Since we can't ship too much out of Atlanta or of Baltimore, this actually is two limited resource constraints.

Now, there's actually one more set of constraints in this problem, but they aren't limited resource nor minimum performance constraints; they belong to the third common kind of constraint, which I call a conservation constraint. A conservation constraint basically says that if you count the same thing two different ways, you have to get the same answer. How does this apply to us?

Well think about the traffic around Chicago. Chicago isn't a source, nor is it a final destination. As a consequence, the number of cars that are coming into Chicago has to be exactly equal to the number of cars that are leaving it. In brief, total in equals total out. And, of course, the same goes for Dallas. So our program looks like this. Minimize shipping cost in dollars subject to the constraints. You must at least meet demand in the three western cities; that's three constraints. You can't exceed the supply in either of the two eastern cities; that's two constraints. And whatever goes into a transfer city is what comes out; that's two more. As you can see, we haven't written down any numbers yet. Numbers are the last thing to go into a formulation.

OK, what's left? Decision variables. What do we have direct control over? Remember that we can get the decision variables by starting with some quantity in the problem and repeatedly asking, on what does that depend? So, the one measurable quantity that we have a firm grip on is our objective, shipping cost. What does it depend on? Well, obviously, on the shipments you make. Good. But remember that we need measurable quantities, number of this, number of that. So, number of cars that you move from place to place. OK, better, but still not enough. Why? Because to figure out my cost, I need to know more than how many cars moved. I need to know where they moved from and where they moved to, because each inter-city link in our map, each edge on our graph, has its own cost. And if I do know the traffic on each link, I can figure the shipping cost as well as checking that I don't violate my constraints. In short, if I know the traffic on each link, I know everything. So, preliminary analysis done. It looks like this.

Minimize the number of dollars spent on shipping subject to the constraints; number of cars reaching Eugene has to be at least what Eugene needs, and the same for the number reaching Fresno and Great Falls; number of cars leaving Atlanta can't be more than Atlanta has, and the same for the number leaving Baltimore; and number of cars going into Chicago has to be the same as what comes out, and the same for the number of cars entering and leaving Dallas.

A few things to notice: First, I relaxed a bit the language for my constraints. Limited resource constraints say "number used less than or equal to number available," but the number of cars used in Atlanta is actually the number of cars that leave the city. Similarly, minimum performance constraints say "number obtained greater than or equal to number required," but in this context, obtaining a car in Eugene means the car reaches Eugene.

Once again, I'd like you to notice that we haven't done anything with the numbers yet. It's all been measurable quantities. If you formulate your linear program correctly, you should be able to read every constraint like a sentence.

Okay, how about decision variables? A moment ago, we said we controlled how many cars we ship along each of the inter-city links, each of the arrows on our map. How many go from Atlanta to Chicago, and how many from Chicago to Eugene, and so on. I'm going to have a decision variable for each of these. Now I could call them w, x, y, z, and so on, but I prefer to use names that remind me what the variable stands for. In this problem, since no two cities start with the same letter, I can unambiguously refer to a link by giving the letter of its origin city and letter of its terminus. So I'll define AC = number of cars going from Atlanta to Chicago. And I'll define decision variables for all the other links in a parallel fashion. So, for example, BF will represent the number of cars that use the direct link from Baltimore to Fresno. Our graph has 10 arrows, so we have 10 decision variables.

The more formulations you do, the better you get at them, because you start noticing patterns. In any transportation/transshipment problem, for example, you're going to expect that you'll have a limited resource constraint for each supply, a quota constraint for each demand, and a conservation constraint for each transshipment point, and a decision variable for each link in the network. You may have some extra bells and whistles added to a particular problem, but the skeleton of the problem is likely to be that. This is true for other problem types, too. Once you work one or two, you quickly develop an intuition on what the program will look like.

OK, we're finally ready to write the program. Let's start with the objective. Since the program includes 10 variables, we have 10 variable coefficients to consider, as well as the constant term. Let's get the constant term out of the way. Earlier in the lecture, we said that to do this, we ask what the quantity of interest, here, our shipping cost, would be if all of the variables were set

174

to zero, minimize number of dollars of shipping cost. Well, the variables tell you how many cars you ship on each link. If they're all zero, you ship nothing, and so your shipping cost is 0; that's the constant term. Good, that was easy.

How do we find all the coefficients of the 10 variables? Just ask, "if this variable goes up by one and we make no other changes, by how much does the quantity of interest increase?" Again, the quantity of interest is the shipping cost, and our variables all represent the traffic on a given inter-city link. If the variable goes up by one, we get one more car of traffic on that link, with no other changes, and so the cost increases by whatever it costs to ship one car on that link. So my objective is just this: 72 AC + 78 AD + 70 BC... and so on, because it costs $72 to ship a unit from A to C, $78 to ship from A to D, and so on. These numbers come from our map.

OK, now let's tackle the first constraint. Number of cars reaching Eugene is greater than or equal to number of cars needed in Eugene. The right-hand side of the constraint is just a number, since we're told that Eugene needs 150 cars. But the left hand side depends on our shipping pattern. Obviously, if we set all the shipping variables equal to 0, we ship nothing, and so Eugene gets nothing. That means the constant term for the left-hand side of our constraint is 0.

How about the variable coefficients? What happens to the number of cars reaching Eugene if I increase the amount of traffic on only a single link in the system? Because that's what it means to increase a variable by 1 here, to send one more car on a single link and leave all the other links alone. So we care about what cars reach Eugene.

Suppose I make a single change of sending one more car on the Baltimore-Fresno link. That doesn't help Eugene at all. Face it, I'm only changing one variable, the traffic on one link, and if that link doesn't end in Eugene, then Eugene could care less about the change. So, mentally, I'm standing on the city limits of Eugene and keeping an eye on the two tracks that do enter it, one from Chicago, and one from Dallas. One more car on either of these tracks will give Eugene one more car. So the constraint's going to look like

CE + DE ≥ 150, because 150 is the number of cars needed by Eugene, and CE + DE totals the traffic on the two rails that enter that city. Once you see that, the other two demand constraints for Fresno and Great Falls are clear.

You expect this kind of demand constraint in any transshipment or transportation problem. There are three ways into Fresno, so Fresno's constraints look like this, BF + CF + DF ≥ 200. There's only one way into Great Falls, so the constraint for them is CG ≥ 100. Count the cars on the links that lead directly into the city, and you have to get at least what the city demands. So, how about the supply constraints, like, number of cars leaving Atlanta ≥ number of cars available in Atlanta? Well, it's very similar. Again, the right-hand side is a known quantity; we're given the number of cars in Atlanta as 300. The constant term for the left hand side? Well, if we set all the decision variables to 0, you're getting this, the number of cars leaving a city is 0, so the constant term for that city is 0. And logic parallel to the one that we used for the demand constraints shows that the only variables that can contribute, in and of themselves, to a car leaving Atlanta are those cars that start in Atlanta, those who begin with the letter A, namely, AC and AD.

If AC goes up by 1, it means we ship one more car from Atlanta to Chicago, and this makes the number of cars leaving Atlanta increase by 1. So, 1 is the coefficient of AC on the left-hand side. Same for AD. So our constraint for the supply leaving Atlanta looks like this. AC + AD ≤ 300, and similarly, the supply constraint for Baltimore is BC + BD + BF ≤ 200. Add up the traffic on all links originating in Baltimore, and it can't be more than Baltimore has.

Now we have the conservation constraints for Chicago and Dallas, but the work we've already done makes this easy. To find out what's going into Chicago, take the traffic on each link that ends in Chicago and add them up, AC + BC. What's leaving Chicago? The sum of all the traffic that's going out of Chicago, CE + CF + CG, all the variables that start with C. So the Chicago constraint is AC + BC = CE + CF + CG. And the same logic gives the Dallas constraint, AD + BD = DE + DF. There is no DG, since Dallas can't ship to Great Falls.

So, are we done? Well, not quite. Most problems have a non-negativity constraint or two, our requirement that the variables in the problem can't be negative numbers. That's the case here, we can't send a negative number of cars along a link. So we actually have 10 extra constraints, each one saying that one of the 10 variables is greater than or equal to zero. Constraints like this are very common, and they're so simple that they're often called trivial constraints. The rest of the constraints, like our supply-and-demand constraints, are called nontrivial. So we need to specify that all of our variables are nonnegative, too. Here, then, is the finished linear program. That is, minimize our objectives subject to several inequalities. Note, the words "subject to" just mean "here come the constraints," which in this case are three minimum-performance constraints, two limited-resource constraints, two conservation constraints, and ten trivial constraints.

Was it worth all of the work? Well, it was to Union Pacific. Their problem was huge, with around 10,000 variables and 1000 constraints, but once their computers were told how to set up the formulation, they were perfectly capable of doing so, setting it up and solving it in 35 to 45 seconds. The model's schedules generally showed cost reductions of 10 to 15%, when compared to the old system of assignment by humans, and the improved schedules and staffing reductions made possible by this generated 35% return on investment.

Today, we've covered the basics of linear programming formulation and given you an example of what the power of linear programming can do. And frankly, once you've seen one transshipment problem like the one we did, you're in good shape to tackle any other. When we get together next time, we're going to expand your repertoire, looking at multi-period problems that unfold over time. We'll see that we can use the tools that we developed in our discussion today to better schedule activities, control inventory, even help determine an investment plan that can give us the money that we need, when we need it. And for our example, we'll leave behind the Union Pacific Railroad in America and travel south to sunny Brazil and to Holambra, the City of Flowers. Exotic foreign lands and linear programming! Hoo-hoo! What more could you want?

Scheduling and Multiperiod Planning
Lecture 9

Multiperiod planning problems are a class of problems in which planning extends over time. In a multiperiod planning problem, we generally need to make essentially the same kind of decisions repeatedly over time. Typically, with multiperiod planning problems, the situation we face at any given point in time depends on earlier decisions, such as with scheduling problems, inventory issues, and personal financial planning. In this lecture, you will learn how to approach problems like these with linear programming.

Multiperiod Planning

- What many **multiperiod planning** problems have in common is a reliance on **conservation constraints**, which are constraints that say that if you count the same thing in two different ways, you have to get the same answer in both counts. In network models, we often take this to mean that total in equals total out. Another way to think of a conservation constraint is as the definition of an **auxiliary variable**.

- To get a handle on what this means, let's think about a production and inventory problem. We have a business where we make a product each month. There's a demand for our product each month, too, and we'll sell whatever we have on hand to satisfy as much of that demand as we can. If we have more product than customers want, what we don't sell goes in inventory.

- There might be more involved, of course—sometimes we're contractually required to meet demand, sometimes we pay a penalty if we can't, or we may be making multiple products with the same equipment, and so on. But let's think about the simple transaction that occurs in any given month, in terms of total in equals total out.

- We start with our leftover inventory from last month. Then, this month, we make products and add them to the pile. That's our total in. Where do the products go? Some of them get sold to meet demand, and whatever's left, we stick into inventory for next month. So,

  old inventory + current production = current sales + new inventory.

- But, if we subtract current sales from both sides, we get

  old inventory + current production − current sales = new inventory.

- This new equation says the same thing as the original but can be looked at as an equation that tells you how to compute new inventory from the other three quantities. Of the four quantities in this equation, only three of them are independent. If we know the values of any three of them, we can figure out the value of the fourth. So, we don't really need all four variables.

- We say that the fourth variable is a dependent variable, or an auxiliary variable. It's one you might choose to use, but you don't really need it, in the sense that its value is completely determined by the values of the other variables. You can make up these "unnecessary variables" anytime you want to, but the caveat is that every one of them has to appear in an equality constraint that essentially defines it, just as new inventory was defined in the example.

Scheduling Problems
- With scheduling problems, the situation might involve making sure that you have enough staff to cover your needs at different times. This kind of problem isn't usually considered a multiperiod problem, in spite of the fact that it evolves through time. Most people would class it as a covering problem—making sure that you have enough coverage over each region in the domain of your problem. The easiest way to see how to formulate a problem like this is often with a timeline.

- Suppose that we run a 24-hour emergency room, and we've decided that we need a certain number of nurses on duty during each 4-hour period through the day. And suppose that each nurse works an 8-hour shift that starts either at midnight, 4 am, 8 am, noon, 4 pm, or 8 pm.

12 am–4 am 5 nurses	4 am–8 am 3 nurses	8 am–noon 6 nurses	noon–4 pm 8 nurses	4 pm–8 pm 9 nurses	8 pm–12 am 11 nurses

Figure 9.1

- Our goal is to make sure that we have sufficient staffing at all times of the day but to not pay more than we have to. We could add all kinds of complications to this problem—such as people being paid more for working the less-desirable shifts, or nurses getting a break for a meal, and so on—but for this example, let's keep it simple.

- In that case, spending as little money as possible on staffing means minimizing the number of nurses used. So, our goal is to minimize the number of nurses used. What are our constraints? There are six **quota constraints**. Every 4-hour interval has to have at least as many nurses on duty as the timeline requires.

- What are our decision variables? We have to decide how many nurses to work in each of the possible shifts—how many start at midnight, at 4 am, at 8 am, and so on. So, let's make variables for that. We'll call the shift that starts at midnight shift 1, the one that starts at 4 am shift 2, and so on. Each shift is 8 hours long, so a nurse working shift 6 comes on duty at 8 pm and works until 4 am on the next day. Let x_i be the number of nurses working shift i, as follows.

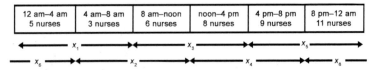

Figure 9.2

- There are x_1 nurses working on shift 1, which runs from midnight to 8 am, and so on. And shift 6 wraps around from one end of the timeline to the other, because it starts at 8 pm today and ends in the wee hours of tomorrow, at 4 am.

- Our objective is the total number of nurses hired. This means that the objective is just the sum of these six variables.

  Minimize $x_1 + x_2 + x_3 + x_4 + x_5 + x_6$

- The coefficients of all of the variables are 1, which just means hire 1 more nurse for any shift and the total number of nurses hired goes up by 1. And the constant term is zero, because if you set all of the variables to zero, you haven't hired any nurses.

- What about the constraints? The first constraint talks about the first 4-hour interval, and it says the following.

  number of nurses working during the 12 am to 4 am interval $\geq$
  number of nurses needed for that interval

- This is a straightforward quota constraint. The right-hand side is the number of nurses that we need from midnight to 4 am, and we're told that this is just 5, from our timeline. What about the left-hand side, the number of nurses on duty from midnight to 4 am?

- If we increase a single variable by 1, such as x_1, that means that we're hiring 1 more nurse for that shift. And 1 more nurse working shift 1 will affect the number of nurses working during the 12 am to 4 am interval, because shift 1 runs from midnight all the way to 8 am.

- So, in particular, the nurses working shift 1 are on duty during the midnight to 4 am interval. On the other hand, the nurses working shift 2 don't start until 4 am, so increasing x_2 by 1 doesn't help us at all in terms of staffing the midnight to 4 am interval. That means that x_2, the number of nurses working shift 2, doesn't appear on the left-hand side of this constraint.

- The timeline shows that only two groups of nurses are on duty during the midnight to 4 am window: shift 1 and shift 6. Midnight to 4 AM is the first 4 hours of shift 1, and it's the last 4 hours of shift 6. So, the midnight to 4 am constraint is as follows.

$$x_1 + x_6 \geq 5$$

- Similarly, each constraint says the following: Add the number of nurses on the two shifts that include this 4-hour interval. That total has to be enough to cover the demand for that 4-hour interval. So, the following is the whole formulation.

Minimize $x_1 + x_2 + x_3 + x_4 + x_5 + x_6$

subject to

$$x_6 + x_1 \geq 5$$
$$x_1 + x_2 \geq 3$$
$$x_2 + x_3 \geq 6$$
$$x_3 + x_4 \geq 8$$
$$x_4 + x_5 \geq 9$$
$$x_5 + x_6 \geq 11$$

and all variables ≥ 0

- In summary, minimize the total number of nurses hired, but require that the number of nurses working during any 4-hour window is at least what we need for that window.

- Why are we using "greater than or equal to" in these constraints? Why not just use "equal to"? Why use more nurses than you need? We're not going to use more nurses than we need—that's what the objective is all about—but that doesn't mean that some 4-hour window might not be overstaffed.

- Imagine changing our staffing requirements for a moment. Imagine, for example, that we needed only 1 nurse in the first 4-hour window,

4 nurses in the second window, and only 1 nurse again in the third window. We can certainly have only 1 nurse on duty during the first and third windows, but those nurses are the only people that could possibly be on duty during the second window, because each nurse works 8 hours—2 successive windows. So, if we have 4 nurses in interval 2, we have to have more nurses than we need in interval 1, or interval 3, or both. Our **program** allows for this possibility.

Important Terms

auxiliary variable: Also called a dependent variable. A variable whose value is completely determined by the values of the other variables in the problem. Formulations may be simplified by using auxiliary variables but must include an equality constraint for each one, which essentially defines how its value is computed.

conservation constraint: An equality constraint in a linear program that generally requires that counting the same thing in two different ways must give the same answer. It can often be interpreted as "total in = total out."

multiperiod planning: Problems in which the same pattern of decisions must be made repeatedly over time.

program: Mathematically, an optimization model. The term corresponds to the idea of a program as a schedule, as in television programming, because many early optimization models were scheduling problems.

quota constraint: The second-most-common kind of linear programming constraint. It says that the amount obtained of something is at least as much as is required. Also called a minimum performance constraint.

Suggested Reading

Caixeta-Filho, van Swaay-Neto, and Wagemaker. "Optimization of the Production Planning and Trade of Lily Flowers at Jan de Wit Company."

Ragsdale, *Spreadsheet Modeling & Decision Analysis*.

1. People often expect that in an optimal solution to a linear program, all limited resources will be completely consumed. This is untrue. To generate a counterexample, imagine that you produce and sell baked goods (and only baked goods) and have found along with the rest of your inventory a jar of dill pickles. (There is nothing actually preventing you from making dill pickle cookies.)

   Answer:

   Using all of the resources available would require using the pickles, too, and because you are in the baking business, that means making some baked good with dill pickles. This product is unlikely to be popular, so profits from it would be low. Worse, the other ingredients of your pickle cookie (or whatever) would no longer be available for the more-profitable products. So, it's likely you could make more money with a solution that let the pickles go unused.

2. The total cost of a firm is its fixed cost plus its variable cost. As the name suggests, the fixed cost of a firm, including such things as the costs of its buildings, can't be varied in the short term. Suppose that two analysts create a linear program for the short-term decisions of a company, using the same constraints. One has the goal of minimizing total cost while the other has the goal of minimizing variable cost. Why would their optimal solutions give the same values of the decision variables? (It's because of this that we are comfortable ignoring fixed costs when solving such linear programs.)

   Answer:

   Fixed costs, by definition, can't be changed over the time horizon of the problem. So, because total cost = fixed cost + variable cost, total cost will be at a minimum precisely when the variable cost is at a minimum.

Scheduling and Multiperiod Planning
Lecture 9—Transcript

There's a lot of money to be made in the flowering plant business, if you know what you're doing. Dealing with living plants causes its own share of risk, of course. They only grow in certain places, and in some of those places, only at certain times of the year. Leave them untended, and they may die. You can't get them to grow faster by working overtime. And if you cut them, they're perishable. You can't just stick them in a warehouse until someone wants to buy them.

The Jan de Wit Company knows all about the challenges of raising and selling lilies. Buying bulbs in Holland and growing lilies in Brazil, they raise and sell about 420,000 pots and 220,000 bunches of lilies per year, making them Brazil's largest grower of oriental and Asian lilies. No, they're not the same thing. In a country that really loves its flowers, Brazilian consumers spend about $1.1 billion a year on them.

But with an operation so large, Jan de Witt faces a serious planning problem. Their rich plot of land is relatively small, only about five acres, but on this they have to plant different batches of bulbs; bulbs come in many different kinds, and different bulbs need different kind of greenhouses, different kinds of soil, different spacing, different amounts of time to grow. And cut flowers are handled differently than potted flowers. Jan de Wit needs to get the right kind of flowers in the right quantities, processed in the right ways, to market at the right time, all with the obvious restriction that they can't use the same patch of earth to grow two different batches of lilies at the same time. So how should they set their production levels and schedule their planting and harvesting if they want to maximize their profit?

This is a wonderful example of a problem where planning extends over time. It's also a scheduling problem. In this case, the job is to figure out what batches of bulbs should be planted, and where and when they should be planted. Obviously, the situation Jan de Wit faces at any given point in time depends on its earlier decisions. This is typical of a whole class of problems called multiperiod planning problems. In multiperiod planning problems, you generally need to make, essentially, the same kind of decisions repeatedly over and over, over time. What do I plant today, and where, and what do I harvest and sell?

You don't have to be in lily businesses to care about multiperiod planning. Think of any firm that deals with production and inventory. It has a certain production capacity over time, perhaps somewhat supplemented by some possibilities of overtime, and it can produce units for the market. But it's possible that the firm may be able to produce more units than the market wants, and so the extra units go into inventory, for later sale. But storing units costs money too, of course. If nothing else, it ties up your capital. So, how can you decide the best balance between production and inventory costs?

Or consider staff issues. Your situation could be as simple as needing a certain number of people to be working during each of a collection of time windows, or it could be much more complicated. You have a number of people who work for you, maybe with differing skill sets, and need to assemble a schedule so that the teams of people working at the same time cover all of the requisite skills. Catholic Diocese of West Texas had a problem like this, scheduling clergy in light of the language needs of the various churches. And believe it or not, even they used linear programming on the problem to handle it. Or how about personal financial planning? Given a number of different financial instruments and an anticipated schedule of inflow and outflow of cash, how can you plan an investment strategy that leaves you in the best position for retirement?

In today's lecture, we're going to look at how we can approach problems like these with linear programming. One thing that a lot of them will have in common is a reliance on conservation constraints. You'll remember we talked about these in the last lecture as constraints that say that if you count the same thing in two different ways, you have to get the same answer in both counts. For our network models, we often took this to mean, total in equals total out. And this perspective will be useful today, too.

Another way to think about a conservation constraint, though, is as the definition of an auxiliary variable. To get a handle on what this means, let's think about a production and inventory. We have a business where we make stuff each month. There's a demand for our product each month, too, and we'll sell whatever we have on hand to satisfy that much of that demand as we can. If we have more stuff than they want, what we don't sell goes in inventory. Now, there might be more involved, of course; sometimes we're contractually required to meet demand, sometimes we pay a penalty if we can't, and we may be making multiple products with the same equipment, and so on. But let's think about the simple transaction that occurs in any given month, in terms of total in equals total out.

I start with my leftover inventory from last month. I take from the stuff that I make this month, and, and I'll add it to the pile. There, that's my total in. Where does it go? Well, some of it gets sold to meet demand, and whatever's left, I stack up, and I put into inventory for next month. So, old inventory plus current production equals current sales plus new inventory.

But if I subtract my current sales from both sides, I get, old inventory plus current production minus current sales equals new inventory—which says exactly the same thing, but can be looked at as an equation that tells you how to compute new inventory from the other three quantities. If you like, of these four quantities in this equation, only three of them are independent. If you tell me the values of any three of them, I can tell you the value of the fourth. So, I don't really need all four variables. We say that the fourth variable is a dependent variable, or an auxiliary variable. It's the one that you might choose to use, but you don't really need it, in the sense that its value is completely determined by the values of the other variables. You can make up these unnecessary variables any time you want to, but the caveat is that every one of them has to appear in an equality constraint that essentially defines it, like new inventory was defined in our example of a minute ago.

The production-and-inventory kind of problem is part of what Jan de Witt has to worry about, in fact, they have two phases of it. For the purchasing department, "making" really turns into "buying," and so they have to make sure that their bulb inventory satisfies the demand for the bulbs required by their planting operation. Then, of course, there is the plant business itself, making sure you grow enough plants of each type to meet whatever contractual agreements that you've committed yourself to for the wholesaler, but not so many that you have a lot of excess inventory. I don't know how well Jan de Wit could store potted plants, but I think that maintaining an inventory of cut flowers, at least, would be problematic.

Of course, Jan de Wit had another kind of multiperiod concern, too, a scheduling problem. For them, it was setting up a planting and harvesting schedule that didn't dedicate the same plot of land to different batches at the same time. In other scheduling problems, the situation might be exactly reversed, making sure you have enough staff to cover your needs at different times. This kind of problem isn't usually considered a multiperiod problem, in spite of the fact that it evolves over time. Most people would class it as a covering problem, making sure you have enough coverage over each region in your domain of your problem. The easiest way to see how to formulate a problem like this is often with a timeline. Let me give you an example. Suppose you run a 24-hour emergency room, and we've decided we need a certain number of nurses on duty during each four-hour period through the day, like this. Let's suppose each nurse works an eight-hour shift and starts either at midnight, 4:00 am, 8:00 am, noon, 4:00 pm, or 8:00 pm.

Our goal is to make sure that we have sufficient staffing at all times of day, but not to pay more than we have to to do so. We could add all kinds of bells and whistles to this problem, like people being paid more for working less-desirable shifts, or nurses getting lunch breaks, and so on, but let's keep it simple.

In that case, spending as little money as possible on staffing means minimizing the number of nurses used. So, my goal is to minimize the number of nurses used. My constraints? Well, six quota constraints, every four-hour interval has to have at least as many nurses on duty as my timeline requires. OK, what are my decision variables? Well, I have to decide how many nurses work in each possible shift, how many start at midnight, 4:00 am, 8:00 am, and so on.

So, let's make the variables for that. I'll call the shift that starts at midnight shift 1; the one that starts at 4:00 am, shift 2, and so on. Remember, each shift is 8 hours long, so a shift-6 nurse comes on duty at 8:00 pm and works until 4:00 am on the next day. I'll let x_i be the number of nurses working shift i, like this. So there are x_1 nurses working in shift 1, which runs from midnight to 8:00 am, and so on. You can see that shift 6 wraps around from one end of the timeline to the other, since it starts at 8:00 pm today and ends in the wee hours of tomorrow, at 4:00 am.

Okay, so, what is my objective? The total number of nurses hired, that means that the objective is just the sum of the six variables, $x_1 + x_2 + x_3 + x_4 + x_5 + x_6$. The coefficients of all the variables are 1, which in terms of our question game from last lecture just says, hire one more nurse for any shift, and the total number of nurses hired goes up by 1. And of course, the constant term is 0, since if you set all of the variables to 0, you haven't hired any nurses at all.

OK, how about the constraints? Well, the first constraint talks about the first four-hour interval, and it says, number of nurses working during the 12:00 am to 4:00 am interval is greater than or equal to the number of nurses needed for that interval, which is a straightforward quota constraint. The right side is the number of nurses that we need from midnight to 4:00 am, and we're told that this is just 5, from our timeline.

How about the left-hand side, the number of nurses on duty from midnight to 4:00 am? Well, if we increase a single variable by 1, like x_1, that means that we're hiring one more nurse for that shift. And if you look at the picture, you can see that one more nurse working shift 1 will affect the number of nurses working between midnight and 4:00 am, because the shift 1 runs from midnight all the way to 8:00 am. So in particular, the shift-1 nurses are on duty during the midnight to 4:00 am interval. On the other hand, the shift-2 nurses don't start until 4:00 am, so increasing x_2 by one doesn't help us at all in terms of staffing the midnight to 4:00 am interval. That means that x_2, the number of nurses working shift 2, doesn't appear on the left-hand side of this constraint.

Well, a quick look at the timeline will show you that only two groups of nurses are on duty during the midnight to 4:00 am window, shift 1 and shift 6. Midnight to 4:00 am is the first four-hour of shift 1, and it's the last four hours of shift 6. So the midnight to 4:00 am constraint is, $x_1 + x_6 > 5$. And once you see this, all of the other constraints are very much the same. They each say, add the number of nurses on the two shifts that do include this four-hour interval. That total has to be enough to cover the demand for that four-hour interval. So, here's the whole formulation. To sum it up, minimize the total number of nurses hired, but require that the nurses working in any four-hour window is at least what we need for that window. Make sense?

By the way, you might wonder why I said "greater than or equal to" in these constraints. Why not just say "equal to"? Why use more nurses than you need? Well, we're not going to use more nurses than we need; that's what the objective is all about, but that doesn't mean that some four-hour window, we might not have it overstaffed. Let's imagine changing our staffing requirements for the moment. Imagine that, for example, we needed only 1 nurse in the first four-hour window, 4 nurses in the second one, and only 1 nurse again in the third four-hour window. We can certainly have only one nurse on duty during the first and third windows, but those nurses are the only people that could possibly be on duty during the second window, since each nurse works 8 hours, two successive windows. So, if we have to have four nurses in interval 2, then we have to have more nurses than we need in either interval 1, or interval 3, or both. And our program allows for this possibility.

All right, we've seen a couple of programs now, the train-shipping one from last time, and the nurse-scheduling from this time, where a lot of the coefficients were just 1. I'd be giving you a very false impression if I let you think this was usually the case. So let's look at one more multi-period problem, a financial one, where you can see a little different use of a timeline and develop a more interesting program.

So, imagine this; You've received a half a million dollars, an inheritance from a rich relative, and you want to invest it so that at the end of four years, you've increased its value to as high an amount as you can. Being conservative, you're only going to put money into very safe investments, where the return is, essentially, guaranteed. To keep it simple, let's suppose that you've decided to limit yourself of four investment options.

Here are three of them. At the end of each year, what have you paid? What's this table telling us? Well, we've got investments A, B, and C. For each $100 you put in investment A, you get $2.00 a year from now, another $2.00 at the end of year 2, another 2.00 at the end of year 3, and at the end of year 4, the investment matures, and you are paid back your $100, as well as $10 more. You've made a total $16, spread out over four years. Investment B is less complicated. It's only available right now, but at the end of year 2, it pays $105 for each $100 that you invested in it, your original investment, plus $5.00. Investment C is only available at end of year 2, but a $100 investment in it will pay you $53 at the end of year 3 and $53 more at the end of year 4. The final investment option is the money market, which is available every year, and always pays a fixed 2% interest per year. At the end of the year, the money comes out, but you can do what you want with it. These interest rates aren't great, but remember that you've decided to stick with safe investments.

One more thing about this problem: You want to take a vacation in year 4, so you want to withdraw $10,000 from the investments at the end of year 3 for that purpose. So, how should you allocate your money? A problem like this looks complicated, but you already know everything you need to set it up. By the way, this kind of problem is sometimes turned on its head when people want to settle up a future debt stream by making one lump sum payment now, like a divorce settlement. It lets you figure out how to minimize the current cash needed to generate the payments over time, assuming various investment options are going to be available.

Again, the problem is easier if we look at a timeline. But this one will look a little different from the last one. Take a look. At the top, we've just got the timeline. But the arrows show, for each investment, when the money goes in and when the money comes out; A gives money for all four years, for example, while each money market investment, at the bottom, pays off after only one year. I'll add some numbers to this picture, showing how much money comes out of each point for each $1.00 invested. I'll also show the original cash available for the investment, and indicate the withdrawal at the end of year 3 from the vacation; $500,000 in on the left, and $10,000 coming out at the end of year 3, with all of the other arrows showing you the returns from $1.00 of the investment over time.

So investment C, for example, has you investing at the end of year 2, and gives you 53 cents on the dollar both at the end of year 3 and the end of year 4. The timeline was actually pretty easy to draw, and once you have it, the formulation of this program is a piece of cake. Let's follow our usual procedure. First: What's the goal? Easy, maximize cash on hand at the end of year 4.

Constraints. What rules do we have to obey? Well, we have a limited resource constraint on our original money. We can't invest more than a half a million dollars at the beginning of year 1. On the other hand, a glance at the program shows that it's silly to leave any money uninvested in year 1; it would be better in the money market. So, we have to invest $500,000 at the beginning of year 1. We also need to have $10,000 cash available at the end of year 3. That's a minimum performance constraint; we have to meet that quota. What else?

Well, if you remember the examples from last time with networks, and our discussion today of conservation constraints, you may have a good idea. At the end of each year and the beginning of the next, some funds become available, and they need to be reinvested, since even the money market is better than just stuffing the cash in your mattress. So at each year break, the total cash coming in to that year has to be the total cash going out of it. And really, that's it.

Okay, how about decision variables? What do we get to decide? What do we get to decide? Or, if you like, what does something like our final cash on hand depend on? Well, it all depends on the investments that we choose to make. Since variables are always number of something, let's sharpen that, the number of dollars that we put into an investment opportunity. Looking at our timeline, there are 7 of these, A, B, C, and the four different years of money market funds.

OK, we've fleshed out the model. Now let's do the math. I'll let A be the number of dollars that I invest in investment A; B be the number of dollars invested in B; and so on. My objective is to maximize my cash on hand at the end of the four years. What does that look like mathematically? Actually, a lot of people get this wrong. They include too much. Remember the question we're supposed to ask. Suppose this one variable only goes up by 1. How much does the thing underneath increase? Here, the thing underneath, is the quantity of interest, which is how much cash we have at the end of year 4. Look at the timeline again.

You can see that if I put one more dollar in investment A and do nothing else, I do, indeed, affect how much cash I have in year 4. And investment A dumps $1.10 into the end of year 4 for each $1.00 invested now. Yes, yes, it also drops money in other years, but that money is going to be reinvested. The only direct impact on final cash from Investment A is that $1.10 return per $1.00 invested. And in the same sense, B doesn't directly contribute to the final cash at all. If the only variable I changed was B, there'd be more money at the end of year 2, but it wouldn't go anywhere. That money from B will show up at the end as a reinvestment. Saying it simply, my final cash is just the money associated with all of the arrows that dump out at the end of year 4. So my final cash is, maximize $1.10 A + 0.53 C + 1.02 M4$, and that's it.

If it feels to some of you like I pulled a fast one, imagine that there was an investment that paid no interest. If I put $1.00 in it now, I'd have $1.00 at the end of year 1, $1.00 at the end of year 2, $1.00 at the end of year 3, and $1.00 at the end of year 4. That doesn't mean I now have $4.00 or $5.00, it's just $1.00 that I looked at at different points in time. The only one that really counts toward my final cash is the last one, the one that dumps into the end of year 4. And all of the constraints run on the same reasoning. At any year's break, the total money in has to equal the total money out. So, at the start of year 1, total money in, which is just the half a million dollars, has to equal the total money out.

Where can this money go? To three different places, I can put it in A, I can put it in B, or I can put it in the first-year money market. So, $500,000 = A + B + M1$. Easy, huh? When you see that first equation, that equals sign, it lets you know that you actually have an auxiliary variable, and this is true. If I told you how much of the half a million I put in A and in B, you'd know how much was left, and that all goes in M1.

OK, how about the end of year 1/beginning of year 2? What's this constraint look like? It's this, $0.02 A + 1.02 M1 = M2$. Take a second to see why this makes sense. The left hand side records all of the arrows dumping into the end of year 1. Each dollar in A gives 0.02 dollars of cash at the end of year 1, just two cents. Each dollar in M1 gives you the dollar back, plus two cents more. So the left-hand side tells us how much cash shows up at the end of year 1. And that's equal to the cash going out of year 1. Where can it go? There's only one arrow that starts at the end of year 1, M2.

So the cash that we got from A and M1 at the end of year 1 is the same as the money that we invest into money market for year 2. OK? And once you see this, you have it all. We keep playing the same game. Look at the timeline, and remember total in equals total out, and you get this for the end of year 2. Money coming in from A; money coming in from B, finishing off; and money coming in from money market 2, finishing off. And where can it go? The two investments available at the end of year 2, beginning of year 3. C and then 3. So, let's move on to the end of year 3. This is slightly different, because of the withdrawal of $10,000 for the well-deserved vacation, but that doesn't cause us any troubles. The constraint looks like this. It still just says, total in equals total out. Money is still dribbling in from A, which won't be done for another year; C is making a healthy contribution; and money market 3 is finishing up, giving back each dollar invested in it, plus 2 cents per dollar interest. And this money is going out, and it only includes one possible investment, M4, and the $10,000 that you need for the vacation.

And those are all of our conservation constraints. And, of course, we have the nonnegativity constraints again, as we usually do. You can't invest a negative amount in anything.

This was actually a pretty sophisticated problem, but I hope you see that once you start to understand how linear programs are put together, you can model the problem with almost ridiculous ease. Wait, am I really done? I said before that I had a limited resource constraint: that I can't spend too much. I also said that I had to have at least $10,000 cash available at the end of year 3; that was a quota constraint. What happened to them? Well, you can certainly include them if you want, but believe it or not, we've already covered those bases. For example, the first constraint actually says you have to invest exactly $500,000; $500,000 = A + B + M1. So, we don't need a second constraint saying that we can't invest more than $500,000.

And how about the requirement of at least $10,000 cash on hand at the end of year 3? Well, the end of year 3 constraint was, $0.02 A + 0.53 C + 1.02 M3 = M4 + 10,000$, where the left-hand side is the money coming in, and the right-hand side was the money going out. If the money coming in were less than $10,000, then to make the equation work, M4, over on the right, would have to be negative. But our nonnegativity constraints say that none of our variables can be negative. So our conservation constraints, actually, are all that we need.

This program has a pile of auxiliary variables, four of them, in fact, one for each equality constraint in our program. Another way of saying this is that if you knew the values of A, B and C, you'd know everything, because whatever you don't put into one of these funds gets dumped into the money market. We might not have needed the money market variables; we never need auxiliary variables, but they can make the formulation a lot easier to create and interpret, as they did here. We came out smelling like roses, or, lilies.

Speaking of which, let's get back to Jan de Wit. How did things go for Jan de Wit Company and their operations? Well, let's compare their performance in 1999 to their performance in 2000; 1999 was before the implementation of the linear program solution, and 2000 was right after it. Revenue up 26%, to about $1.6 million; pot sales up 14.8%; bunch sales up 29.3%. And of course, the formulation could be used year after year, just by changing the parameters, like demand. With confidence in the new procedures, Jan de Wit decided in 2001 to depart from its previous market limitations and take on its main competitor in the market for oriental cut lilies.

Of course, the higher revenues don't mean much if your costs are out of control. How did Jan de Wit fare on cost? Costs dropped, from 64% of sales in 1999 to 62% of sales in 2000, and with the expectation of getting to only 54% in 2001. Total sales minus variable cost increased 32% in a single year, from 1999 to 2000. Not bad, especially considering that the Brazilian flower market had considerable excess supply in 2000. And the new system let Jan de Wit better anticipate its upcoming bulb needs, allowing them to sign longer-term contracts with suppliers and lock in better prices. Quality? Well, the percentage of first-quality potted plants stayed the same in both years, an impressive 93%. But cut lilies jumped from 11% best quality in 1999 to 61% best quality in 2000. Bulb quality improved, too, and pre-planting bulb losses were cut to about a third of their 1999 rates.

The linear program did more than determine how to schedule the planting of the lilies that Jan de Wit had already decided upon. It also considered what mix of products would be most productive, given the demands of the market. That is, in addition to where and when to plant each of the batches of bulbs, it also had variables for how many of each bulb type should be planted. Only 6 of the top 10 varieties in 2000 had been on the top-10 list in 1999. The model found both what to do and how to do it.

It was a big problem, with about 420,000 decision variables and about 120,000 constraints, but in spite of this size, the model could be summarized on one sheet of paper. There were 25 different families of constraints and 15 different families of variables. So while the problem was mechanically huge, conceptually, it's quite reasonable. And Jan de Wit's success shows that computationally, it was both possible and well worth doing. They didn't gild the lily. They turned the lily into gold.

I hope I've been able to convince you of the usefulness of linear programming, both that it can be applied to problems of practical importance, and that it can often help find answers that are far superior to those obtained by human trial and error. But maybe you've noticed that our discussion so far has one big gap, namely, how do you solve the darned things?

In these last two lectures, we've seen how a practical problem, given in English, can be reformulated into a mathematical model, a linear program. Hopefully, you're starting to feel comfortable with that. But once you have the formulation, once you have, as it were, translated the original problem into the language of mathematics, then how do you go about finding the best solution to the problem? Before 1947, no one really had a very good answer to that. But in that year, George Dantzig—the guy with the bouillon cube diet, remember?—George Dantzig came up with the simplex method: a fairly straightforward approach that would find the optimal solution to any linear program that had an optimal solution. It's essentially his algorithm that's implemented in many computer environments that can solve linear programs, including spreadsheet programs, like Excel or Calc. We're going to fill in this gap in our discussion of linear programs, the solution part, because I'm going to show you how to enter any reasonably-sized linear program into a spreadsheet and quickly and easily find its optimal solution.

But before we get into this general solution procedure, we're going to take a little detour into an alternative, easier to visualize way of solving linear programs, solving them by, essentially, drawing the appropriate picture. There are some significant limitations to this graphical approach, but the detour is worth making, since one picture really can be worth 1000 words. Most of us process images much better than numbers or equations, and this pictorial approach is going to let us get a more concrete feel for some of the key characteristics of linear programs and their solutions. And surprisingly, the key to solving linear programs is going to be linked to a practical survival skill. Where should you go when the flood waters rise?

Visualizing Solutions to Linear Programs
Lecture 10

I n this lecture, you will learn how to solve a linear program by using a graphical method—by drawing a picture. There are some significant limitations to this graphical approach, but there are also some important advantages. Most people process images better than numbers or equations, and this pictorial approach provides many insights into some of the key characteristics of all linear programs and their solutions. More complicated linear problems may be more difficult to grasp in their entirety, but their solutions rely on the same observations and properties that the simple problems do.

A Graphical Approach to Personal Financial Investment

- Let's say that you have $80,000 to invest in three different investment options. Each investment pays you back after four years. The first option is moderate **risk** and moderate return. It's 2/3 likely to give you a 40% return over four years, but there's a 1/3 chance that instead it will lose you 20% of your investment in that time.

- Even in this linear program, there is a stochastic element, because the objective is to maximize your average return in this situation, also called your expected return. To calculate the expected return, multiply each possible outcome by its probability of occurring and add the results, as follows, resulting in an expected return of 20%.

  $$2/3(40\%) + 1/3(-20\%) = 20\%$$

- The second option is high risk and high return. It's 50% likely to give you a 150% return on investment, but it's also 50% likely to give you only half of your money back. This means that its expected return works out to be 50%.

  $$1/2(150\%) + 1/2(-50\%) = 50\%$$

- The final option is essentially a risk-free investment, and it gives a 15% return, total, over the four years.

- Your goal is to maximize the expected return, or average return, on the money you invest. To limit your downside risk, though, you want to be sure that you lose at most $10,000, even if the risky investments turn sour. You also want a diversified portfolio, so you want no more than $40,000 in any single investment.

- Initially, this may seem like a three-variable problem—how much money do you put in each investment? But because we know that the total investment is $80,000, as soon as we know how much we put in the first two investments, we know how much we put in the third. That is, the third variable is really an auxiliary variable. So, the program is as follows.

Maximize $0.20M + 0.50H + 0.15(80,000-M-H)$

subject to

$M \leq 40,000$
$H \leq 40,000$
$80,000-M-H \leq 40,000$
$0.2M + 0.50H - 0.15(80,000-M-H) \leq 10,000$

and M, H both ≥ 0

- The objective just records the expected, or average, return on each of the three investments: 20% for the medium-risk one, 50% for the high-risk one, and 15% for the risk-free one. We're trying to maximize this expected return, and we get it by multiplying the expected rate of return for an investment by the amount of money that we put into it.

- The first three constraints just say that none of these three investments can have more than \$40,000 put into it. The fourth constraint is our Murphy's law constraint, which says that the money we lose on our investments, even if everything goes wrong, is at most \$10,000. In this worst case, we lose 20% of the medium-risk money and 50% of the high-risk money, and this loss is counterbalanced somewhat by what we make on the risk-free investment. We also have **nonnegativity constraints**, because we can't invest negative dollars in any option.

- We can clean up the algebra by multiplying out parentheses and gathering like terms. We can also rewrite the problem in terms of thousands of dollars, rather than just dollars, to get rid of all the triple zeroes at the ends of everything. The following program might look different at first glance, but mathematically, nothing has changed.

Maximize $0.05M + 0.35H + 12$

subject to

$M \leq 40$	No more than \$40,000 in moderate-risk
$H \leq 40$	No more than \$40,000 in high-risk
$M + H \geq 40$	No more than \$40,000 in risk-free
$0.35\,M + 0.65\,H \leq 22$	Must lose no more than \$10,000

and M, H both ≥ 0

Finding Possible Answers

- How do we solve the problem? We have two decision variables, M and H, and we'll give each one its own axis in a Cartesian plane. Then, we'll add the constraints one at a time to see what points in that plane satisfy each one. A point that satisfies every single constraint is going to be a **feasible** solution, a possible answer. We'll be able to see easily exactly which points—which combinations of M and H—these are.

- Then, we're going to have to decide which of these possible answers is the best one—for this problem, which one gives the largest expected return. And this is going to work because of the geometric character of linear functions.

- Linear relations between two variables graph as straight lines. More generally, linear relations among a collection of variables graph as lines, or planes, or higher-dimensional equivalents called **hyperplanes**. The crucial point about these objects is that they are flat. They don't wiggle, or bend, or kink, or jump. So, everything we draw in this problem is going to be a straight line, thanks to the linearity of our program.

- We graph all of the constraints to see what's possible. The points in the shaded region in **Figure 10.1** satisfy every one of our six constraints: the four **nontrivial constraints** and the nonnegativity constraints. That's true of the points on the border of the shaded region, too. Conversely, any point outside of the shaded region violates at least one of the constraints; it's on the wrong side of a line, so it doesn't represent a possible answer to our question. The shaded region is called the **feasible region** because it contains precisely the points that satisfy every single one of the constraints.

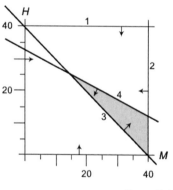

Figure 10.1

Finding the Best-Possible Answer
- Among the points in the feasible region, which is best? To get at the answer, we're going to pick a point in the feasible region and figure out what its expected return would be. For example, suppose that we pick the point with coordinates $M = 40$, $H = 0$, which

means putting $40,000 in the medium-risk investment and $0 in the high-risk investment. The other $40,000, then, will go in the risk-free investment.

- What expected return do we get? The $40,000 in the risk-free return is guaranteed to give us 15%, which is $6000. The $40,000 in the medium-risk investment on average pays 20%, so it gives us $8000 on average. That's a total of $14,000 in expected return. We're going to do the same thing with every other point in the feasible region.

- This is a linear program, so it graphs as straight lines, planes, or hyperplanes, and the defining property of these things is that they're flat. Our objective—expected return—is linear, and that means that the three-dimensional graph of our expected returns is going to be a flat plane. It might be tilted, but there won't be any dips, bumps, or kinks.

- For this problem, a flat triangle is formed, and no matter how it's tilted, one of the corners is going to be a highest point—it's going to be as high or higher than everything else. And that's the point we want. This is the extreme point theorem of linear programming: If there's an optimum point, there's one at a corner of the feasible region. So, we could just check all three corners to find the optimum. But we can also take a different approach.

- Imagine that we take the triangular field and flood it to a certain depth. When the water gets high enough, some of the points will be submerged, but others will still peek above the surface. Let's use an aerial view, looking straight down. We see the triangular field, the feasible region. And cutting across it is the line separating the submerged part of the field from the part still peeking above the water.

- This "flood line" is usually called an **objective function line (OFL)**. In our flooded field, all the points along it are right at the water's edge, so they all have the same height. In terms of our original problem, this means that each of the points along the OFL gives the same expected revenue. That is, it gives a higher expected return than the level represented by the water.

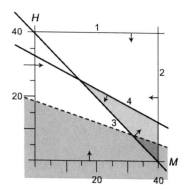

Figure 10.2

- If you want a shot at a higher expected return, send some more water into the picture. The water level will rise, and the points that are still not submerged make you more money on average.

- How do we know how to draw the flood line? We just take the objective function and set it equal to whatever number pleases us. The number specifies the expected return we're looking for—it's the depth of the water. When we set the objective equal to a number, we get a linear equation, so it graphs as a straight line. That's why the water's edge is a straight line in the aerial view.

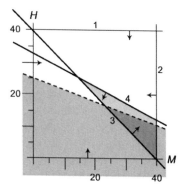

Figure 10.3

- If you think of this as a map, you'll avoid a common mistake. In the picture, the coastline between flooded and not flooded separates water in the southwest from land to the northeast. As the water rises, the coastline covers more land and moves northeast. But, of course, this wouldn't have to be so. The Nile River flows north on the map, but that doesn't mean it's flowing uphill.

- In general, when you draw the first objective function line, you need to take a moment to figure out which side of the coastline is water and which is dry land. The highest land is not always in the northeast.

- To find the highest point, we just keep the water coming. The coastline keeps creeping in—in this case, from the southeast—and eventually only one point in our shaded triangle is still above water, which is the upper-left corner of the triangle.

- Where is it? From the graph, we can tell that it's somewhere around $15,000 in the moderate-risk investment and $25,000 in the high-risk investment, which would leave about $40,000 for the risk-free investment.

- We can get precise values for the coordinates by taking the two constraint equations that meet at the optimal point and solving them simultaneously. This involves more basic algebra.

- Solving constraint equations 3 and 4 simultaneously gives us that $M = 13\ 1/3$ and $H = 26\ 2/3$. That is, the best we can do is to put $13,333 in the moderate-risk option, $26,667 in the high-risk option, and the remaining $40,000 in the risk-free option. Doing so will give an expected return of $22,000 on your $80,000 investment, or a 27.5% expected return.

$$0.2(13,333) + 0.5(26,667) + 0.15(40,000) = 22,000$$

- That's the unique, best answer to this problem. Any other solution either gives a lower return or breaks a constraint.

feasible: Possible. In a mathematical program, feasible solutions are those that satisfy all of the constraints.

feasible region: The set of decision variable combinations that correspond to feasible solutions to the problem. In a two-variable linear program, the feasible region, if it exists, will be one contiguous convex region bordered by straight lines.

hyperplane: The higher-dimensional equivalent of a plane. Think of a flat wall. Don't worry—most people can't visualize them.

nonnegativity constraint: A constraint requiring a particular decision variable to be at least zero. Also called a trivial constraint.

nontrivial constraint: Any constraint other than a nonnegativity constraint.

objective function line (OFL): In a graphically solved linear program, a line of constant objective function value. OFLs act like contour lines on a topographic map. In problems with more than two variables, OFLs are replaced with planes or hyperplanes of constant objective function value.

risk: Variability in the desirability of the possible outcomes of a situation involving uncertainty. Decision makers may be risk neutral, meaning that they choose whatever course of action gives the highest expected payoff. Alternatively, they may be risk averse or risk loving. Risk-loving individuals are inclined to accept a gamble even if, on average, it does not benefit them. Risk-averse individuals are inclined to refuse a gamble even if, on average, it does benefit them.

Suggested Reading

Ragsdale, *Spreadsheet Modeling & Decision Analysis*.

1. In this lecture, we addressed only maximization problems, but an entirely equivalent procedure exists for solving minimization problems. If the word "maximize" in a linear program is changed to "minimize," what effect, if any, would this have on the feasible region? The original objective function line? The direction that the "flood line" moves? How would you characterize the optimal point for the "minimize" problem in terms of our flooded field?

    Answer:

    The feasible region would not change at all, nor would the original objective function line. The flood line would move in the opposite direction, though. We could imagine withdrawing water from a partially flooded field, rather than adding it. The optimal point would be the very last point of the feasible region revealed as the floodwaters recede.

2. The original objective function line that is drawn is generated by setting the objective equal to an arbitrary constant, setting the "depth of the water" in our flood interpretation. Show mathematically or by logic that if two different objective function lines are drawn for the same problem, they must be parallel. (Hint: Remember that they're straight lines, because the objective is linear.)

    Answer:

    The algebra is straightforward, the simplest method probably being to put both equations in slope-intercept form and verifying that they have the same slope. Logically, it's even easier. Each objective function line sets the objective equal to some fixed number. Suppose that you have objective function lines obtained from setting the objective equal to two different numbers, such as 10 and 20. If they are not parallel, then they intersect, and at the point of intersection, the objective function would then have two different values—both 10 and 20. Clearly, this is impossible.

Lecture 10: Visualizing Solutions to Linear Programs

Visualizing Solutions to Linear Programs
Lecture 10—Transcript

The comedian Steve Martin had a monologue on *Saturday Night Live* that went like this, "You can be a millionaire, and never pay taxes! You say, 'Steve, how can I be a millionaire and never pay taxes?' First, get a million dollars. Now…"

It might seem that I've been doing much of the same thing, maybe flipped around, in the last two lectures. My proposal for how to solve complicated problems also has two steps. First, translate the problem from English into mathematics, creating a linear program. Second, solve the linear program. We've been doing great on the first step. So far on the second part, the actual solving, we got nothing.

But we can fix this, and that's what we're going to be doing in the next two lectures. And it may surprise you to learn that, really, we've already done the hard part. You've already got the million dollars! The hard work is in writing the program, like we've been doing; the rest is remarkably mechanical.

I just love this about mathematics, being able to crack a specific problem in life is always gratifying. But with mathematics, you've got all these problem types that we know how to solve in generality. Linear programs are one such type. Turn George Dantzig's simplex method loose on a linear program, any linear program, and it will find the optimal solution to it, or tell you that it doesn't have an optimal solution. And it will do so by a series of purely mechanical steps.

It isn't the only way to solve linear programs. There are faster solution procedures for some specialized problems, or, for extremely large problems, there are some clever ways that are really quite surprising, but all of them are algorithmic. That means that solving a linear program is just mechanical cranking, the kind that a computer will do without complaint. I'm going to show you how to set this up yourself, using a spreadsheet, in the next lecture. But today, my focus is a bit different.

The simplex method is an algebraic approach, and that's great. But for most people, looking at algebra doesn't provide a lot of intuition about what's going on, or why being linear is such a big deal. So today, we're going to solve a linear program by using a graphical method, by drawing a picture, if you like. The approach has its limitations, but it has some important advantages, too.

Limitations? Yeah. The technique requires one dimension of space for each decision variable in the problem. So a one-variable problem could be solved in one dimension, on a number line. Two variables require two dimensions, both width and height, so you can solve them on a plane, like a piece of paper. Three variables? Hmm, three dimensions, height, width, and depth. You can do a three-variable problem if you're good at making perspective drawings, or you could have a piece of software do that for you. But, just like we saw in regression and data mining, once you have more than three variables, you're entering the Twilight Zone. You need more than three dimensions. The math is fine, but the visualization is pretty rough.

But we've already seen that interesting examples of linear programming can involve thousands of variables. So, why bother with a graphical approach? Well, for the same reason that we did it for regression and data mining; doing so will provide a lot of insights into all linear programming problems. More complicated problems may be harder to grasp in their entirety, but their solutions rely on the same observations and properties that the simple ones do.

So, let's take a look at a problem. This one's about personal financial investment. Let's say that you've got $80,000 that you're going to invest it in three different investment options. Each investment pays you back after a four-year period. The first option is moderate risk and moderate return. It's $2/3$ likely to give you a 40% return over the four years, but there's a $1/3$ chance that, instead, you'll lose 20% of your investment in that time. You might have noticed that even in this linear program, I'm sneaking in a bit of a stochastic element, since our objective is going to be to maximize your average returns in this situation, also called your expected return. We'll have a lot more to say about expected values later in the course, but the calculation here is easy; take each possible outcome, multiply it by the probability of occurring, and add the results. So we get this: $^2/_3 (40\%) + ^1/_3 (-20\%) = 20\%$; $^2/_3$ of the time, you gain 40%; $^1/_3$ of the time you lose 20%, and so your expected, or average return is 20%.

OK, the second option, high risk, high return. It's 50% likely to give you 150% return on investment, that's $250 back for each $100 you invest; unfortunately, it's also 50% likely to give you back only half of your money; that means the expected return works out as, one half of 150, plus one half of −50, is 50%. Like I said, high risk, but high average return.

OK, the final option, it's essentially no risk. OK? It has a 15% return total, over the four years. Your goal is to maximize your expected return, the average return, on the money that you invest. To limit your downside risk, though, you want to be sure that you lose, at most, $10,000, even if the risky investments turn sour. You also want a diversified portfolio, so you want no more than $40,000 in any single investment. At first blush, this may sound like a three-variable problem. How much money do you put in each of the three investments? We could do it that way, but since we know that the total investment is $80,000, as soon as we know how much we put in the first two investments, we know automatically how much we put in the third. That is, the third variable is really an auxiliary variable. So, if we put M dollars in the medium risk-investment and H dollars in the high-risk one, then we put 80,000 minus M, minus H in the risk-free one. So we can get away with two variables. OK, here's what the program looks like.

The objective just records the expected, or average, return on each of the three investments, 20% for the medium-risk one, 50% for the high-risk one, and 15% for the risk-free one. We're going to try to maximize this expected return, and we get it by multiplying the expected rate of return for each investment by the amount of money that we put into it.

The first three constraints just say that none of these three investments can have more than $40,000 put into it. Remember that 80,000 minus M, minus H is just the money that we dump into the risk-free investment. Finally, constraint 4 is our Murphy's Law constraint. In English, it says that the money that we lose on our investments, even if everything goes wrong, is at most $10,000. In this worst case, we lose 20% of the medium-risk money and 50% of the high-risk money, and this loss is counterbalanced somewhat by what we make on the risk-free investment. As usual, we also have nonnegativity constraints, since we can't invest negative dollars in any option.

In light of our work in the last two lectures, I hope that this formulation looks sensible to you. If so, that's all we need for now. While we use math to solve the problem graphically, we'll be talking in English as we interpret what we're doing. Before we start, I'm going to clean up the algebra a bit by multiplying out the parentheses and by gathering like terms. I'm also going to rewrite the program in terms of thousands of dollars, rather than dollars to get rid of all of those pesky triple zeroes at the end of everything.

It looks somewhat different at first glance, but mathematically, nothing has changed. For example, look at the original constraint 3 and its makeover. The original is most easily read as saying you can't put more than $40,000 in the risk-free investment. The new math is most easily read as saying that you have to put at least $40,000 in the other two investments combined. But since we are starting with $80,000, these mean exactly the same thing.

OK, so how do we solve the problem? Well, we have two decision variables, M and H, and we'll give each one of them its own axis in a Cartesian plane. Then we'll add some constraints, one at a time, and see what points in the plane satisfy each one. A point that satisfies every single constraint is going to be a feasible solution, a possible answer. We'll be able to see easily exactly which points, which combinations of M and H, these are.

Then we're going to have to decide which of these possible answers is the best answer—for this problem, the one that gives the largest expected return. We're going to come up with a rather clever way of doing this. And all of this is going to work because of the geometric character of linear functions. Remember that linear relations between two variables graph as straight lines, hence, linear. More generally, linear relations among a collection of variables graph as lines or planes or higher-dimensional equivalents called hyperplanes. The crucial point about these objects is that they are flat. They don't wiggle, bend, kink, or jump. So everything we draw in this problem is going to be a straight line, thanks to the linearity of our program.

OK, we start with the constraints to see what's possible. First, the easiest ones: the nonnegativity constraints. As you can see, they aren't much. I've drawn little arrows pointing from the two axes into the first quadrant. The value of M, the money we put in the moderate-risk fund, can't be negative, and neither can H, the money put in the high-risk fund. So the best answer is going be somewhere in the flue, shaded region, the points that satisfy these two requirements. It's usually called the first quadrant. OK? Now let's add constraints 1 and 2. They're not much either. They just say that M and H can't be more than $40,000, 40 on our graph. A vertical line through 40 on the M axis shows all of the points [where] M is equal to 40, so we have to be on or to the left of that vertical line. The same idea for the second constraint. It says that H has to be less than or equal to 40, so we need to be on the horizontal line, H = 40, or below it. Remember, M is measuring our horizontal position on this graph, and H is measuring our vertical position. Again, you can see the little arrows on the lines to show their "good sides," where the constraints are true. The points that satisfy all of the constraints so far are shown in the blue square.

We're on a roll. Let's go for constraint 3. It's a little harder than the ones so far, but not much, since we know that everything that graphs is a straight line, like everything does in a linear program. You can find the line with simple graphing techniques, but let's do it logically. The constraint says that we have to have a total of at least $40,000 in the moderate-risk and high-risk investments together. To find the border line, push the constraint to its limits; make it exactly 40,000. How could we make such investments?

Well, one obvious way would be to put 40,000 in the moderate-risk investment and nothing in the high-risk one. The second one would be to put $40,000 in the high-risk investment and nothing in the more moderate one. On our graph these solutions correspond to M = 40, H = 0 on the horizontal M axis; and M = 0, H = 40 on the vertical H axis. And connecting them gives the line for constraint 3, showing all of the other ways to split $40,000 between them. Any point above this line gives more than $40,000 invested in those two together, as the constraint requires. That's what the little arrow on constraint line 3 says; any point on or above this diagonal line satisfies constraint 3, but if I wanted to satisfy all of the constraints I've introduced so far, I'm restricted to points inside the triangle. Constraints 1 and 2 only let me go so far.

In the same way, we can throw in constraint 4, which says we lose at most $10,000 on all of this. That limits how much you can put in the risky investments, especially the high-risk one, which is why constraint 4's arrow points down and to the left. You could plot this line using the slope-intercept form from basic algebra, or a lot of other ways. And if you're a bit rusty on that, don't worry about it. Remember, our whole reason for doing this graphically is to strengthen your intuition. Just get the feel of what's going on here.

So, that's all of the constraints. Now, look at the blue region. The points in that region satisfy every one of our six constraints, the four nontrivial constraints and the nonnegativity constraints. That's true of the points on the border of the blue region, too. Conversely, any point outside this blue region violates at least one of the constraints; it's on the wrong side of the line, and so it doesn't represent a possible answer to our question. The blue region is called the feasible region, because it contains precisely the points that satisfy every single one of the constraints. If you like, it's the OK Corral, since every point inside that region is OK for all the constraints.

OK, so we now know what answers are possible. The question is, among those points in the feasible region, which is best? Well, here's one way to get at the answer. I'm going to pick a point inside the feasible region and figure out what its expected return would be. For example, suppose I pick this point here in the lower-right corner of the triangle. It has coordinates M = 40, H = 0, which means that I put 40K into the medium-risk investment and 0 into the high-risk investment. The other 40K, then, will go into the risk-free investment.

What expected return do I get here? We'll, the 40K in the risk-free return is guaranteed to give me 15%, which is $6000. The 40K in the medium-risk investment, on average, pays 20%, so it gives $8000 on average. That's a total of 14K in expected return over the two years, not bad.

I'm going to record this information in a rather bizarre way. I'm going to lay my feasible region down on the table top, then pull out a piece of extra-long uncooked spaghetti. I'll will cut it to exactly 14 centimeters long to represent the $14,000 that we get from this point. Then I'll take a drop of crazy glue, put a dab right on that corner, and glue the spaghetti right on top of that point. I'm going to do this with every other point in the feasible region. Pick the point, find the coordinates, plug them into the objective function, get an answer, cut a piece of spaghetti to that length, and glue it onto the point. So when I'm done, I'm going to have a crop of spaghetti growing straight up out of my feasible region field.

It's going to take a really long time, but I'm going to do it anyway. Why? Because I have no social life. But also, because what this field looks like is really important because we're looking for the best expected return on the investment; that's what we're looking for, and that means finding the tallest piece of spaghetti in this field.

So here's the question. I want you to imagine what you'd see when you looked over the tops of this field of spaghetti. Because it seems to me that there are three possibilities. The first one is that they'd be a patternless mess, sticking up like a punk-rocker's hair. The second possibility is that the tops of the spaghetti would form a set of gently rolling hills. And the third possibility is that the tops of the spaghetti would define a surface that's flat, not necessarily level, but flat, as if a samurai took a sword and went hiyaaa! What do you think?

I'll give you a hint. This is a linear program. And that makes all the difference. Remember, linear things graph as straight lines, planes, or hyperplanes, and the defining property of these things is that they're flat. Our objective, expected return, is linear, and that means the tops of our spaghetti, or more properly, the three-dimensional graph on our expected returns, is going to be a flat plane. Like this, tilted maybe, but with no dips or bumps or kinks. And knowing this is terrifically important, because again, we're looking for the single, tallest piece of spaghetti. And for our problem here, the spaghetti tops form a flat triangle. Now, I don't know how it's tilted yet, maybe like this, or this, or this. But I hope that you see that, no matter how it's tilted, one of the corners is going to be a highest point; it's going to be as high or higher than everything else. And that's the point we want. What we've just discovered is called the extreme point theorem of linear programming, a fancy name that just means that if there's an optimum point, it's going to be at one of the corners of the feasible region. So we could just check all three corners and find the optimum.

But to get a feel of how we're going to proceed later in the course, I'm going to take a different approach. Imagine I take my triangular field of spaghetti and flood it to a certain depth. When the water level gets high enough, some of the stalks will be submerged, but others will still peek out above the surface. Now let's look at an aerial view, looking straight down. We see the triangular field in the feasible region, cutting across it, the line separating the submerged part of the field from the part still peeking above the water. Like this.

The flood line is usually called an objective function line, or OFL. In our flooded field, all the points along it are right at the water's edge, so they all have the same height. In terms of our original problem, this means that each points along the OFL gives the same expected revenue. Every point still peeking out of the water has spaghetti higher than the water. That gives a higher expected return than the level represented by the water. If you want a shot at a higher expected return, send some more water into the picture. The water level will rise, and the points that are still not submerged make you more money on average. Like this.

You can probably see where this is going, and I bet you can tell me where the optimal point will be without any more help from me. Before I confirm your answer, though, I'd like to point out a couple of little matters. First, how did I know where to draw the flood line? I just took the objective function and set it equal to whatever number pleased me. The number specifies the expected return I'm looking for; if you like, it's the depth of the water. When you set the objective equal to a number, you get a linear equation, so it graphs as a straight line. That's why the water's edge is a straight line in the aerial view.

For my original data, I decided, rather arbitrarily, to set the depth of the water at $18,500. So my original dotted line was obtained by setting the objective function equal to 18.5, like this, and then, graphing the result, which is a nice, linear function.

If you think of this as a map, you'll avoid another common mistake. In our picture, the coastline between flooded and not flooded separates the water into a southwest land and the northeast. As the water rises, the coastline covers more and more land as it moves northeast. That's because in this picture, the spaghetti toward the upper right is taller than the spaghetti toward the lower left. But of course it wouldn't have to be that way. The Nile River flows north on the map, but that doesn't mean it flows uphill. In general, when you draw a first objective function line, you need to take a moment to figure out which side of the coastline is water and which is dry land. Dry land is not necessarily on the northeast!

OK. You see where this is going, right? To find the highest point, we just keep the water coming. The coastline keeps creeping in, and eventually only one point of our blue triangle is still above water, which where the upper-left corner of the triangle is in this case. Where is that? Well, you can see from the graph it's somewhere around 15,000 in the moderate risk, and 25,000 in the high-risk investment, which would leave about $40,000 for the risk free. We can get precise values for the coordinates by taking the two constraint equations that meet at the optimal point and solving them simultaneously. More basic algebra.

Solving constraints 3 and 4, the two that meet, simultaneously gives us $M = 13 \frac{1}{3}$ and $H = 26 \frac{2}{3}$. That is, the best you can do is put $13,333 and change into the moderate-risk option, and $26,666 and change into the high-risk option, and the remaining $40,000 goes into the risk-free option. Doing so will give you an expected return of 0.2 (13,333) + 0.5 (26,667) + 0.15 (40,000), or, $22,000. That's an expected return of $22,000 on your $80,000 investment, or a 27.5% return expected for the two years. That's the unique, best answer to this problem. Any other solution either gives a lower return or breaks a constraint.

Okay, we've solved our first mathematical optimization problem. But I'd like to point out that we've gained some considerable insights in doing so. Let's look at a few. First, before you draw your first objective function line, the original flood line, you don't know where the optimal point is. The optimal point wasn't the best because it was closest to the top edge of the picture; remember this is an aerial view; top is not necessarily up. It was the best because it was the last point to be flooded, because the spaghetti was highest at that point. People often think that for a maximize problem, the good answers are always as far to the upper right as you can get, or as far from the origin as you can get, and neither of these is true in general. In our problem, the best answer was actually the one closest to the origin.

In fact, if you changed this problem a little bit, you'd get a completely different optimal point. Remember that the medium risk investment had a $\frac{2}{3}$ chance of giving you a 40% over the four years, and a $\frac{1}{3}$ chance of losing 20%. If it were 95% likely to make the money and only 5% likely to lose it, the feasible region would remain unchanged, but the slope of the objective function line would lessen. The resulting picture would look like this. Hey, look at the flood line now. Our old optimal solution is no longer optimal! Now, as the water creeps up from the lower left, the last point to be submerged is the upper-right corner of the feasible region at M = 40,000, H = 12.31 thousand.

But let's go back to our original problem and see what else we can see. I've shown the original solution as a red star. So here's a question. Suppose that you're actually a bit more adventuresome in your investments than this picture suggests. But the reason that you imposed the conditions that you did was because your spouse doesn't like risk. Suppose that you felt that with a bit of persuasion, you could get him or her to relent on one of the four nontrivial constraints. Which would you focus on?

Well, you probably can't say for sure, but you should be able to narrow it down to one of two. Either the requirement for no more than $40,000 in the risk-free option, constraint 3, or the guaranteed loss of no more than $10,000, constraint 4. Why? Because it's these two constraints, constraints 3 and 4, that are keeping you from getting a better answer. They're just barely satisfied, right now you have exactly $40,000 in the risk-free fund, and your maximum possible losses [are] exactly $10,000. You can tell because the lines of these two constraints actually run through the optimal point. We say that these are the binding constraints. They have no left-overs; they're pushed to their limits.

But, if either of these constraints were eased, the effect on the graph would be that constraint line would back up, keeping its same slope, but making the feasible region larger. And when this happened, the part of this larger feasible region would, again, appear above the flood line, meaning that the new highest point would make you more average money than before.

Surprisingly, you actually want to invest more in the risk-free fund, if you were allowed to do so. Constraint 3 would move to the lower left, expanding the feasible region and giving you a better optimal expected return. This is because more money in the risk-free investment frees you up to put more money in the high-risk investment too. This change might be an easier sell to your spouse than raising your maximum loss limit. On the other hand, the constraint that you can't put more than 40,000 in the moderate-risk investment, constraint 2, is nonbinding; it doesn't pass through the optimal point. The reason you are making the investments you are has nothing to do with that constraint. You want to invest less than $40,000 in the moderate-risk fund.

And look at constraint 1! Here, the situation is even more extreme. You'll notice that constraint line 1 doesn't even form a side of the feasible region. What that means is, if you satisfy the other constraints in the program, you automatically satisfy number 1. Saying it another way, you couldn't put 40,000 in the high-risk fund if you tried. You'd always break a maximum loss constraint.

A constraint like this is called redundant. In this problem, the nonnegativity constraints are redundant, too. Take them away, and the feasible region doesn't change. You can safely ignore redundant constraints unless the other constraints in the program change. In fact, I actually left one constraint out of our original formulation, that you can't invest negative money in the low-risk fund. But no harm done. It's a redundant constraint. Why? Well, there's $80,000 to be invested, and whatever doesn't go into the moderate or high-risk investments gets dumped into the low-risk one. But with constraints that say that at most 40,000 goes into each of those risky funds, it's impossible to make the low-risk fund investment negative. Just remember, binding constraints are satisfied with none to spare, and it's the binding constraints that are going to limit the quality of your solution. Redundant constraints are the least of your worries; they add no additional requirements at all.

OK, I want to finish up here today with three other matters that arise in linear programs. We've worked on an example which had one and only one optimal solution, a unique optimal solution. But there are three other things that can happen when you tackle linear programs. A program may be infeasible, it may have alternative optima, or it may be unbounded. An infeasible program is one that doesn't have a feasible region. The constraints are mutually contradictory.

For example, suppose I told you that you had to spend at least $10 on food and at least $5 on beverage, but your total bill couldn't be more than $11. Well, you can do any two of these, but all three of them together put you in an impossible situation. Infeasible programs don't have optimal answers; they don't have any feasible answers. So they don't have a best one! In real life, if you get an infeasible problem. If you get one, you're going to need to relax or remove some of the constraints, or you're doomed.

The second possibility is alternative optima, and that happens when the OFL is tilted at just the right way, so that the optimal objective function line passes through, not a single point, but an entire side of the feasible region. Like this. Here, the optimal OFL would align perfectly with constraint 4, and so all of the feasible points on constraint line 4, the entire top of the triangle, would all be optimal points. This would happen in our problem if the moderate-risk investment had just under a 90% chance of making money.

Another example, easier to see, would be a restaurant that has 5 waiters and waitresses available, 5 waiters, 5 waitresses, but needs only 4 servers. Assuming men and women are equally good at the jobs and are paid equally, then using 4 men, or 4 women, or 3 men and 1 woman, and so on are all equally good answers. It's nice to have alternative optima, since it gives you some flexibility in optimal solution.

OK, the last possibility is that the program is unbounded. This is kind of the opposite of the situation to infeasible programs. Infeasible programs have no optimal solutions because the constraints are too binding. Unbounded programs have no optimal solutions because the constraints aren't restrictive enough. No matter what feasible answer you find to an unbounded program, you can always find another one that's even better. If you got rid of constraints 2, 3, and 4 in our program, the only constraint left, other than nonnegativity, would say that you can't put more than \$40,000 in the high-risk fund. But you could put trillions and trillions into the risk-free fund and make as much money as you like.

To have an unbounded program, you have to have an unbounded feasible region, that is, the feasible region that goes on forever and ever in some direction, but that's not enough. If I tell you that the only rule is that x has to be at least 0, you have an unbounded feasible region, the entire upper half of the number line. The problem maximize x is unbounded, since you can always pick a bigger value. But with the same feasible region, minimize x isn't unbounded. The best answer is $x = 0$. All of those bigger numbers are still available, ad infinitum, but for this problem, minimize x, you don't want to use them.

In brief, if you have an OFL moving in the direction that makes the objective get better, and if moving it that way never leaves the feasible region, then you've got an unbounded program. In real life, an unbounded program almost always means you've missed something. Real life quantities can't approach infinity. If you think about what stops you from making your decision variables infinitely big, you'll probably find a constraint you forgot to add to the program.

OK. We've seen the ideas behind how to find an optimal solution in a linear program, and done so in the relatively simple case of two decision variables. Along the way, we've come across some key ideas, feasible, infeasible, bounded, unbounded, binding, nonbinding, redundant, nonredundant. But it was also a lot of hands-on labor, and what do you do when you have more than two variables? The answer is just what we're going to do next time. You put your program in a computer, and you use the power of George Dantzig's simplex method to crank out the information that you need.

Solving Linear Programs in a Spreadsheet
Lecture 11

In this lecture, you will learn how to find the optimal solution to virtually any linear program that has one by using a spreadsheet program on your computer. The example in this lecture will give you a good idea of what goes on behind the scenes once a problem is formulated and what its answer may look like. The technique that the spreadsheet uses is essentially George Dantzig's simplex method. The example will be done in Calc, which is part of OpenOffice, a freeware alternative to Excel in Microsoft's Office suite. The interfaces of both programs are quite similar, but there are some variations between the two.

The Simplex Method

- Suppose that you have seven different investment options available for your $500,000 nest egg: investments A, B, and C and four money-market funds. Let's say that you want to accrue as much wealth as possible at the end of the fourth year, but you also want to withdraw $10,000 at the end of the third year for a vacation. The program is as follows.

Maximize $1.10A + 0.53C + 1.02M_4$	*Maximize cash at end of year 4*
subject to	*subject to*
$500,000 = A + B + M_1$	*(beginning of year 1)*
$0.02A + 1.02M_1 = M_2$	*(out of year 1 = into year 2)*
$0.02A + 1.05B + 1.02M_2 = C + M_3$	*(out of year 2 = into year 3)*
$0.02A + 0.53C + 1.02M_3 = M_4 + 10,000$	*(out of year 3 = into year 4)*
and all variables ≥ 0	

- Each constraint says that the money that becomes available at the end of one year is equal to the money that's disbursed at the beginning of the next year, and the objective tells us how much money becomes available at the end of year 4. In some ways, it's amazing that so much information boils down to such a small formulation, isn't it? That's part of the charm and power of the language of mathematics.

- The first step is to clean up the constraints by putting them in what is referred to as **standard form**, in which the left-hand side of the constraint is just a linear expression with no constant term and the right-hand side is just a constant term—no variables.

- The first constraint needs to be flipped around end for end, as follows.

$$500,000 = A + B + M_1 \rightarrow$$
$$A + B + M_1 = 500,000$$

- And the variables on the right-hand sides of the other constraints need to be subtracted, to get them over on the left-hand sides.

$$0.02A + 1.02M_1 = M_2 \rightarrow$$
$$0.02A + 1.02M_1 - M_2 = 0$$

$$0.02A + 1.05B + 1.02M_2 = C + M_3 \rightarrow$$
$$0.02A + 1.05B + 1.02M_2 - C - M_3 = 0$$

$$0.02A + 0.53C + 1.02M_3 = M_4 + 10,000 \rightarrow$$
$$0.02A + 0.53C + 1.02M_3 - M_4 = 10,000$$

- A method called the matrix representation is a straightforward approach that allows you to accurately and correctly enter any linear program into a spreadsheet. "Solving" a linear program means finding the best values for all of the variables, so we're going to start off with a row that just lists the names of all of the variables in the program. Directly under those, we're going to have cells that will—eventually—hold their best values. To keep things clear, these so-called decision variable value cells are bordered with dotted lines. Next comes the objective: to maximize final cash on hand. Add a new row for the objective, aligned with the variables we've already entered, and type in the objective's coefficients.

	A	B	C	D	E	F	G	H
1		A	B	C	M_1	M_2	M_3	M_4
2								
3								
4	Maximize	1.10	0.00	0.53	0.00	0.00	0.00	1.02

Figure 11.1

- The objective is just $1.10A + 0.53C + 1.02M_4$, so we just recorded those coefficients—1.10, 0.53, and 1.02—underneath their corresponding variables. The other variables get zero entries in this row, because they aren't in the objective.

- The spreadsheet is going to need to calculate how much money this expression works out to be. And that, of course, depends on how much money we put in each investment. Those numeric values are going to go in the cells with the dotted lines, and although we obviously don't know how much money to put in each fund, we're for the moment just going to put any numbers in those cells, as placeholders.

	A	B	C	D	E	F	G	H
1		A	B	C	M_1	M_2	M_3	M_4
2		300	200	100	1000	1000	1000	2000
3								
4	Maximize	1.10	0.00	0.53	0.00	0.00	0.00	1.02

Figure 11.2

- Not only are these values not optimal—they're not even feasible. Remember that we have to invest $500,000 right now, and clearly we aren't. But it doesn't matter. When we get to it, the spreadsheet's **solver** will find the best values for these cells.

- Sticking with these numbers, we want to know how much profit is supposed to be generated by them. The objective is $1.10A + 0.53C + 1.02M_4$, so the answer is as follows.

$$1.10(300) + 0.53(100) + 1.02(2000) = 2423$$

- In the spreadsheet, the decision variables' values in the boxes with the dotted lines are lined up with the objective function coefficients in the gray box. Then, corresponding terms are multiplied: 300 times 1.10, 200 times 0, 100 times 0.53, and so on. Finally, all of these products are added. Both Excel and Calc have a built-in function for doing this with two rows of numbers. It's called **SUMPRODUCT**. First, you do a bunch of products, and then you do a sum.

- To use it, you just type "=SUMPRODUCT(", followed by the first range you want—in this case, the cells with the dotted lines in B2 to H2—and then the second range that you want—in this case, the gray cells from B4 to H4. Between these two ranges goes a separator. In Calc, it's a semicolon, and in Excel, it's a comma. At the end, you close the parentheses.

=SUMPRODUCT(B2:H2;B4:H4) (Calc)

=SUMPRODUCT(B2:H2,B4:H4) (Excel)

- When you enter this, the spreadsheet calculates the correct value.

	A	B	C	D	E	F	G	H	I	J
1		A	B	C	M_1	M_2	M_3	M_4		
2		300	200	100	1000	1000	1000	2000		
3										
4	Maximize	1.10	0.00	0.53	0.00	0.00	0.00	1.02	2423	= final cash

Figure 11.3

- This is dynamic, of course—change the numbers in the cells with the dotted lines and the spreadsheet automatically computes the new "final cash."

- Next, we're going to play the same game with the left side of the constraints.

	A	B	C	D	E	F	G	H	I	J
1		A	B	C	M_1	M_2	M_3	M_4		
2		300	200	100	1000	1000	1000	2000		
3										
4	Maximize	1.10	0.00	0.53	0.00	0.00	0.00	1.02	$2423.00	= final cash
5										
6	subject to									
7									LHS	
8	Now	1	1	0	1	0	0	0	1500	
9	Year 1 end	0.02	0	0	1.02	−1	0	0	26	
10	Year 2 end	0.02	1.05	−1	0	1.02	−1	0	136	
11	Year 3 end	0.02	0	0.53	0	0	1.02	−1	−921	

Figure 11.4

- Column I is labeled LHS, which stands for "left-hand side." When you multiply each entry in the row with dotted lines by the corresponding element in a gray row and then add all of those products together, the LHS entry tells you the result.

	A	B	C	D	E	F	G	H	I	J	K
1		A	B	C	M_1	M_2	M_3	M_4			
2		300	200	100	1000	1000	1000	2000			
3											
4	Maximize	1.10	0.00	0.53	0.00	0.00	0.00	1.02	$2423.00		= final cash
5											
6	subject to										
7									LHS		RHS
8	Now	1	1	0	1	0	0	0	1500	=	500,000
9	Year 1 end	0.02	0	0	1.02	−1	0	0	26	=	0
10	Year 2 end	0.02	1.05	−1	0	1.02	−1	0	136	=	0
11	Year 3 end	0.02	0	0.53	0	0	1.02	−1	−921	=	10,000

Figure 11.5

- But a constraint has two sides, and we still need the right-hand sides. In this particular program, all of our constraints are conservation constraints—equality constraints—so let's indicate that and include the right-hand sides in the spreadsheet as well.

- For a feasible answer, all of the constraints have to be satisfied. Because all of our constraints are equality constraints, the LHS and RHS in each row have to be identical, or we don't have a feasible solution. So, our current proposed solution is way off base. None of these constraints is satisfied.

- We could play around with trying different numbers in the cells with dotted lines and seeing if we could find a way to make all of the constraints true, but that would not be the best-possible investment strategy. We don't need to play around with these numbers, because that's the job of the spreadsheet and the **simplex method**.

- Everything on the spreadsheet that isn't a number is actually going to be irrelevant to the spreadsheet's solver. We have to tell the solver where the objective is, where the decision variable values are, and where the constraints are.

- Excel has a Solver Add-in that comes on the Office disk. If you ask Excel to install the add-in, you'll see it appear on the menus or ribbons on the top of your Excel screen. For Calc, which is what we're using, the add-in is preinstalled. It's accessible from the "Tools" menu. While the popup windows in the two applications look slightly different, they want the same information and work in the same way. The following is the process for Calc.

Figure 11.6

225

- The "target cell" is the cell that holds the objective function. In this case, that's year 4 cash on hand, in cell I4. We want to maximize it, so we leave the radio button selecting "maximum" alone. And the "changing cells" are the cells that contain the values of our decision variables—the cells with the dotted lines, B2 through H2.

Figure 11.7

- That leaves the constraints. You can enter each constraint as its own row in Solver, but if a number of consecutive constraints have the same relational operator—≤, =, or ≥—you can do all of those in one row. Because all of our constraints are equalities, we can do all of them at once, by saying that the four left-hand sides, from I8 to I11, have to equal the four right-hand sides, from K8 to K11.

- We have only one more concern, if we're using Calc. We forgot the nonnegativity constraints on all of the variables. This is easily remedied by clicking "options" and then checking the box by "Assume variables as nonnegative." Then, get out of "options" by clicking "OK," and you're back in the Solver.

- In Excel, assuming that all variables are nonnegative is the default. Also, if you're using a version of Excel that includes "Premium Solver," you'll also want to tell it to use "Standard LP/Quadratic" as the "Solving Method."

Figure 11.8

- As soon as we press the "solve" button in the bottom-right corner, the spreadsheet will find the best-possible investment strategy out of all of the myriad possibilities. For problems of even considerable size, the simplex method for linear programs is blindingly fast. The best investment strategy is as follows.

	A	B	C	D	E	F	G	H	I	J	K
1		A	B	C	M_1	M_2	M_3	M_4			
2		5000	0	20200	0	10000	0	10706			
3											
4	Maximize	1.10	0.00	0.53	0.00	0.00	0.00	1.02	$571,626.12		= final cash
5											
6	subject to										
7									LHS		RHS
8	Now	1	1	0	1	0	0	0	500000	=	500,000
9	Year 1 end	0.02	0	0	1.02	−1	0	0	0	=	0
10	Year 2 end	0.02	1.05	−1	0	1.02	−1	0	0	=	0
11	Year 3 end	0.02	0	0.53	0	0	1.02	−1	10000	=	10,000

Figure 11.9

- According to the boxes with the dotted lines, the best way to invest the money is to put all of it in investment A originally. After 1 year, you'll have $10,000 come back from A, and this gets funneled into money-market 2 for a year. At the end of that time, it comes out as $10,200. Also, A gives another $10,000, so the total of $20,200 gets plowed into investment C. Investment C, after a year, gives 53% of this back, or $10,706. We also get another $10,000 from investment A, which is just enough to pay for the vacation, so the $10,706 gets one more year in the money-market fund. It matures, as does the money from investment A, along with another 53% of the money from investment C, and we end up with $571,626.12 on hand at the end of year 4. With these lousy interest rates, it's difficult to make a lot of money, but the solution we found was the best-possible one.

simplex method: An algorithm invented by George Dantzig for finding the optimal solution to any linear program that has one. For most linear programs, the simplex method finds the optimal solution with surprising speed.

solver: Generally, a piece of software for solving a class of problems. Specifically, the add-in in Excel or Calc capable of finding optimal solutions to linear and nonlinear programs.

standard form: A linear constraint in standard form has all of its variable terms on the left side of the constraint and its constant term on the right side. $2x + 4y \leq 7$ is in standard form.

SUMPRODUCT: A function in Calc and Excel that takes two row or column arrays of equal length, multiplies corresponding elements, and then adds the results. In vector calculus, this is the dot product or scalar product of two vectors.

Suggested Reading

Anderson, Sweeney, Williams, Cam, and Cochran, *Quantitative Methods for Business*.

Ragsdale, *Spreadsheet Modeling & Decision Analysis*.

Strayer, *Linear Programming and Its Applications*.

Questions and Comments

1. Why should a manager focus his or her attention on those constraints that are binding on the optimal solution?

Lecture 11: Solving Linear Programs in a Spreadsheet

Answer:

Because any change in a binding constraint will probably chang[e] quality of the optimal solution. On the other hand, a small change i[n] nonbinding constraint will never affect the optimal solution, nor can a chance in a nonbinding constraint ever improve the optimal solution.

2. Imagine beginning with a linear program that has an optimal solution. We change this program by removing a constraint. Argue that the optimal objective function value may stay the same or may improve, but cannot grow worse. Argue that the program may become unbounded, but cannot become infeasible. (Hint: Think about what, if anything, happens to the feasible region.)

Answer:

Any solution that worked in the original program (that is, was feasible) is still feasible. We know that it originally satisfied all of the constraints, so we can be sure that it satisfies the smaller set with one constraint removed. Thus, either the old optimal solution is still the best-possible solution (which is what would happen if the removed constraint were nonbinding on the optimal solution), or a new, better solution is now possible. Because the old optimal solution is still feasible, the new program is feasible. But the new program may now be unbounded. As an example, consider the program "maximize x," subject to $x \leq 5$. With the constraint, the optimal solution is $x = 5$. Without the constraint, x may become arbitrarily large.

is the ultimate exercise in deduction. Each step
before it by carefully established theorems, creating
from premises to the conclusions. Actually doing
, is a very different matter. Mathematicians rely to
on their intuitions when trying to guess what a right
conclusion m... , and what twists and turns that unbreakable chain might
take before it links that conclusion to the premises. Intuition, in this sense,
is a practiced way of seeing things, getting a feel for how things usually fit
together and play off of each another.

That's what the last lecture was about, really. The graphical method of
solving linear programs is in some ways an almost useless thing, with more
than two variables, it's problematic, with more than three, it's basically
impossible. And it's labor intensive, with all of that line drawing and region
shading and the like. The value of the graphical method is that it gives us a
feel for some important ideas, like the best solution being determined by the
constraints that are binding on that solution, the ones that are pushed to their
limits at the optimal point. We'll use these ideas repeatedly, especially in the
next lecture.

And now you're ready to learn that you can easily find the optimal solution
to virtually any linear program that has one, using nothing more than the
spreadsheet program on a home computer. If you're eager to try what you've
learned in this course so far on problems of your own, this will be right up
your alley. But even if that's not for you, what we do will give you a good
idea of what goes on behind the scenes once a problem is formulated, and
what an answer may look like. We'll solve some of the problems we've
already looked at, as well as presenting a new, richer application that will
give you a better idea of what linear programming can do for you.

The technique that the spreadsheet is going to use is essentially Dantzig's
simplex method, and now that you can visualize a linear programming
problem from the last lecture, I can tell you how it works. I'll even show

you a visual example of what I'm talking about at the same tim[e] problem that has three decision variables, just to give you a sense o[f] that looks like.

What is a linear program with three variables going to look like, graphically? It's going to look very different from linear regression with three variables. First off, with three variables, we again need three dimensions, so our representation isn't going to live in a flat plane, like a video screen. It's going to be hanging in space. In two dimensions, the constraints and the objective function lines graphed as lines. Now, we're a dimension higher up, so constraints are going to graphing as planes—unbending, flat surfaces.

The feasible region is bounded by constraints, so now it will look like a faceted gem, a polyhedron—something like this. And the OFL, the objective function line, separated the flooded part of our picture from the un-flooded part, that was the water line. Now it's no longer a line, it's the water level, cutting through the feasible region at some angle, like this.

For this particular problem, raising the water level, that is, increasing the objective function value, means moving the plane on an upward diagonal direction. To make it easier to talk about, let's tilt our heads so that the water level is in a horizontal plane, as it would be on Earth, there. The point that we're looking for is the one way at the top, the last one that would go under water when the objective function's water level rises. Here's how the simplex method finds it. It begins by finding a corner of the feasible region, any corner, like here.

Then it looks along all of the edges, leaving that corner, to see if moving along any of those edges makes the objective function get better. Since we've tilted our heads, "better" in this picture means "up." I'll call such an edge an improving edge. If the corner doesn't have any improving edges coming out of it, then you're already at the best corner, and you're done. But if there is an improving edge, then the simplex method picks the one that improves the objective most rapidly and moves along it until it reaches another corner.

…e, with a
…f what

…v making the OFL plane horizontal, the edge with the
…ng objective is just a complicated way of saying, the
…Our starting corner has two improving edges, so we
…Then repeat the process. In our example, the steepest
…second corner takes us to the peak, and since every
…corner worsens the objective function—in our tilted
…e're done. That's the simplex method. Conceptually,
…pretty darned simple.

By the way, you might remember that with two decision variables, you can always find an optimal point defined by where two constraints, two lines, meet at a corner. Those are the two that are binding on the optimal solution. With three variables, you can see that the optimal point is where three constraints, three planes, meet. Just like two walls and a floor define a corner of a room. This pattern continues into higher dimensions. If the program is infeasible, of course, you can't get a starting point. There is no feasible region to find a corner of. And if a program is unbounded, the procedure never ends. Imagine a feasible region that was like a silo, stretching up to the stars. But otherwise, you'll eventually reach a corner that has no improving edges, and that point is your optimal solution.

You can see how mechanical this process is. And that's the beauty of it. Because although the mathematics involved in moving from corner to corner is arithmetically tedious, it's a lot like solving a family of simultaneous linear equations for a bunch of unknowns, conceptually, it's straightforward. In other words, it's exactly the kind of thing that computers are wonderful at. So all we have to do is to get the hay down where the horses can eat it. And that's pretty easy.

The demonstrations you'll see today will all be done in Calc. Calc is part of OpenOffice, a freeware alternative to Excel in Microsoft's Office Suite. For my own work, I usually use Excel, but anyone can get Calc for free, so I thought it a good choice for this series. Feel free to use whichever you have. Their interfaces are quite similar, and I'll point out the few small variations between the two as we go. Okay, let's see how we find our best answers!

Let's start with our financial planning problem from two le... may recall we had 7 different investment options available for million dollar nest egg, investments A, B, and C, and four money... funds. And we wanted to accrue as much wealth as possible at the end... year 4, although we wanted to withdraw $10,000 at the end of year 3. Don't worry about the details. Let's get the program back, so that we can have it for easy reference, there.

Each constraint says that the money that becomes available at the end of one year is equal to the money that's disbursed at the beginning of the next year, and the objective tells us how much money becomes available at the end of year 4. In some ways, it's amazing that so much information boils down to such a small formulation, isn't it? That's part of the charm and power of the language of mathematics.

Okay, before we get started, I'm going to clean up my constraints a little bit by putting them in what's referred to as standard form. In standard form, the left-hand-side of the constraint is just a linear expression with no constant term, and the right-hand-side is just a constant term. No variables. The spreadsheet can actually handle things that differ a little from standard form, but doing this is still a good idea; it makes the interpretation of some things we'll be doing later easier to understand. So, the first constraint needs to be flipped around end for end, like this. And the variables on the right-hand sides of the other constraints need to be subtracted off to get them over on the left-hand sides, where they belong, like this. Later in today's lecture, I'm going to show you how you can pretty up a problem like this, but right now, I'm going to show you a straightforward approach that will let you accurately and correctly enter any linear program into a spreadsheet. I call it the matrix representation.

Solving a linear program means finding the best value for all of the variables, so we're going to start out with a row that just lists the names of all of the variables in the program. Directly under those, I'm going to have cells that will, eventually, hold their best values. To keep things clear, I always border these decision variable value cells in red, all right? Next comes the objective,

...hand. Here's how I do this. Add a new row for the
...e variables that I've already entered, and type the
...ito that line, like so.

...e sure you see what's going on here. The objective
...3 C + 1.02 M4, so I just recorded those coefficients,
underneath their corresponding variables. The other
...ntries in this row, because they aren't in the objective.
Well, ...or a human reading this, but the spreadsheet is going to
need to calculate how much money this expression works out to be. And
that, of course, depends on how much money we put in each investment. I
told you that those numeric values are going to end up in the red cells, and
although I obviously don't know how much money to put in each fund, for
the moment, I'm going to put in any numbers in those cells, as place-holders.
I'll be creative, and just stick them in.

Not only aren't these values optimal, they aren't even feasible! Remember
that we had $500,000 to invest right now. Well, clearly, we aren't. But I don't
care. When we get to it, the spreadsheet's Solver will find the best values for
these cells. But for now, I'm going to stick with these numbers, and I want
to know how much profit is supposed to be generated by them. Well, the
objective is 1.10 A + 0.53 C + 1.02 M4, so the answer is 1.10 (300) + 0.53
(100) + 1.02 (2000), which works out, if you do it, to $2423. I could have
included all the other variables too, of course, but they're all multiplied by
0 for their objective function coefficients, so it wouldn't change my answer.

What I'd like you to look at is the geometry of how this calculation is done in
the spreadsheet. The decision variables' values in the red boxes are lined up
with the objective function coefficients in the gray box, then corresponding
terms are multiplied, 300 × 1.10, 200 × 0, 100 × 0.53 and so on. And finally,
all of these products are added up. Happily, both Excel and Calc have built-

in functions for doing this with two rows of numbers. It's called sumproduct.
First you do a bunch of products, multiplying, then you do a sum.

To use it, you just type, "=SUMPRODUCT(", followed by the first range
that you want, for us, that the red cells in B2 to H2, and then the second
range that you want, for us that's the gray cells from B4 to H4. Between

234

these two ranges goes a separator. In Calc, that's a semicolon, while in Excel, it's a comma. And at the end of it all, you close the parentheses. So we get this. And when you enter this, voila! The spreadsheet calculates the correct value. This is dynamic, of course, changes the numbers in the red cells, and the spreadsheet automatically computes the new final cash. I've labeled the calculation as "final cash," as you can see.

We're going to play the same game with the left-hand side of the constraints: After all, they're the same kind of expressions that what we tackled in the objective, just number time variable, plus number times variable, and on and on, all the way through. For ease in reference, I'll give each constraint an English label, to remind us what the constraint is talking about.

Column I is labeled left-hand side, as you can see, which just stands for left-hand side, LHS. When you multiply each entry in the red row by the corresponding element in the gray row, then add up all of those products, the LHS entry tells you the result. For example, the Now row just computes (1) (A) + (1) (B) + (1) (M1) to get 300 + 200 + 1000, or 1500. This is the total amount being invested right now, given the numbers that I typed into the red cells. Just remember that those investment numbers were bogus, though. I made them up. Again, all of the entries in that LHS column are just these SUMPRODUCT formulas, the sumproduct of the red numbers, with the numbers in that row. OK, well, great.

But a constraint has two sides, and we still need the right-hand sides. In this particular program, all of our constraints are conservation constraints, equality constraints, so let me indicate that and include the right-hand sides in my spreadsheet as well. And that's looking pretty good. OK, let's step back and see what we have, and what we don't. For a feasible answer, as you know, all of the constraints have to be satisfied. Since our constraints here are equality constraints, the LHS and RHS in each row have to be identical, or we don't have a feasible solution. So our current proposed solution is way off base; none of the constraints is satisfied.

Well, we could play around with trying different numbers in the red cells and seeing if we could find a way to make all of the constraints true, but even then, that would only be a possible investment strategy, not the best possible one. And the truth is, we don't need to play around with any of these numbers. That's the job of the spreadsheet and the simplex method. I'd love to tell you that we could just tell the spreadsheet "go!" at this point and get the answer, but I'm afraid that's not true. Everything in this spreadsheet that isn't a number is actually going to be irrelevant to the spreadsheet's Solver. We have to tell the Solver where the objective is, where the decision variables are, and where the constraints are. Fortunately, that's easy.

Excel has a Solver add-in that comes on the Office disk. If you ask Excel to install the add-in, you'll see it appear on the menu or ribbons on the top of your Excel screen. For Calc, which is what I'm using here, the add-in is pre-installed. It's accessible from the Tools menu. While the popup windows in the two applications look slightly different, they want the same information, and they work in the same way. Here it is for Calc.

The target cell holds the objective function. For us, that's year 4-year cash on hand, in cell I4. We want to maximize it, so we leave the radio button selecting maximum alone. And the changing cells are the cells that contain the values of our decision variables, the red cells, B2 through H2. That leaves the constraints. You can enter each constraint on its own, in its own row in Solver, but if a number of consecutive constraints have the same relational operator, all <, or all >, or all =, then you can do all of those in one row. And since all of our constraints are equalities, you can do them all at once by saying that the four left-hand sides, I8 to I11, have to equal the four right-hand sides, K8 to K11. Like this.

And that's almost all that there is to it. We have only one more concern, if we're using Calc. We forgot the nonnegativity constraints on all of the variables. Easily remedied; click options and then the check box by "Assume variables as nonnegative." OK, get out of Options by saying OK, and you're back in the Solver. By the way, in Excel, assuming that all the variables are nonnegative is the default, and you have to tell Excel if you want variables to be allowed to take on negative values. As we've seen, we usually you don't. Also, if you're using a version of Excel that includes "Premium Solver," you'll also want to tell it to use the "Standard LP/Quadratic" as the "Solving Method."

Okay! All set. Relax a while as our spreadsheet finds the best investment alternative from of all the myriad of possibilities. It'll start as soon as I press the little "Solve" button. Done. Rest's over. It wasn't a very long, was it? As I've been telling you, problems of even considerable size, and the simplex method, end up solving problems blindingly fast. And here is the best investment strategy. What's it say? Look at the red cells.

The best way to invest money is to put all of it in investment A originally. After one year, you'll have $10,000 come back from A, and this gets funneled into money market 2 for a year. At the end of that time, it comes out as $10,200. Also, A gives another $10,000. So the total, $20,200, gets plowed into investment C. Investment C, after a year, gives 53% of this back, or $10,706. We also get another $10,000 from investment A, which is just enough to pay for the vacation, so the $10,706 gets one more year of money market. It matures, as does the investment A money, and another 53% of the investment C money, and we end up with $571,626.12 on hand at the end of year 4. With these lousy interest rates, it's hard to make a lot of money, but the solution we found here was the best possible one.

On the other hand, suppose that you decide that you want to skip the vacation, and instead, make a home improvement that will cost $50,000 in year 4. Now what happens? Again, change the right-hand side of the last constraint and rerun the program. Now things are quite different! We can't put all of the money in investment A anymore; it won't leave enough for the renovation. So, about $480,000 goes in A, but the rest goes in B. When the money comes out of either of these investments, it's immediately sunk back into the money market or withdrawn for the house renovation. Now C is not used at all.

I'd like you to notice that when we made this change, a $50,000 withdrawal at the end of year 3, the solution we saw not only changed in its particulars, like how much money was left at the end of the year-4 money market, it changed its character. A different set of investment options were being used, and some of the previously used options were abandoned. Keep this in mind, because knowing how big an alteration will cause this kind of sea change is important in assessing the sensitivity of a solution to changes in the existing situation. We'll look at this issue in much more detail in our next lecture. But for now, let's look at the solution of another of our problems and see what else we can learn from the spreadsheet.

Remember The Great Courses Railroad, our first linear program? We do the same thing; write constraints in standard form, use SUMPRODUCT to compute the objective and left-hand sides, and tell Solver the same kind of stuff we told it before, except this time we'll hit the radio button for MINIMIMUM, since we're trying to spend as little money shipping as we can.

Also, we have two ≤ constraints, followed by two = constraints, followed by two ≥ constraints, so the constraints have to be entered in three blocks instead of one. When we run this, here's what we get. Which tells us what? Actually, quite a bit. Let's start with the optimal solution. Ship 100 cars from Atlanta to Chicago and onto great falls, and 150 from Atlanta to Dallas and onto Eugene. All 200 of Baltimore's cars go to directly to Fresno, like this. The total shipping cost is $90,800.

We can also see that if this pattern of supplies and demands were constant over time, we're kind of foolish to have an office in Dallas at all. We never send a car through Dallas. Of course, in this problem, we'd probably have different supplies and demands on different days. When we got the new information, we could quickly rerun the program. That's what Union Pacific does in determining how to ship its cars. So what else can we see?

Well, remember our discussion of binding constraints in the last lecture? A binding constraint is one that is satisfied with none to spare. The constraint is pushed to the limit. Graphically, a constraint was binding on a solution point if the constraint actually passed through the solution point. But they're just as easy to find here. The binding constraints are the ones where the number in the left-hand column and in the right-hand column are the same. In this solution, every constraint is binding, except for the Atlanta supply constraint.

With the conservation constraints at Chicago and Dallas, that's no surprise; they're equalities, they have to have the two sides equal. But the other ones being binding suggests that changing their right-hand sides will probably impact your total cost, while changing the nonbinding constraint's right-hand side by a small amount, won't. You can see the latter. We have 300 cars in Atlanta, but are only shipping 250. If we had less cars in Atlanta, we wouldn't care. It wouldn't hurt our optimal cost, unless we they dropped to less than 250 cars available in Atlanta.

But how about the binding constraints? Well, the optimal solution is, push each binding constraint to the limit; that's what makes it a binding constraint. This suggests that making the constraints harder to satisfy would probably worsen our optimal cost, while making them easier to satisfy would probably make our optimal cost better. Since we've got the linear program, we can do a little what-if experimentation to see if it's so. We can change a number in the linear program, rerun it, and see what happens. For example, it seems to make sense that if western cities need more cars, it's going to cost more money, and that needing less cars will need less money. But how much?

Let's try it; let's increase the demand at Fresno by 1, to 201. What happens? Costs do climb, by $234. Atlanta needs to send one more car to Dallas, and then on to Fresno. On the other hand, if Fresno needed one less car, 199, then rerunning this program shows that this would only save $138. So the cost of shipping, not surprisingly, is dependent on the demand at Fresno, but in a rather interesting way. In fact, if you keep on rerunning this program with different demands and record the total cost each time, it looks like this.

And you can see that as you change the demand at Fresno, the effect is linear, almost. Every car below 200 demand saves you $138. Every car above 200 demand costs you $234. The graph is piecewise linear, which just means it's linear in pieces. But the behavior changes from one linear relation to another at exactly 200 cars demanded by Fresno. It would have been nice to have known that without having to rerun the program a bunch of times. And happily, there is a way to know about that, and that's what we're going to talk about next lecture. For now, just keep in mind this picture, a piecewise linear relationship between the right-hand side of a constraint and the objective function value, here, between Fresno demand and total shipping cost.

I want to spend the rest of the lecture today showing you a different way to approach setting up a linear program in a spreadsheet that might actually be easier for some of you. Certainly, if it's done well, it can be easier for those who would use your spreadsheet. I'm going to create a fairly complicated manufacturing example, a multi-period model that's like the Jan de Wit lilies problem in some ways, deciding what products to produce, how many to produce, and when to produce them. It's a variant on a nice problem by Wayne Winston and Chris Albright in their excellent text, *Practical Management Science*.

We've got a factory that can turn cloth into clothes, but we only have so much cloth and only so much production capacity each month. We're making summer clothes, so we have a three-month time horizon for the problem. Each month, we have a production capacity to make 2400 items of clothes, a shirt or a pair of pants each count as an item. Pants take more cloth than shirts. We have some pants and shirts already in stock and some cloth as well, although we'll be getting additional cloth at the beginning each of the months 2 and 3. We have signed a contract with a wholesaler to deliver a certain number of shirts and pants each month for the next three months. They have to be delivered on time, not early, not late. We've already agreed on the selling price for this order, so our goal is to minimize the total cost of production and storage to get the job done.

I'm going to set this up in a spreadsheet, but not using our matrix approach. My goal, instead, is to organize the sheet in a way that's easy to understand, easy to modify, and easy to create. So first, let's make a section for the item parameters, how much cloth each garment uses, how much it costs to make, and how much it costs to store. The blue boxes prompt the user that these values are user-definable. If the values change, they can simply enter the new values in the spreadsheet, and it will still work. Linear programs are often used in this way, over and over.

Now, the production schedule. As you can see, there are some more parameters here, the garments currently in inventory and the monthly demands, but the red cells will be my decision variables, how much of each type of garment to produce in each month. Like last time, I've filled these with some bogus, made-up values. If we know all of these quantities, we can track inventory. I've left blank the cells for those calculations; we'll do them in a minute. Note how the structure of the sheet takes advantage of the parallelism in the problem. We have essentially identical processes going on for shirts and pants in the left-right division of the sheet, and the layout highlights that. Similarly, the activities of one month echo those of another, so each month gets its own column.

In the third section, we're going to keep track of our resources, labor and cloth. We use the same layout as in the previous section. The numbers here are, again, provided by the user, and could be changed as the user desires. We have a bunch of currently blank cells, which we're going to compute, how much cloth and labor capacity we use, as well as keeping tabs on our inventory of cloth. And after all of that, we get to the money. How much money does doing all of this cost us? Again, I'll track this month by month, breaking out the money that I spend on production and on inventory. All these cells are going to need to be computed. While there is no one layout appropriate for every problem, spending a bit of time thinking about how the pieces of your problem fit together can result in a much more intuitive sheet.

OK, now let's put in the math. The first part is production. The relations are what we've talked about before for production and inventory. Take the old inventory, add the new production, subtract the sales, and you've got the new inventory. Notice that the end of the month inventory for one month becomes the new inventory for the following month. You can see that my made-up numbers for the red cells are going to give an infeasible solution, since I have −80 shirts left over at the end of month 2. That's okay while we're creating the program. Solver will fix it.

Let's move on to resource usage. Unused cloth can be carried over from month to month, so it behaves exactly like the inventory work we've just been doing. Add up what's on hand, subtract off what we use, and that's the new inventory. To compute fabric used in this part, just multiply how many items we make times the amount of cloth an item takes, which was recorded up in the parameters section, then add these up. That's cloth.

And labor is actually a bit easier. Every item takes one unit of labor, so we just add up the total items made each month. Those were our decision variables in the production section, remember? Unused production capacity is lost, so we'll just enforce the labor restrictions with limited resource constraints. Notice that I've shown the inequality constraints explicitly in the spreadsheet, to help the user understand the program requirements.

So, all we have left are costs. And again, this is pretty straightforward. Multiply the amount that each item costs to produce by the number of the item produced to get production cost. Similarly, take the monthly storage cost of an item times the number of items that are stored each month to get the storage costs. The storage and production costs per item are in the parameters section. Finally, we find the total monthly cost, then add all of the total monthly costs to get the total cost, and finally put it in green to make it easy for the user to find it.

All that's left is to identify the information to Solver and solve. We know the objective cell and the decision variable cells. But what constraints do we need? Well, we certainly need the limited resource constraints on a production capacity for each month, as shown on the sheet. The only other conditions we have to enforce is that the end-of-month inventory can never be negative, not for pants, not for shirts, not for fabric. All of the other relations, like the conservation constraints on inventory and production, are already part of the spreadsheet.

And, here's the answer when we run the program. Here's the optimal solution! The red cells show us the production levels that we should set for shirts and pants each month to reach the minimum cost. Of course, the report shows a lot more than that, too, since we created it to be easy to read. For example, we can see that it's never a good idea to store shirts, we produce them just in time, but we do carry an inventory of pants at the end of months 1 and 2.

From the cost section , we see that the optimal cost is $37,700. And from the resources section we see that the fabric supply is never a binding constraint; we have extra fabric left over at the end of each month. So as a manager, micromanaging fabric is not a good use of my time. On the other hand, we use all available labor in months 2 and 3. So, if you're going to service your equipment or give vacations to workers, month 1 would be the time to do it.

There's both an art and a science to laying out a sheet like this, but I think you'll agree that the final product is considerably easier to both use and interpret. I also hope that you're beginning to get an idea of how much practical problem-solving power lies within the domain of linear programs. Not every problem can be described by a linear program, but there's a remarkable breadth of problems that can be approached in just this way. And when one can, the simplex method, even if implemented by something as simple as a spreadsheet, can allow us to find the very best answer with surprising ease and speed.

Of course, to get good answers, we need good information. In the real world, we can sometimes be handed surprises on some of these things. What if labor holds out for more money, or some of the cloth that we're supposed to receive is delayed? The usual answer is that, if the changes aren't too great, the answer that we've already found is left unchanged, or changes is a small and predictable way. Too big a change, though, and we may need to rework the whole solution.

In our next lecture, we'll look more closely at exactly this issue, a way to anticipate if, and how, the best answer changes when we find that things aren't quite as we imagined they were, and how far we can push things before dramatic changes occur. It's called sensitivity analysis, and it nicely augments the power of what we've done so far. Among other things, it will allow us to determine, not the market price of a resource, but something considerably more important—how much that resource is worth to you. See you then.

Sensitivity Analysis—Trust the Answer?
Lecture 12

W   ith sensitivity analysis in operations research, we take a problem for which we have found the best answer and then ask how sensitive this best answer is to perturbations in the problem. How much can we change a program parameter before we affect the best solution? If changing a parameter does affect the optimal solution, can we anticipate and characterize the nature of this change? And if we can, when, if ever, does this rule break down? These questions for linear programs will be addressed in this lecture.

Sensitivity Analysis: The Railroad Problem

- Suppose that we want to ship freight cars by railroad from the source cities of Atlanta and Baltimore, through the transshipment cities of Chicago and Dallas, and on to the destination cities of Eugene, Fresno, and Great Falls. Each source has a limited number of cars available (that's the first two cities), and each destination has a demand for cars that has to be met (that's the last three cities). And the constraints for Chicago and Dallas say that for those two cities in the middle, what goes in goes out.

- If you look at the matrix representation of this linear program, you'll see that all of the problem **parameters**—the numbers given in the problem—appear in the gray boxes. The boxes with the dotted lines are the best values of the variables as found by the spreadsheet. The black cell with the white writing is the objective. The LHS ("left-hand side") column holds the left sides of the constraints, computed by using the decision cells (dotted lines) and the gray cells beneath them. So, the entry in cell L8 tells us that 250 cars leave Atlanta, and the entry in cell L14 tells us that 100 cars come into Great Falls.

- **Sensitivity analysis** asks the following question: What happens when we change one or more of the numbers in gray? These might be changes resulting from deliberate choices, or they might be changes imposed on us by some random variation.

	A	B	C	D	E	F	G	H	I	J	K	L	M	N
1		AC	AD	BC	BD	BF	CE	CF	CG	DE	DF			
2		100	150	0	0	200	0	0	100	150	0			
3														
4	Maximize	72	78	70	277	136	213	214	138	206	156	$90,800		= final cash
5														
6	subject to													
7												LHS		RHS
8	Atlanta	1	1	0	0	0	0	0	0	0	0	250	≤	300
9	Baltimore	0	0	1	1	1	0	0	0	0	0	200	≤	200
10	Chicago	1	0	1	0	0	-1	-1	-1	0	0	0	=	0
11	Dallas	0	1	0	1	0	0	0	0	-1	-1	0	=	0
12	Eugene	0	0	0	0	0	1	0	0	1	0	150	≥	150
13	Fresno	0	0	0	0	1	0	1	0	0	1	200	≥	200
14	Great Falls	0	0	0	0	0	0	0	1	0	0	100	≥	100

Figure 12.1

- In the top gray row, changing a number would mean changing how much it costs to ship one car along a link from city to city. In the right-hand column, it would be changing how many cars are available in a city, or how many are needed, or allowing cars to be added or removed in Chicago or Dallas. Changing the numbers in the big gray rectangle to the lower left doesn't make good physical sense in this problem.

- We're doing math, and the power of math is generality. In this problem, the objective is cost, and the constraints are talking about freight cars. In a different problem, the numbers in the shaded regions might represent quantities that aren't even close to these. But if it's a linear program, the kinds of effects when we change a gray cell are always going to be the same.

- Changing the right-hand side—the constant term—of a **nonbinding** constraint, such as the number of cars available in Atlanta, doesn't impact the optimal solution. At least, it doesn't impact the optimal solution unless you change the RHS ("right-hand side") so much that you chew through the **slack**—the cars that we are currently leaving unused in Atlanta.

- In the optimal solution, after we ship 250 cars from Atlanta, 50 of Atlanta's 300 cars are left in the train yard—that's the slack. If you lost some or all of those 50, or if you got more cars in Atlanta, you really wouldn't care. You'd just have more or less surplus. If you lost more than 50, of course, you'd have eaten through your slack, and the supply constraint would be broken. You'd need to change your shipping schedule, and it would probably cost you more money.

- Let's try this same game with a constraint that is originally **binding** in our optimal solution. Let's look at the bottom one, the demand for Great Falls.

- It wants at least 100 cars, and it's getting exactly that. No slack. Binding. Now, what do you think will happen to optimal cost if we change the amount of cars required by Great Falls, either up or down? If they need more cars, cost will go up, and if they need fewer cars, cost will go down.

- The following shows a one-car increase in demand for Great Falls—101 cars, as shown in the bottom cell of the RHS column—and the cost did increase by $210. The old cost was $90,800.

	A	B	C	D	E	F	G	H	I	J	K	L	M	N
1		AC	AD	BC	BD	BF	CE	CF	CG	DE	DF			
2		101	150	0	0	200	0	0	101	150	0			
3														
4	Maximize	72	78	70	277	136	213	214	138	206	156	$91,010		= final cash
5														
6	subject to													
7													LHS	RHS
8	Atlanta	1	1	0	0	0	0	0	0	0	0	251	≤	300
9	Baltimore	0	0	1	1	1	0	0	0	0	0	200	≤	200
10	Chicago	1	0	1	0	0	−1	−1	−1	0	0	0	=	0
11	Dallas	0	1	0	1	0	0	0	0	−1	−1	0	=	0
12	Eugene	0	0	0	0	0	1	0	0	1	0	150	≥	150
13	Fresno	0	0	0	0	1	0	1	0	0	1	200	≥	200
14	Great Falls	0	0	0	0	0	0	0	1	0	0	101	≥	101

Figure 12.2

- The following shows a decrease in the Great Falls demand by 1— to 99, as shown in the bottom cell of the RHS column. Again, the original cost changes by $210, but dropping this time.

	A	B	C	D	E	F	G	H	I	J	K	L	M	N
1		AC	AD	BC	BD	BF	CE	CF	CG	DE	DF			
2		99	150	0	0	200	0	0	99	150	0			
3														
4	Maximize	72	78	70	277	136	213	214	138	206	156	**$90,590**		= final cash
5														
6	subject to													
7												LHS		RHS
8	Atlanta	1	1	0	0	0	0	0	0	0	0	249	≤	300
9	Baltimore	0	0	1	1	1	0	0	0	0	0	200	≤	200
10	Chicago	1	0	1	0	0	-1	-1	-1	0	0	0	=	0
11	Dallas	0	1	0	1	0	0	0	0	-1	-1	0	=	0
12	Eugene	0	0	0	0	0	1	0	0	1	0	150	≥	150
13	Fresno	0	0	0	0	1	0	1	0	0	1	200	≥	200
14	Great Falls	0	0	0	0	0	0	0	1	0	0	99	≥	99

Figure 12.3

- There's nothing really mysterious about this figure of $210 in both cases. Solver did look at all possible routings for the modified problem, but it found out that it was best to continue to use all of Baltimore's cars for Fresno, so another car to Great Falls has to go from Atlanta to Chicago to Great Falls. Atlanta to Chicago costs $72, and Chicago to Great Falls costs $138. That total is $210. And it isn't just one car, of course. Every time we need another car in Great Falls, the answer is to ship it from Atlanta to Chicago to Great Falls, and pay $210.

- The key point here is that each time we change the right-hand side by 1, we alter our optimal solution according to the same recipe. If demand in Great Falls increases by 1 car, we'll send 1 more car from Atlanta to Chicago to Great Falls, and pay $210. If they want another, we do it again. This is a linear pattern, but it doesn't hold forever. Eventually, we run out of cars to send from Atlanta. This happens when Great Falls demand passes 150.

- What we've done with this example isn't intended to be impressive; it's intended to be instructive. What's important is that what we've done here *always* works.

- When you change the RHS of a nonbinding constraint, it doesn't affect your objective or your schedule, until you change the RHS so far as to make it become binding—you "use it all up." If you change the RHS of a binding constraint, it generally changes both the optimal schedule—what you're supposed to do—and your objective value—how well you do—but it does so in a predictable, linear way, such as how one more demand in Great Falls means one more car from Atlanta to Chicago to Great Falls and a $210 increase in cost. This linear pattern holds for a while, but it may break down at some point.

- So, change the demand at Great Falls from 100 cars to a new value, and it'll change your total cost by $210 for every 1 car you add or subtract from that demand. This $210 is called the **shadow price** of the constraint, although it's also known as the **dual price**. In general, the shadow price of a constraint tells you how much the optimal objective value increases (or decreases) each time that you increase (or decrease) the right-hand side of the constraint by 1 unit—at least for a while.

- We can't play this game forever. If the demand at Great Falls gets too high or too low, that $210-per-car cost no longer applies, because our "ship a car from Atlanta to Chicago to Great Falls" recipe breaks down.

- However, for a certain range of right-hand-side values, the recipe works great, and in that range, you can trust the shadow price. This range where everything is behaving is called the **right-hand-side range** of the constraint (RHS range).

- Using our new terminology, we could summarize our findings from this example by stating the following: The shadow price of the Great Falls constraint is $210 per car, and the RHS range is from

0 to 150 cars. That is, as long as the number of cars Great Falls needs stays between 0 and 150 cars, each increase or decrease in its demand increases or decreases the total cost by $210. Also, the shadow price of the Atlanta constraint is $0 per car, and the RHS range is from 250 cars to infinity. That is, as long as the supply of cars in Atlanta stays at 250 or above, increasing or decreasing Atlanta supply does not change total cost.

Important Terms

binding: A constraint is binding on a solution if it has zero slack in that solution.

dual price: See **shadow price**.

nonbinding: A constraint is nonbinding on a solution if it has positive slack in that solution. (Negative slack means that the constraint is violated.)

parameters: The values of the constants relevant to a problem, such as the quantity available of a resource. Models are often created so that the parameters can be changed to reflect different situations.

right-hand-side range (RHS range): The range of values over which the constant term in a constraint may vary without changing the shadow price of any constraint.

sensitivity analysis: An investigation of how an optimal solution changes as parameters of the problem are varied from some set of original values. Sensitivity analysis can be done with most optimization techniques, such as linear programming or decision theory.

shadow price: Also called dual price. The amount that the optimal objective function value increases when the right-hand side of a constraint is increased by 1. Shadow prices may change when one leaves the right-hand-side range of a constraint.

slack: For any inequality constraint, the slack is the side of the constraint claiming to be bigger minus the side claiming to be smaller. Hence, the slack in $4x + y \geq z$ is $4x + y - z$. The slack in $8x \leq 12$ is $12 - 8x$. A constraint that is satisfied always has a slack of 0 or more.

Suggested Reading

Ragsdale, *Spreadsheet Modeling & Decision Analysis*.

Winston and Albright, *Practical Management Science*.

Questions and Comments

1. The video lecture referred to the 100% rule for right-hand-side (RHS) ranging. It is applied to determine if multiple simultaneous changes in the right-hand sides of constraints will leave shadow prices unchanged. To apply it, divide each increase made in an RHS by the allowable increase to that RHS. Divide each decrease made to an RHS by the allowable decrease to that RHS. Add the results. If the total is 1 or less, then shadow prices are unchanged. If the total is more than 1, then the 100% rule fails, and shadow prices may change or may stay the same.

   Example:

   In the video lecture, we looked at increasing month 1 pants demand by 100 and decreasing month 3 pants demand by 100. The allowable increase for month 1 pants demand was 572, and the allowable decrease in month 3 pants demand was 300. The 100% rule would then have us calculate $100/572 + 100/300$. Because this total is less than 1, shadow prices remain unchanged by the combined RHS changes.

2. The shadow price of a nonbinding constraint is always zero. Provide an argument why this should be so in the case of a nonbinding limited resource constraint.

Answer:

If a limited resource constraint is nonbinding, then the quantity of the resource used in the optimal solution is less than what is available—that is, there is some leftover. Then, adding more of that resource cannot improve the optimal solution. If more of the resource made the objective value better, the current optimal solution would not be leaving leftovers of the resource to begin with. On the other hand, the current solution leaves some leftover in the constraint. If the quantity available of the resource is reduced by only a fraction of this leftover, then the old optimal solution is still a feasible solution, so the change in RHS did not make the optimal solution any worse. So, a small enough change leaves the optimal value of the objective unchanged. Therefore, the constraint has a zero shadow price.

Sensitivity Analysis—Trust the Answer?
Lecture 12—Transcript

Universities often have breaks during the semester, where the students can get out and clear their minds. So let's send you on a little vacation. Your destination is a resort spot out in the middle of the Nevada desert, which the internet map programs tell you is 250 miles away from your present location. You have only 10 gallons of gas in your car according to your gas gauge, but you get 30 miles per gallon, which gives you an effective 300-mile range. You're eager to get to the resort, but with the unfamiliar roads, you decide to drive at a constant and leisurely 50 miles per hour. Well, 250 miles at 50 miles per hour means it's going to take five hours to complete the trip. Piece of cake.

But, how about if your gas gauge is a bit off, and you actually only have nine gallons of gas, not 10? No problem. At 30 miles per gallon, your range is still 270 miles, so you'll reach your destination of 250 miles in the desert in five hours, same as before. It's a difference that, effectively, makes no difference. But if the gauge is off a bit more, say, if you only have eight gallons in your tank, then your range is only 240 miles, and you run out of gas with 10 miles of desert between you and your destination, and your trip time is going to be longer; depending on the time of day and how much traffic is on this desolate wasteland road, maybe much, much longer.

On the other hand, if I went back to the 10 gallons of gas but made the trip 260 miles long, rather than 250, the travel time would change, but not by much. An extra 10 miles at 50 miles per hour would mean $\frac{1}{5}$ of an hour, 12 minutes. That's true for every additional 10 miles of trip distance, up till we reach our 300-mile limit; 10 more miles, 12 more minutes; 10 fewer miles, 12 fewer minutes. And that, of course, is a linear relationship between the travel distance and the travel time. But it doesn't last forever. If I make the trip more than 300 miles, this 12-minutes-for-10-miles rule no longer works. You're back to hiking the desert again, and the linear relationship breaks. And the break point in this case is the trip of 300 miles.

This simple analysis actually gives you a pretty good idea of what's meant by the term sensitivity analysis in operations research. We take a problem for which we have found the best answer, and then ask how sensitive this best answer is to perturbations in the problem. How much can we change a program parameter, like the gas in the tank, before we affect the best solution? If changing a parameter does affect the optimal solution, like changing our distance to our destination, can we anticipate and characterize the nature of this change? And if we can, like the the rule 12 minutes for each added 10 miles, then, when, if ever, does this rule break down?

Sensitivity analysis can be done with a variety of analytical techniques any time we want to know how sensitive our solution is to variations in the parameters that characterize the problem. Today, we're going to look at these questions for linear programs. It'll be easiest to see what's going on in the matrix representation of a linear program that we worked out last time, so, let me pull up one that's already familiar with to you, The Great Courses Railroad. We're shipping freight cars from the source cities of Atlanta and Baltimore, through the transshipment cities of Chicago and Dallas, and on to the destination cities of Eugene, Fresno, and Great Falls. Each source has a limited number of cars available; that's the first two constraints. Each destination has a demand for cars that has to be met; that's the last three constraints. And the constraints for Chicago and Dallas say that, for those two cities, in the middle, what goes in, goes out.

If you look at this matrix representation, you'll see that all of the problem parameters, the numbers given in the problem, appear in gray boxes. The red boxes are the best values of the variables as found by the spreadsheet. The black cell with the green writing is the objective. The left-hand side column holds the left sides of the constraints, computed by using the red decision cells and the gray cells beneath them. So right here, we can see 250 cars leave Atlanta, and here, we can see that 100 cars come into Great Falls.

Sensitivity analysis asks, what happens when we change one or more of the numbers in gray? These might be changes resulting from deliberate choices, or they might be changes imposed on us by some random variation. In the top gray row, changing a number would mean changing how much it costs to ship one car along a link from a city to city. In the right-hand column, it would mean changing how many cars are available in a city, or how many are needed, or allowing cars to be added or removed in Chicago or Dallas. Changing the numbers in the big, gray rectangle in the lower left doesn't make good physical sense in this problem, so we'll come back to that later.

Well, we're doing math, and the power of math is generality. In this problem, the objective is cost and the constraints are talking about freight cars. In a different problem, the numbers in the shaded regions might represent quantities that aren't even close to these. But if it's a linear program, the kinds of effects that we see when we change a gray cell are always going to be the same.

Last time we saw that changing the right-hand side, the constant term, of a nonbinding constraint, like the number of cars available in Atlanta, doesn't impact the optimal solution at all. At least, it doesn't as long as the amount that you change the right-hand side isn't so much that you to chew through the slack, the cars that we are currently leaving unused in Atlanta. Take a look. In the optimal solution, after we ship 250 cars from Atlanta, 50 of Atlanta's 300 cars are left in the train yard, that's the slack. If you lost some or all of those 50, or if you got more cars in Atlanta, you wouldn't really care. You'd just have more or less surplus, more or less slack. If you lost more than 50, of course, you'd have eaten through your slack, and the supply constraint would be broken. You'd need to change your shipping schedule, and it would probably cost you more money.

But let's try this same game with a constraint that's originally binding in our optimal solution. I'm going to look at the bottom one, the demand for Great Falls. It wants at least 100 cars, and it's getting exactly that. No slack. Binding. Now what do you think will happen to optimal cost if we change the amount of cars required by Great Falls, either up or down? You probably figured that, if they need more cars, the cost will go up, and if they need fewer cars, the cost will go down. Exactly right.

Here's a one-car increase in the demand for Great Falls, 101 cars. And yet you can see cost did go up by $210. The old cost was $90,800. And here's decreasing the cost of Great Falls demand by 1. And again, the original cost changes by $210, but dropping this time. There's nothing really mysterious about this figure of $210 in both cases. Solver did look at all possible routings for the modified problem, but it found that it was best to continue to use all of Baltimore's cars for Fresno, so another car to Great Falls has to go from Atlanta to Chicago to Great Falls. Atlanta to Chicago costs $72; Chicago to Great Falls costs $138. Total: $210. And it isn't just one car, of course. Every time we need another car in Great Falls, the answer is to ship it from Atlanta to Chicago to Great Falls, and to pay $210.

They key point here is that each time that we change the right-hand side by 1, we alter our original solution according to the same recipe. If the demand in Great Falls increases by 1 car, we send 1 more car from Atlanta to Chicago, one more from Chicago to Great Falls, and we pay $210. If they want another, we do exactly the same thing, a nice, linear pattern. But like in our Nevada resort example, it doesn't hold forever. Eventually, we run out of cars to send from Atlanta. This happens when the Great Falls demand passes 150. What we've done in this example isn't intended to be impressive, it's intended to be instructive. Let me summarize what we said here, because, what's important is, what we've seen here always works.

When you change the right-hand side of a nonbinding constraint, it doesn't affect your objective or your schedule, until you change the right-hand side so far as to make it become binding; you use it all up. If you change the right-hand side of a binding constraint, it generally changes both the optimal schedule, what you're supposed to do, and your objective value, how well you do, but, it does so in a predictable, linear way, such as one more demand in Great Falls means one more car from Atlanta to Chicago to Great Falls and $210 increase in cost. This nice, linear pattern holds for a while, but it may break down at some point.

So change the demand at Great Falls from 100 cars to a new value, and it'll change your total cost by $210 for every one car you add or subtract to that demand. This $210 is called the shadow price of the constraint, although it's also known as the dual price. In general, the shadow price of a constraint tells you how much the optimal objective function value increases or decreases every time that you increase or decrease the right-hand side of the constraint by 1 unit, at least for a while. I have to say at least for a while, because we saw that we can't play this game forever. If the demand at Great Falls gets too high or too low, that $210 per car number is no longer applicable, because our ship-a-car-from-Atlanta-to-Chicago-to-Great-Falls recipe breaks down.

But, for a certain range of right-hand-side values, the recipe works just great, and in that range, you can trust the shadow price. This range where everything is behaving great is called the right-hand-side range of the constraint, RHS range, for short. So, using our new terminology, I could summarize our findings from the last few minutes by saying this. Ready? The shadow price of the Great Falls constraint is $210 per car, and the right-hand side range is from 0 to 150 cars. Translation: As long as the number of cars Great Falls needs stays between 0 and 150 cars, each increase or decrease in its demand increases or decreases total cost by $210. I could also say that the shadow price of the Atlanta constraint is $0 per car, and the right-hand side range is from 250 cars to infinity. That is, as long as the supply of cars in Atlanta stays at at least 250 cars or above, increasing or decreasing Atlanta supply does not change total cost at all; that's the 0 shadow price. Getting it?

Okay our analysis so far has been something you could almost certainly have done on your own. Now let me try to convince you that all of this is worthwhile by looking at a harder problem. I'm going to put our problems from last time, the three-months, pants-and-shirts, inventory and production problem, in this same kind of matrix form, and look at the same kinds of questions. And here is the program. Well, yuck! The first thing to notice is that this program looks a lot less comprehensible than my user-friendly representation from last time, huh? Last time, I had six variables, for shirts and pants production in each of the three months, but in the matrix approach, it was easier to create a bunch of additional auxiliaries, for inventory of shirts and pants and cloth at the end of each month. While I could do a sensitivity analysis with the program in the user-friendly form that we created last time, this matrix form will actually be easier for us to see what's going on today.

And don't let all of these numbers put you off. To do the work that we're going to be talking about, we only need to know a few things. First, what the objective is, which here, is minimizing the cost in dollars. Second, what the right-hand side of the constraints are telling us. These right-hand sides, the gray cells in the last column, record, for each of the three months, the demand for pants and shirts, the amount of cloth that becomes available, and the production capacity available from labor.

We're going to change one of these right-hand-side numbers and see what it does to the optimal solution. This time, it isn't so easy to intuit the changes that will occur to our solution when we change the right-hand side of a binding constraint. For example, suppose that we could get 10 additional units of production capacity in one of the three months. Which would you choose?

It's far from simple to see how the optimal production schedule would change in response to this extra capacity, but with the right tool, you can find this information out from the spreadsheet, without rerunning the program. While Calc in OpenOffice doesn't have this feature at the present time, you can get such a report from Excel or from a free linear programming solver, like WinQSB. In Excel, you simply ask for the sensitivity report in the window that Solver pops up after it has solved the linear program. The report it generates covers most of a page, but I'm only going to show you a portion of it here.

This part talks about right-hand side ranging constraints numbers 10, 11 and 12. These are the constraints that say that you can't use more production capacity from labor than what's available in any of the three months. The next column is labeled Final Value, but really it just tells you the left-hand side value of each constraint. Here, that means how much capacity you actually used in each of the months. We'll skip the shadow price column for a second. The Constraint R.H. Side column, not surprisingly, tells you the right-hand side of the constraints. Here, it says that 2400 units of production capacity were available each of the three months.

So the top row says that a production capacity is a nonbinding constraint in month 1. It says that we're only using 1780 units of capacity, but that there are 2400 units of capacity available. But in the next two rows, you can see that that capacity is binding in months 2 and 3; 2400 used equals 2400 available. Final value = constraint right-hand side. All this is all old news. We could have read these values off of our matrix formulation.

But now comes the new part. The report also shows the shadow price of each constraint for months 1, 2, and 3. The shadow prices for production capacity are given as being 0, −3, and −6 respectively. Remember, shadow price tells you what happens to the objective, here, cost, when the right-hand side of the constraint increases by 1. So, the 0 tells us that an additional unit of capacity is useless in month 1, which makes sense, since we already have unused capacity in month 1. But an additional unit of capacity in month 2 would reduce the costs by $3, that's a −3 shadow price, while an additional unit of capacity in month 3 would reduce costs by $6. We can make a similar statement about capacity decreases, too. If you're going to have people going on vacation, you don't want them doing it in month 3, since each 1-unit decrease in capacity increases cost by $6.

A few moments ago, I asked, if you had to get 10 more units of production capacity for just one month, which month do you want it in? Well, now we can answer that, month 3, where it should be worth 10 × (that −6 dollar shadow price) = −60, a $60 cost reduction.

Let's see if it's true by rerunning the program with that 10 additional units of capacity in month 3. You can see the change I made in available capacity in the last row, and look at the objective. Sure enough, the costs dropped from $37,700 to $37,640, exactly the $60 drop that we predicted for the shadow price. Of course, that's not a very big savings. How about if we could get not 10 more units of capacity, but 400 more units of capacity in month 3. Now we're talking!

Since each additional unit of capacity is worth $6 savings, according to the shadow price, we should save 400 × 6 = $2400! We predict the costs will drop from $37,700 to $35,300. Let's see if we're right by rerunning the program. Change the month 3 capacity from 2400 to 2800, and, and, it didn't work. We predicted a savings $2400, and we're only saving $2100. We were $300 off. The shadow price lied. Well, actually, no, it didn't. Remember that we've said that shadow prices only apply over a limited range, the right-hand side range? It's like a coupon from a store that lets you buy light bulbs for $1 each, limit six. You can get six light bulbs for $6, but who knows what you're going to have to pay for a seventh one. How big this range is for a given constraint is given in the sensitivity report, too, under Allowable

Decrease and Allowable Increase. These columns tell you how much you can increase or decrease the right-hand side of the constraint without getting into trouble, like the shadow prices changing and the like.

Here, take a look at the bottom row of the report on the screen, which talks about the capacity in month 3. You'll see that the allowable increase is only 300. So the first 300 units of labor capacity that we add will each reduce the optimal cost by $6. But after the first 300, it's anybody's guess. We'd have to rerun the program with the new right-hand side to find out the new shadow price. It turns out to be $3 per unit for each additional unit of capacity between 300 and 400, although you wouldn't know that. It's the old diminishing-returns idea from economics, though; as a resource becomes more plentiful, a unit of it becomes less valuable.

Now the changes I've been talking about so far involve changing only one right-hand side, but you can use shadow prices even when changing more than one right-hand side. As we said, that would generally mean changing either the amount of available something, or the amount required of something. You have to be a little careful with what happens when you change more than one right-hand side at a time; you have to use something called the 100% rule, which is a bit beyond our topic today. But, as long as the changes that you make in the right-hand sides are relatively small compared to what each of the ranges allows, you can still trust the shadow prices to be accurate. So with that in mind, how could a manager use this report?

I'm showing you more of the report this time, the range information on all of the constraints. You are the manager. What constraints do you want to focus on? Pants? Shirts? Cloth? Labor? Well, look at the shadow prices. We can forget about cloth. Constraints 7 through 9 show getting or losing a bit of cloth supply doesn't matter; the shadow prices are all zero. OK. How about the demand for pants and shirts? Here the shadow prices are all positive, so increasing demand for any things in these months is going to increase your costs. But you can see that this is true for pants more than for shirts, higher shadow prices, and the later we get in the production cycle, the more costly each additional garment becomes. The top row shows that an additional pair of pants demanded in month 1 increases costs by $6, while an additional pair in month 3 would bump cost by twice as much, $12.

Hmmm, which means that if we could get a wholesaler to take early delivery on some of those pairs of pants, we could shift demand from month 3 to month 1 and save money. Let's try it. The allowable decrease on constraint 3 is 300 pairs of pants. Let me play it safe and use only 100. So I'll reduce month-3 demand by 100, and simultaneously increase month-1 demand by 100. Month 1 allows an increase of 572 demand, so 100 isn't pushing the constraint very hard. Our analysis says we should save $12 × 100 = $1200 by reducing month-3 demand, and increase costs by $6 × 100 = $600 by increasing month-1 demand. The net effect should be that we save $600. Does it actually work? Let's rerun the program. You can see in red the changes I made in the demands in the two months. And checking the costs? Costs are down from $37,700 to $37,100. Sure enough. $600 savings. And of course, you could play similar games with other aspects of production. For example, we see from the shadow prices that getting the wholesaler to accept pants or shirts earlier can reduce our costs. Pretty handy stuff.

There's a second kind of ranging that we can do, that's equally useful. So far we've been changing the right-hand side of a constraint, and these quantities generally represented either the quantity available of something or the quantity required of something. But how about the gray cells across the top of our matrix? They're the objective function coefficients. In our problem today, the objective has been cost, so they represented the cost per unit of each decision variable. Changing these numbers always changes how much one unit of the variable is worth, here, how much cost.

So suppose, for example, that the production cost of pants drops from its original $6. Let's drop the pants all the way down to a little more than half price, then, $3.01 per pair. Should I make more pants at this price? A lot more? A few more? No more at all? Let's see. I'll change the cost of making pants on month 1 down to $3.01 and rerun the program. And as you can see, and this was really amazing to me when I first saw it, you don't change production at all. All of the decision variables in the red cells stay exactly the same. We reduce costs because we're saving $2.99 on each of 700 pairs of pants that we were already making in month 1, but we don't shift production to take advantage of this bargain; we don't shift it at all. Our optimal production schedule is preserved.

But something interesting happens if I can get this month 1 pants production cost just two more cents down to $2.99. Take a look. Whoa! Look at this! Production shifts dramatically. We make lots more pants in month 1, with a corresponding reduction in pants in month 2. Of course, this requires making less shirts in month 1 and more in month 2 to make up for the deficit. All the labor comes into play in month 1, and is now free time in month 2. And these changes, of course, alter cloth reserves and storage costs. But it's worth it, because the costs do go down.

Take a second to think again about what just happened here, because it's kind of startling. As the cost of making pants in month 1 starts to drop, while there's no impact on the production schedule for a while, this is actually true all the way down to a cost of $3 per pair. But the instant we drop below this figure (!) the solution changes dramatically. This is typical of what happens when we change an objective function coefficient in any linear program. It's called objective function coefficient ranging, or OFCR. If the change is small enough, you keep doing exactly what you were doing before. If it's big enough, you see a fairly radical change in activity, a change in which the constraints that are binding and are nonbinding can shift. And the sensitivity report shows you where this happens. Let me show you a different part of the sensitivity report.

This is the objective function coefficient range information, the OFCR info. The top row of the report says that we're currently making 700 pairs of pants in month 1, and that the objective coefficient, that's the current cost, is $6 per pair. The increase and decrease columns are applied to this figure of $6 per pair. The decrease is $3, which drops the $6 per pair down to $3 per pair. The report is telling you that beyond this point, production will probably change. Before that point, it definitely won't. That's what we just saw. The increase column tells you something new; it indicates that if month 1 pants production costs go above $7, that's the $6 plus the $1 allowable increase, then the production is probably going to change then, too.

And you can guess how! The only thing we changed was making month-1 pants more expensive, so if we change production because of this, it's going to be to cut back on month-1 pants production. The report says that $7 per pair is the price point that initiates this kind of change. Let's test it. If pants in month 1 cost $6.99, there's no change in our production from our original optimal solution; the red cells haven't changed. But if we go just above $7 price point? Yep, sure enough: Pants production in month 1 drops, with month 2 picking up the slack, and all other aspects of the optimal production schedule shift around to accommodate it. Note that these changes are quite complex and ripple through the entire factory, but that we knew what price was going to be the breaking point by a quick glance at the sensitivity report. As long as the cost of month-1 pairs of pants stay between $3 and $7; we can't do better by changing schedules. Outside that range, the old answer is probably no longer the best answer.

And what happens when the cost of a pair of pants in month 1 is exactly $3, or exactly $7? Well, we get alternative optima. As we crossed the end of the objective function coefficient range, we moved from a place where the old solution was best to a place where the new solution was best. Exactly at the breakpoint, they're both equally good. This always happens.

It's just like two trains leaving the same station, travelling in the same direction along parallel tracks. Imagine the slow train leaves a half an hour earlier. For short trips, to arrive at your destination as soon as possible, you'll want to take this earlier but slower train. But if you watch the two trains, you'll see the fast one leaves the station, eats through the lead that the slow train built up, and eventually overtakes the slow train. For any longer trip than that, you'd want to be on the fast train. But, if your destination happened to be at the exact point where the fast train caught up with the slow train, you wouldn't care which one you'd ridden. You have alternative optima.

In the same way, if we're inside the objective function coefficient range, what you do is not going to change. Outside of the range, what you're going to do probably does change. And at the end of the range, both the old and new solutions are going to be equally good. And again, you can change multiple coefficients without changing the optimal schedule, as long as you don't change them too much. But, don't think that if the change stays inside each one of its own ranges, you can count on the same schedule.

Think about this; you and I sell the same product in side-by-side booths at a market to 100 people. Your price is $6, and mine is $7.50. Obviously, they all want to buy from you, and this would be the case even if you raised your prices by $1. Similarly, if I lowered my prices by $1, they'd still all buy from you. The optimal schedule wouldn't change. But let us both make those changes, where your extra $1 puts you at $7 and my $1 discount puts me at $6.50, it's a whole new ball game. Again, if you want to look at multiple objective coefficient changes, keep them small compared to the allowable changes, or learn about the 100% rule for objective function coefficients.

Earlier in the lecture, I talked about changing the numbers in the gray boxes in the matrix representation of a linear program, and so far, I've only talked about changing them in the last column, right-hand-side ranging, and changing them in the first row, objective function coefficient ranging. What about the big gray box? Those are the numbers that appear on the left-hand side of the constraints. This would be, if you like, left-hand side ranging.

Well, the truth is, left-hand-side ranging is much trickier. The reason, once again, is the issue of linearity. When you change one or more right-hand side values in a linear program, the effect on the solution is a linear one, as long as you don't push it too far. Changing one or more objective function coefficients has the same kind of effect, as long as you don't push it too far. The linear impact is on shadow prices, rather than on the optimal schedule, but it's still a linear effect. But changing a coefficient on the left-hand side of a binding constraint generally produces a nonlinear change in the optimal solution, much more complicated. From a practical standpoint, it's often easiest to just rerun the program multiple times with different values substituted for the left-hand side coefficient, just to get an idea of the impact of varying it.

Well, earlier we talked about [how] applying the idea of optimization and how it has relevance beyond the formal mathematical models that we've discussed. What are you trying to accomplish? What do you control? What rules do you have to obey? Important questions in a lot of life's situations. And that's also true about some of the ideas we've developed today.

Objective function coefficient ranging shows us that, in a linear program, behavior doesn't change until the payoffs are moved beyond some tipping point, the end of the objective function coefficient range. It's interesting to apply this line of thought to public policy issues, such as environmental policy. If companies are engaging in behavior destructive to the environment, legislatures often enact laws to penalize them, such as fines. This was the case, for example, in West Virginia in 1939, when the state enacted penalties for failing to adequately reclaim land after strip mining.

If you've been to West Virginia, you may guess what happened. It turned out to be cheaper to pay the fine than to reclaim the land, so the land was left scarred and barren. More recent pieces of legislation have substantially increased the penalties, including loss of mining permits, and behavior has changed. A sufficiently large penalty should rarely, if ever, have to be assessed.

It may be cynical, but when I see a piece of legislation proposing fines to curb, say, pollution, I'm always interested to see how the polluters respond to the legislation. If the fine is large enough, outside the firm's objective function coefficient range, they'll cut back on pollutants. If not, the legislation does nothing to address our pollution problem, but it may do a great job of raising government revenues from fines and providing political cover for politicians.

Determining appropriate sanctions can be tricky, but this part, at least, is pretty simple. If you actually want to change behavior, you need a penalty that's severe enough that the violator prefers making the change to paying the penalty. Sensitivity analysis can help you find the tipping point.

Integer Programming—All or Nothing
Lecture 13

U p until this point in the course, we've mostly been assuming that variables could take on any value, subject to the constraints. But in linear programming, add the harmless-sounding additional requirement that your answer needs to involve only integers, and you can make a train wreck out of an otherwise terrific solution technique. In real-life problems, we often have no use for fractional answers. No one will buy three-fourths of a car for three-fourths of the sticker price, for example. Programs where variables have to be integers are called integer programs and are the subject of this lecture.

Linear Programs with Integer Variables

- The following is a linear program with the objective of maximizing $8x - 3y$, solved graphically.

- The arrow on the dotted objective function line (OFL) tells us that the further one moves in the direction of the arrow, the larger the objective function value is growing. If you're maximizing $8x - 3y$, large x's are good and large y's are bad.

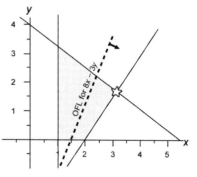

- In the image, the optimal point is marked with a star. What makes it optimal? With graphical solutions, we imagine the feasible region being a flat but tilted field that is gradually being flooded. In this interpretation, an OFL is the line marking the water's edge after some degree of flooding. The arrow points to the side of the flood line that's still above water.

Figure 13.1

- As we let in more and more water, the OFL creeps to the lower right, keeping the same slope as it covers more and more of the feasible region. The last point above water is the optimal point, and that's right at the star. The simplex method would get there by a different technique, walking along the edges of the feasible region and always going uphill, but the result is the same. We've solved this linear program.

- But what if it's an **integer program**? For example, maybe x is the number of bulldozers we sell and y is the number of fueling stations we buy. We can't sell 1/3 of a bulldozer or buy 0.77 fueling stations. Such things have to be integer quantities. And you can probably see from the graph that the optimal solution is a little to the right of $x = 3$ and a little below $y = 2$. In other words, its coordinates aren't integers.

- To be an acceptable answer to this integer linear program, both coordinates have to be whole numbers. We can show this on the graph as follows.

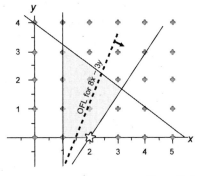

Figure 13.2

- The crosses in the graph form a lattice—a regular array of points. And these lattice points are the only points that are permissible answers to this integer programming problem. They're the only points where both coordinates are integers. And you can see clearly that the optimal solution to our linear program isn't on a lattice point. So, what do we do?

- A natural guess is to pick the lattice point nearest the linear program's optimal solution. This essentially corresponds to rounding each decision variable in the program to the nearest whole number. In this case, that's $x = 3$ and $y = 2$. But the problem with this approach is that the point is outside of the feasible region, so we don't get a feasible solution.

- What about finding the nearest lattice point to the linear program's optimal point that's inside the feasible region? Mathematically, finding which point that is isn't quite as simple as you might think, when the number of variables gets high—but we can deal with that.

- The real problem is that this doesn't necessarily give the best answer. The nearest integer solution that's feasible in our problem is at $x = 2$, $y = 2$, almost directly to the left of the linear program's optimal point. The objective in this problem is to maximize $8x - 3y$. That means that the point $(2, 2)$ gives an objective value of $8(2) - 3(2)$, which is 10. But the point $x = 2$, $y = 0$, which is much farther away from the linear program's optimal point, does better: $8(2) - 3(0) = 16$. In fact, for the integer program, $(2, 0)$ is the optimal point. The optimum integer point is marked with a star on the graph. (See **Figure 13.2**.)

- In this particular problem, there really aren't that many feasible lattice points to check—there are only seven crosses in the shaded region. But in a more complicated problem, this number can quickly explode.

- So, integers can cause us problems, even when the underlying program is linear. What do we do about it? We have a few options. The first thing we can do is ignore the problem and hope that it goes away. That's not as flippant a suggestion as it sounds. Suppose that we try to solve an integer programming problem by ignoring the fact that the variables have to be integer. This linear program is called the linear programming (LP) relaxation of the linear integer program.

- Linear programs can generally be solved very quickly, so maybe we'll get lucky. Maybe the solution to the **LP relaxation** will have all integer values. In some cases, it's not as unlikely as you might think. There are families of problems, such as one-product transportation problems, where you can be sure that you'll always get integer optimal solutions.

- However, most linear programs have nonintegral optimal solutions. So, when you aren't lucky, what's next? One possibility is to fake it—meaning adjusting the answer given by the LP relaxation by a little bit. We could poke around the feasible lattice points in the vicinity of the LP optimum and choose the one that's best. The problem with this idea is that the integer optimum might not be in that neighborhood. In fact, it might be quite far away.

- In some practical cases, though, you might not care. For example, suppose that you solve the LP relaxation and it says that the minimum cost is $900,000, and then by rounding some of the numbers in this solution, you get a feasible integer solution that costs $903,000. You're within $3000, or about 0.3%, of the LP optimum. Maybe that's good enough for you.

- But suppose that it isn't, and you insist on finding the best integer answer. How are integer problems solved? A common technique is called **branch and bound**, and we can get an idea of how it works by using our graphical example.

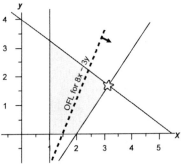

Figure 13.3

Branch and Bound
- Start with the LP relaxation of the problem. Solving our graphical LP, we'll find that the optimal point is at about $x = 3.14$, $y = 1.71$, and the optimal objective value is 20.

- If we're going to have an integer solution, x can't be 3.14. It's either going to be 3 or less, or it's going to be 4 or more. So, we take our one original integer program and replace it with two: one with a new requirement that x has to be 3 or less, and one with a new requirement that x has to be 4 or more. Graphically, it looks like the graph in **Figure 13.4**.

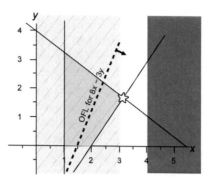

Figure 13.4

- We're breaking the picture up into two regions: the striped one where $x \leq 3$ and the dark gray one where $x \geq 4$. That's the "branch" part of "branch and bound." Whichever of these two different programs gives the better integer answer will provide the best integer answer to the original problem.

- But how do we solve these two branches? The dark gray region doesn't have any feasible points in it, so we can get rid of it and focus on the striped region. We got lucky— if not, we'd have to continue our work with both regions.

- In the striped region, $x \leq 3$, so we add that constraint to the picture and solve the new LP relaxation. You can see in **Figure 13.5** that this constraint cuts off the right tip of the feasible region, which makes the old LP optimal solution slide down and to the left. This new solution is $x = 3$, $y = 1.5$. Obviously, this is not an integer solution.

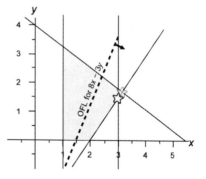

Figure 13.5

- So, now what? Well, y can't be 1.5 in an integer solution. It either has to be 1 or less, or it has to be 2 or more. So, again, we create two regions, as shown in **Figure 13.6**.

- Imposing the linear program's constraints on these two big regions gives us two regions to search for the best answer. We get a striped triangle where $y \geq 2$ and a dark gray trapezoid where $y \leq 1$.

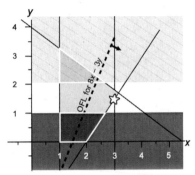

Figure 13.6

- These are two new problems. And when we solve these LP relaxations, we find that the top star is at $x = 2.75$, $y = 2$, giving an objective of 16. The lower star is at $x = 2.666$, $y = 1$, giving 18.333 for the objective.

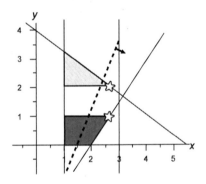

Figure 13.7

- Neither of these is an integer solution. But they both have x between 2 and 3. So, for each one, either x is less than or equal to 2, or x is greater than or equal to 3. We divide each of these problems in two, playing the same kind of game we've already played twice.

- And we lucked out again. We can see from **Figure 13.8** that x can no longer be 3 or more, so we know that x is 2 or less. That adds this one more constraint, $x \leq 2$, to our picture and shrinks the feasible regions of each of the two ongoing problems.

- For the striped problem, the new best answer is at $x = 2$, $y = 2$—an integer solution. The objective is $8(2) - 3(2) = 10$. For the dark gray program, the new optimal solution is at $x = 2$, $y = 0$—again, an integer solution. Its objective is $8(2) - 3(0) = 16$, which is better than 10. So, the bottom star at $(2, 0)$ marks the optimal integer solution to the original problem.

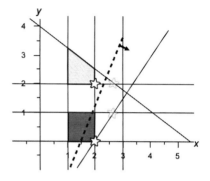

Figure 13.8

- The "bound" part of "branch and bound" is a winnowing technique. It just says that if we found one integer solution, we don't need to consider derived problems whose best answers have no chance of doing better than that one. It's kind of like saying that if you're looking for the oldest person in town and you've found a 90-year-old, you can skip the local school where you know that everyone is 65 or younger.

Integer Programs versus Linear Programs

- The previous example was an easy and short example of branch and bound. This procedure typically involves an absurd amount of work. And that's what makes integer programs in general so much more difficult to handle than linear programs. And it's part of why some real-life integer problems require so much ingenuity to solve.

- In spite of this, for problems of a reasonable scope, you're going to be perfectly capable of handling integer and mixed-integer programs in either Calc or Excel. Given the amazing amount of work required to solve an integer program as compared to a linear one, it's almost embarrassing how little is required from you to turn a linear program into an integer program in Excel.

Important Terms

branch and bound: A technique for finding optimal solutions to integer linear programs by repeatedly replacing a linear integer program with two variants, each of which differs from the original by the inclusion of one additional constraint.

integer program: A program in which all of the decision variables are required to be integers—that is, positive or negative whole numbers or zero. If only some of the variables have this requirement, the program is termed "mixed integer."

LP relaxation (of an integer program): The linear program that results from suspending the integer requirements of an integer linear program.

Suggested Reading

Cornuejols, Trick, and Saltzman, "A Tutorial on Integer Programming."

Winston and Albright, *Practical Management Science*.

Questions and Comments

1. In the video lecture, we said that if *ANNEX* was a 0/1 variable for "We build the annex" and *CONTRACTOR* was a 0/1 variable for "We hire a contractor," then the requirement "If we build the annex, we hire a contractor" is represented by the constraint *ANNEX* $\leq$ *CONTRACTOR*. Verify this.

    [Hint: Recall that a variable is 1 if and only if the corresponding statement is true and 0 otherwise. Under what circumstances is *ANNEX* $\leq$ *CONTRACTOR* false, when *ANNEX* and *CONTRACTOR* are either 0 or 1? Under what conditions is the statement, "If we build the annex, we hire a contractor" false?]

Answer:

Consider the following statement. I promise, "If I like you, I will shake your hand." How could I be found a liar? If I don't like you, I'm not a liar no matter what I do. The statement is only a promise of what I'll do if I *do* like you. If I like you and shake your hand, I am clearly not a liar. But if I like you and yet fail to shake your hand, then I *am* lying.

This is true of if-then statements in general. Such a statement is false if and only if the premise ("if" part) is true and the conclusion ("then" part) is false.

Now look at the relation $A \leq B$, where both A and B are either 0 or 1. When can it be false? Clearly, only if $A = 1$ and $B = 0$, because $1 \leq 0$ is false. So, "If A, then B" is false only if A is true and B is false, and $A \leq B$ is false only if A is 1 and B is 0. The correspondence between the logic of if-then and the arithmetic of $A \leq B$ is perfect.

2. Suppose that *PROJ1* is a binary variable that is 1 if project 1 is being worked on at the plant this weekend and 0 if it is not. Define *PROJ2* and *PROJ3* similarly, for project 2 and project 3. Let *OPEN* be a 0/1 variable that is 1 if the plant is open this weekend and 0 otherwise.

Verify that the constraint $PROJ1 + PROJ2 + PROJ3 \leq 3OPEN$ exactly enforces the requirement that if any of the projects are being worked on this weekend, then the plant must be open this weekend.

Answer:

Just look at the possible cases. If *OPEN* = 0, then the right-hand side is 0, so the constraint says $PROJ1 + PROJ2 + PROJ3 \leq 0$. Because each of the *PROJ* variables is 0 or 1, this works only if they are all 0. Hence, if *OPEN* = 0, then all of the *PROJ* variables are 0, too. On the other hand, if *OPEN* = 1, then the constraint says $PROJ1 + PROJ2 + PROJ3 \leq 3$, which is no restriction at all. The three 0/1 variables obviously can't add to more than 3.

Conclusion: If the plant is closed, then no project can be worked on. Or, equivalently, if a project is worked on, the plant must be open.

You can also reason more directly. If any of the *PROJ* variables equals 1, then the left-hand side adds to more than 0. This means that $3OPEN$ must be more than 0, which means that *OPEN* must be 1. On the other hand, if $OPEN = 1$, it puts no restriction on any of the *PROJ* variables. We reach the same conclusion.

Integer Programming—All or Nothing
Lecture 13—Transcript

Remember the Pythagorean Theorem from way back when? The first time I heard something like it was when I was a kid, because, in *The Wizard of Oz*, the Scarecrow recites a fairly mangled version of it to impress everyone that he has a brain. I didn't get all of it because I was cowering behind the couch at the time, hiding from the flying monkeys.

What Scarecrow should have said was, for a right triangle with legs of length a and b and hypotenuse c, you always have $a^2 + b^2 = c^2$. It's quite a workhorse in mathematics. We use one version of it or another all the time, like we did when we defined Euclidean distance in n-dimensional space in our data mining work.

But, if I was surprised that the Scarecrow didn't know the theorem after all, what surprised me even more was when a friend of mine in grad school took a survey of how many math grad students or math faculty could prove the Pythagorean Theorem. He found that almost none of us could, including me.

So, in case you want to slam dunk a mathematician sometime, here's a cute little geometric proof. Start with two squares, side by side, like this. The blue square is 12×12, and the red square is 5×5. Glue them together, and then make a right-angle cut, like this, to split off two congruent triangles. Since my original squares were 12×12 and 5×5, the legs for these triangles have lengths of 5 and 12. Now, you just swing the triangles around, and it's easy to verify that what you get is a square, and obviously it has the same area as the original two squares. So its total area has to be $5^2 + 12^2$, which is 169. And that means that each side of the square is 13 units long, since $13^2 = 169$.

Actually, this proof works with any original squares of any size, so you get $a^2 + b^2 = c^2$ for any right triangle, which I think is really cute. On the other hand, the fact that the resulting big square that you get is a whole number of units long, that's something different. If you pick an arbitrary pair of squares to start with, you're almost certain to get a larger square whose side is an irrational number. If you start with a 6×6 and a 10×10 square, for example, then when you pivot them, you get a bigger square, but each side of it is about 11.661904 units long. Most people seem to know that 3, 4, and 5 will work. The next smallest solution in whole numbers is 5, 12 and 13, like we just did here. If you ask a high school student to use the Pythagorean Theorem, they can generally do it. But if you restrict them to solutions where all three legs have to be whole numbers, so-called Pythagorean triples, that's a much tougher challenge.

There's a point here that I don't want us to miss. A lot of mathematical problems become much more challenging when you impose the seemly minor requirement that the variables in the solution have to be integers. In the 3rd century B.C., the Hellenistic mathematician Diophantus of Alexandria was looking for solutions to such equations, which are actually now called Diophantine equations, in his honor. In 1900, many such questions were still open, and the brilliant mathematician David Hilbert challenged his peers to determine whether there was any general algorithm that could determine, in a finite amount of time, whether a given polynomial equation has an [integer] solution. The answer was found 77 years later. It's no. You can solve equations like $a^2 + b^2 = c^2$, but you can't solve them all in a general algorithm.

In fact, even figuring out which problems you can solve can be nightmarishly difficult. A 358-year-long nightmare was started back in 1637 by the French mathematician Pierre de Fermat. It may not be easy to find Pythagorean triples, but there are an infinite number of them. Fermat wanted to think about higher powers. Can you find a solutions to $a^3 + b^3 = c^3$, or $a^4 + b^4 = c^4$, and so on. Nice question.

Fermat wrote in the margins of one of his books that he had discovered a truly marvelous proof, that there aren't any positive integer solutions to $a^n + b^n = c^n$, if n is bigger than 2, but that there wasn't room in the margin to write it down. This result is the famous Fermat's Last Theorem, which is actually a misnomer in a couple of ways. First, he died about 30 years later, and he did a lot of good math in the intervening years. And, given how intractable later mathematicians found the problem to be, it's almost certain that Fermat's proof had holes in it. So it was only a conjecture, not a theorem. But nobody published a proof or disproof of this simple-sounding integer problem until 1995, 358 years later. When the problem was finally solved, and Fermat was right, there aren't any solutions, it was heralded by mathematicians and the popular press alike. That wasn't the moon landing. It took much longer than that.

The point of all of this is that, up to this point in this series, I've mostly been assuming that variables could take on any value they please, subject to the constraints. You want a fraction? No problem. And we've got some great solution procedures.

But in a linear program, add the harmless-sounding additional requirement that your answer involve only integers, and you can make an absolute train wreck out of an otherwise terrific solution technique. And in real life problems, we often don't have any use for fractional answers. Face it, no one will buy three quarters of a car for three quarters of the sticker price.

Programs where variables have to be integers are, sensibly enough, called integer programs. Take a linear program and require that all of the variables be integers, and you got an integer linear program. If its got both integers and real variables, it's called a mixed integer linear program. To save breath, I'm just going to call them all integer programs unless it becomes important to distinguish among them.

So, let's restrict ourselves to linear programs with integer variables. How much trouble are we buying for ourselves with the integer restriction? The best way to give you an idea is to return to our graphical approach to a linear program. So here's a linear program with the objective of $8x - 3y$, solved graphically. The arrow on the dotted line, that's the objective function line, tells us that the further one moves in the direction of that arrow, the larger the objective function value is growing. And that makes sense, $8x - 3y$. If you're maximizing, big x's are good, and big y's are bad, and that's what the arrow is saying.

In the image, you can see that the optimal point is marked with a star. What makes it optimal? Well, when we talked about graphical solutions, we imagined the feasible region is being a flat, but a tilted field, that was gradually being flooded. In this interpretation, the OFL line here is marking the water's edge after some degree of flooding. The arrow points to the side of the flood line that's still above water. And as we let more and more water in, the OFL line creeps to the lower right, keeping the same slope as it covers more and more of the feasible region. The last point above water is the optimal point, and that's right at the red star.

The simplex method would get there by a bit different technique, walking along the edges of the feasible region and always going uphill, but the results are the same. We've solved this linear program. But what if it's an integer program? For example, maybe x is the number of bulldozers that we sell, and y is the number of fueling stations that we buy. We can't sell $1/3$ of a bulldozer or buy 0.77 fueling stations. Such things have to be integer quantities. And you can probably see from our graph that the optimal solution is a little to the right of $x = 3$ and a little below $y = 2$. Its coordinates aren't integers. To be an acceptable answer to this integer linear program, both coordinates have to be whole numbers.

We can show that on a graph like this. It looks like the Cartesian plane has come down with chicken pox. All those red x's form a lattice, a regular array of points. And these lattice points are the only points that are permissible answers to this integer programming problem, because they're the only points where both coordinates are integers. And you can see clearly that the optimal solution to our linear program isn't on a lattice point. So what do we do? Well, a natural guess is to pick a lattice point nearest the linear program's optimal solution. This essentially corresponds to rounding each decision variable in the program to its nearest whole number, for us, that's $x = 3$, $y = 2$. But you can see the trouble with that immediately. That point is outside of the feasible region, so you don't get a feasible solution.

OK, how about this, Find the nearest lattice point to the linear program's optimal point that's inside the feasible region. How does that do? Well, mathematically, finding what point that is isn't quite as simple as you might think, and when the number of variables gets high, it can be worse, but, we can deal with that. The real problem is that this doesn't necessarily give you the best answer. The nearest integer solution that's feasible in our problem is there at $x = 2$, $y = 2$, almost directly to the left of the linear program's optimal point. Now, you'll remember that the objective function for this problem was $8x - 3y$, maximize that. That means that the point $(2, 2)$ gives an objective value of $8(2) - 3(2)$, which is 10. But the point $x = 2$, $y = 0$, which is much further away from the linear program's optimal point does better: $8(2) - 3(0)$; that's 16. In fact, for this integer program, $(2, 0)$ is the optimal point. I've marked the optimum integer point with a green star on the graph.

Now, in this particular problem, there really weren't that many feasible lattice points to check. You can see that there were only seven red crosses inside that blue region. But, for a more complicated problem, this number can quickly explode. For example, suppose that my program has 10 variables, each of which are allowed to take on a value from 1 to 10, integer values. Then the feasible region contains 10^{10}, or 10 billon, lattice points. A lattice point for every person on the planet, with a few billion to spare. An exhaustive search just isn't practical.

So, integers can cause us problems, even when the underlying program is linear. What do we do about it? Well, we have a few options. The first thing we can do is ignore the problem and hope that it goes away. That's not as flippant a suggestion as it sounds. Suppose I try to solve an integer programming problem by ignoring the fact that the variables have to be integer? This linear program is called the LP relaxation of the integer linear program. And as you know, linear programs can be solved very, very quickly. So, maybe we'll get lucky. Maybe the solution to the LP relaxation will have all integer values. In some cases, that's not as unlikely as you might think. There are whole families of problems, such as single product transportation problems, where you can be sure that you'll always get integer optimal solutions. The Great Courses Railway problem never had a chance of calling for 13.3 cars to travel from Atlanta to Chicago.

That said, most linear programs have non-integral optimal solutions. So when you aren't lucky, what next? Well, one possibility is to fake it. By this I mean, adjust the answers given by the LP relaxation just a little bit. We could poke around the feasible lattice points in the vicinity of the LP optimum and choose the one that's best. The problem with that is that, as we've seen, the integer optimum might not be in that neighborhood; it might be quite far away.

In some practical cases, though, you might not care. For example, suppose you solve the LP relaxation, and it says the minimum cost is $900,000. And then by rounding off some of the numbers in this solution, you get a feasible integer solution that costs you $903,000. Well, you're within $3000, or about 0.3%, of the LP optimum. Maybe that's good enough for you.

But suppose that it isn't, and you insist on finding the best integer answer. How are integer problems solved? Well, a common technique is called branch and bound, and we can get an idea of how it works by using our graphical example. So, you start with the LP relaxation of your problem. Solving our graphical LP, we'll find that the optimal solution point is at about $x = 3.14$, $y = 1.71$, and the optimal objective value is 20. Now, if we're going to have an integer solution, x can't be 3.14. It's either got to be 3 or less, or it's going to be 4 or more. So, we take our one original integer program and replace it with two, one with a new requirement that x has to be 3 or less, and one with the new requirement that x has to be 4 or more.

Graphically, it looks like this. We're breaking up the picture into two regions, the orange one where $x \leq 3$, and the red one, where $x \geq 4$. That's the branch part of branch and bound. Whichever one of these two different programs gives the better integer answer will provide the best integer answer to the original problem. It's essentially a divide-and-conquer technique.

OK. How do we solve these two branches? Well you'll see that the red region doesn't have any feasible points in it at all, so, we can get rid of it and focus on the orange region. We got lucky, if not, we'd have to continue our work with both regions. But, in the orange region, $x \leq 3$, so we add that constraint to the picture and solve the new LP relaxation. You can see that throwing in this constraint cuts off the right tip of the feasible region, which makes the old LP optimal solution slide down and to the left. This new solution is $x = 3$, $y = 1.5$. Obviously not an integer solution.

So now what? Well, y can't be 1.5 in an integer solution. Either it has to be 1 or less, or it has to be 2 or more. So again, I create two regions, like this. Imposing the linear program's constraints on these two big regions gives us two regions to search for the best answer. We get an orange triangle where $y \geq 2$ and a red trapezoid where $y \leq 1$. Two new problems. And then we solve these LP relaxations. We find that the top one is the orange star at $x = 2.75$, $y = 2$, giving an objective of 16. The lower one is the red star at $x = 2\,^2/_3$, $y = 1$, giving $18\,^1/_3$ for the objective. Here they are. And neither of these are integer solutions. But, they both have x between 2 and 3. So for each one, either x is less than or equal to 2, or x is greater than or equal to 3. Divide each of these problems in two, playing the same kind of game that we've already played twice.

And, we luck out again We can see from the picture that x can no longer be 3 or more, so we know that x is 2 or less. This adds this one more constraint, $x \leq 2$, to our picture and shrinks the feasible regions of each of the two ongoing problems. For the orange problem, the new best answer moves is at $x = 2, y = 2$, an integer solution The objective here is $8(2) - 3(2) = 10$. For the red program, the new optimal solution is at $x = 2, y = 0$, yes again, an integer solution. And its objective is $8(2) - 3(0) = 16$, which is better than 10. So, the red star wins; it marks the optimal integer solution to the original problem.

And that's branch and bound. By the way, the bound part of branch and bound is a winnowing technique. It just says, if we found one integer solution, we don't need to consider derived problems whose best answers have no chance of doing better than that one. It's kind of like saying, if you're looking for the oldest person in town, and you've found a 90 year old already, you can skip the local school where you know that everyone is 65 or younger.

What we just did was an easy and short example of branch and bound. There's a mathematical term for a procedure like this; it's called a pain. Seriously, it's an absurd amount of work. And that's what makes integer programs in general so much more difficult to handle than linear programs. And it's part of why some of the real-life integer examples we've been talking about in the course so far, like that FAA flight routing problem, required so much ingenuity to solve. Because although we didn't have the language to discuss it at the time, a lot of those programs were really integer or mixed integer programs. The FAA could not send half of a plane north of a storm and the other half south. Jan de Witt couldn't plant part of a lily bulb. In some instances, the LP relaxation and adjusting it a bit was good enough. Other ones took insight and cleverness to handle a huge problem of this difficult integer kind.

In spite of this, for problems of reasonable scope, you're going to be perfectly capable of handling integer and mixed-integer programs in either Calc or Excel. Given the amazing amount of work required to solve an integer program, as compared to a linear one, it's almost embarrassing how little is required from you to turn a linear program into an integer program in Excel. Let's look at an example, stock cutting.

Sheet metal is often stored on large rolls, like rolled-up carpet. The roll might be, say, 22-feet wide. But the machines in my factory need rolls that aren't so wide. For this week's work, I need 5 rolls that are 10-feet wide, 5 rolls that are 8-feet wide, 4 rolls that are six-feet wide, and 4 that are 4-feet wide. To get usable rolls, we're going to cut the 22-foot roll like a loaf of French bread into rolls of usable width.

Here's one that has been cut into a 10-foot roll, a 4-foot roll, and an 8-foot roll. Now, management doesn't want us wasting material through inefficient cutting, so cutting two 10-foot pieces out of the 22-foot roll is forbidden, because we'd have this 2-foot piece of useless scrap left. Our goal is to figure out how to carry out the stock cutting so that we get all the cut rolls that we need and use up the fewest possible number of 22-foot rolls.

The trick in stock-cutting problems is to begin by listing all of the different patterns that we could sensibly use in cutting a 22-foot roll into usable pieces. How can we cut a 22-foot roll into pieces that are 4-, 6-, 8-, and 10-feet wide? We've just seen one way, by a 10, an 8, and a 4. But we could also do three 6's and a 4 for 22, or two 8s and a 6. If you keep going, it turns out there are seven possible, sensible cutting possibilities. My decision variables are going to be the number of rolls that are cut in each of these seven patterns, so I have seven variables.

Now, obviously, these have to be integer values. Either I cut up a roll or I don't. The goal is to minimize the sum of these seven variables, since that's the total number of 22-foot rolls that I'm cutting. And I have four constraints, namely that I get at least as many as I need of each of the four roll sizes. I'll show you in a spreadsheet. Easy peasy. It's really pretty simple when you look at it. The red boxes across the top record how many 22-foot rolls we cut in each of the seven patterns; these values are just made up for now. The objective counts the total number of the rolls cut, and we want to minimize that. The constraints just keep track of how many rolls of each size we make, and ensures that we get at least as many as demanded.

When I tell Solver to find the optimal solution to the linear program, we get this. Yep. Nonintegral optimal solution. I only need to cut about 5.9 rolls for it, but it's a nonsensical solution. I can't cut fractional rolls. How do I tell Calc or Excel that all of these variables have to be integers? Very simply. Here's the interface in Calc. As you can see, I just added a new constraint. The cell reference column on the left just gives the locations of all of my decision variables, and the operator pull-down menu, from that, I chose integer. It works the same way in Excel. That's it. When we press Solve, the spreadsheet starts from that branch and bound drudgery that we talked about a minute ago, cranks through it, and a problem of this size, well, it's not too complex, it takes less than a second, and we get the best answer solution, the best integer solution. And here it is.

You can see that it's considerably worse than the LP relaxation. If fractions were OK, we could get away with cutting a little under six rolls. But with integers required, we're going to need to cut up seven, and in the process, we're going to get extra, unneeded rolls of the 8-foot, 6-foot and 4-foot varieties. The price that you'd pay for integer variables. And by the way, don't try to generate and use the spreadsheet's sensitivity report when you're doing an integer program. The fact that only lattice points can be solutions for integer programs means that most of what we interpreted from the sensitivity report in the last lecture, like shadow prices and ranges, goes right out the window. They may be approximately right, but it's safer to investigate program changes by using a tool that resolves the linear program for a variety of parameter values, like we discussed when changing the coefficient on the left-hand side of a constraint.

But from the point of view of creating a program or putting into a spreadsheet, you can see that the integer variable requirement isn't really much of a bother. It just corresponds to adding one constraint into your spreadsheet program. From the point of view of the difficulty of finding the optimal solution, this is a quantum leap in difficulty, though.

But let me talk about a different flavor of integer variables, one that at first blush may sound extremely limited. It's called by a number of names, binary variable, Boolean variable, or 0/1 variable, and, as the names suggest, it can only take on two values, 0 or 1. That doesn't sound like much, but remember that your computer works in binary, and that all that magic comes from shuffling around a bunch of 1s and 0s. One thing that makes binary variables so useful is that, having only two values, they lend themselves to yes/no or true/false quantities. For example, either a worker gets assigned to a shift, or she doesn't.

Binary variables are great for quantities in which there's no middle ground. Either you assign this person to the shift, or you don't. You can't kinda assign them. In fact, we'll often define binary variables with just such a statement, and we understand that the variable is 0 if the statement is false, and 1 if the statement is true. And that coding is very useful, too, because 0 and 1 are about the friendliest numbers you could hope for arithmetically, especially when it comes to multiplication; 1 times anything leaves the number unchanged; 0 times anything is 0.

Because of this, when I multiply a number by a 0/1 variable, I think of that variable as a light switch, turning the number on or off in the problem. Let me show you what I mean. My firm needs at least 80 employees under normal conditions, but if we accept a this new job from a major customer, we're going to need 40 more. So I'll let JOB be a 0/1 variable. It's 1 if we accept the job and 0 if we don't. We want to write the constraint that says we have enough employees for whatever we decide to do. That's going to be this, $EMP \geq 80 + 40\ JOB$, where EMP is the number of employees.

Look at it. When JOB is 0, we don't take the extra work; it just says that $EMP \geq 80$. That's our usual work force. But when $JOB = 1$, we do have the extra work, it says $EMP \geq 80 + 40$, or 120. That is, it says that when we have the extra work, we need 40 more people. The JOB variable turned that 40 on and off. You can get quite clever with this on-off trick. You can even include constraints expressed in propositional logic, statements like, If we decide to build the annex, then we have to hire a contractor. Mathematically, this one becomes, $ANNEX \leq CONTRACTOR$, where ANNEX and CONTRACTOR are 0/1 variables. If $ANNEX = 1$, we build the annex, then CONTRACTOR has to be at least 1 to make the relationship true, and being a 0/1 variable, that means it has to be 1, and hence, true. So, ANNEX implies CONTRACTOR, like we wanted.

Binary variables are everywhere, which is a big part of why integer programming is so important. And happily, they're no more difficult to represent in Excel or Calc than integer variables are. You add a new constraint, the same way that we saw a few moments ago for integer variables, but instead of selecting INT from the pull-down menu, we chose BIN, for binary. And that's it. When binary variables appear in real life, they often appear in great quantity. Let me give you a real-life example.

A number of years ago, I was consulting for a firm that delivered its products from its warehouse to retail stores throughout its distribution region. They had about 200 stores that they were delivering to at the time. The trouble was, they were seeing substantial delays in delivery times. Almost all of the stores made a new order every two weeks, every 10 business days. Orders would come in, be processed in a process that took two days, and then shipped out. But there were some restrictions. The warehouse couldn't handle shipping to more than 35 stores on any given day. They were serviced by five different freight carriers, but they couldn't handle more than two of those freight carriers to come to their warehouse on any given day. And since each store was delivered to by a specific freight carrier, if the carrier didn't come on a particular day, then the order didn't ship on that day.

When I was called in on the problem, the average delay between ordering and delivery was about seven business days, and some stores had been experiencing delays of more than 10 days on the receipt of their orders, which meant that they were actually making their new order before the last one had come in. Let's just say that the retailers could have been happier. So, when I was called in, I asked some questions. It turned out that each store had its own specific order day. Store 116, for example, might order on the 1st and 3rd Tuesdays of the month. Looking at the tangle of order routings, I asked how these days were assigned. The answer? Randomly. When a new store entered the system, they just picked a day.

I suggested that we change order days, making stores that served the same geographic area, by the same freight companies, order on the similar days. I was told, and it was certainly true, that, clearly, I didn't work in with retailers. They were used to the current schedule, and they would not change it happily. So, no change there. We talked about some policies of assigning days to new stores, but the problem at hand was the current collection of 200 stores. So, I decided to solve it by writing an integer program.

Let's rough it out together. Each store gets one delivery day out of the 10 work days in the cycle, that's 10 binary variables per store For each store, do I ship on day 1, or not? On day 2, or not? That's 10×200, or 2000 binary variables. Then there are the 5 shipping companies, which either come a day, or don't. That 5 companies $\times$ 10 days = 50 more variables; hardly worth talking about. But call it 2050 variables.

How about constraints? Well, each store has to ship sometime, that's one more constraint per store, or 200 constraints; no store can ship on a day when a carrier isn't there, that's 200 more constraints; no day can have more than 35 stores, that's 10 more constraints; no day can have more than 2 carriers, 10 more. And the requirement that every order took at least two days to ship I got for free, from the way that I defined my variables. So, 420 nontrivial constraints and about 2000 nonnegativity constraints. And notice something, please. Even though this program is huge compared to our usual examples, conceptually, it's no big deal. We have two kinds of variables, for stores and shippers, and we have 5 kinds of constraints. That's why I've been using small examples in this course. It's a difference of scale, not of kind. Finding the answer to big problems is much more impressive, but the mental effort is often not much more, in terms of the heavy lifting.

OK, anyway. I threw this thing into a mainframe computer and ran it as a linear program, no integer requirements. It came back in less than three seconds with the answer. It was a useless answer, of course, telling me to have $^3/_8$ of a truck at the loading dock, and so on. So I imposed the requirement that all of the variables had to be binary, and reran it. It wasn't done in three seconds this time. It wasn't done in three minutes. It wasn't done in three hours. Let me cut to the chase. I had it running in the background on a mainframe, but I came back three days later, and it still wasn't done.

When I said integer requirements make a problem harder to solve, I wasn't blowing smoke at you. Although, even though it wasn't done, I had found an answer that took the original average delay time of around 7 days and got it down to 2.5. And no store had to wait longer than 5 days for its order. Considering that it was impossible for any order to ship in less than 2 days— that processing time, remember?—2.5 days on average looked pretty good. I terminated the program at that point. I wrote a front-end interface that would take the company's store information and turn it into an integer program automatically, as well as a back end interface that would take the output of the program and convert it into a shipping schedule that the shipping department could read and follow. With that in hand, when the situation did change, they could just rerun the program to get a new answer.

I submitted my solution to the company, and it was met by a lot of happy faces. One of them was an ex-student of mine who brought the problem to my attention to begin with, and who got a nice bonus as a result. In part, that was because of something that happened a few months later. The national organization of the company released a directive to all of its regional offices, insisting that delay times for deliveries had to be no more than 5 days on average by the end of the next year. My client could say, Yeah, we were already on that, and got it down to 2.5 days, right now. More happy faces. It feels good to make things better, you know?

Where Is the Efficiency Frontier?
Lecture 14

Most companies are interested in improving their efficiency. But how can we demonstrate inefficiency? It involves turning inputs into outputs. A technique called data envelopment analysis (DEA) mathematically boils down to a linear program, and that means that we have all of the tools available that we have for such programs. This includes being able to solve them, of course, but it also means that we have sensitivity analysis. As you will learn in this lecture, DEA allows us to consider and compare efficiency in a comprehensive way.

Data Envelopment Analysis

- **Data envelopment analysis** involves the combination of efficiency and linear programming. To learn about DEA, we'll start by comparing just two producers. In standard DEA terminology, these would be referred to as **decision-making units (DMUs)**, but "producers" is a more familiar term and captures the essence.

- There are two different ways to think about their relative efficiency: one focusing on inputs and one focusing on outputs. The output perspective would say that you are more **efficient** than another person if you can make more of every output than the other person does without using more of any input than the other person does. The input perspective would say that you are more efficient than another person if you can use less of every input than the other person does and still make at least as much of every output as the other person does.

- Let's take the latter "input" perspective: Greater efficiency means using less of everything to make at least as much of everything. If you can squeak by and make what another person makes using only 60% as much of each input, then that person's relative efficiency when compared to you is only 60%.

- That's a decent start, but it leaves a lot of ground to be covered. You might use less of everything than the other person, but you might make less of some things, too. In addition, there are more producers in the world than just you and that other person. How does that change things? Let's start by dealing with a special case—the one in which all of the producers have the same supply of inputs.

- Let's say that we have 9 teams of workers in a factory and that each team is doing essentially the same job: making units of the product and shipping them out. Each team has the same inputs. Those inputs might include many things—such as raw materials, and so on—but to keep it simple, we're going to focus only on personnel. Each team has 54 people: 30 trained workers, 20 untrained workers, and 4 supervisors. We'll also assume that they all face the same demand.

- What about outputs? Suppose that we decide that there are two outputs that matter for a team: productivity and rush responsiveness. Productivity is the average number of units that the team produces per week. Rush responsiveness is the average number of rush orders that are delivered on time per week. For each team, we collect data on these two outputs. Because there are only two outputs, we can show them on a scatterplot, where the horizontal position shows the productivity of the team and the vertical axis shows its responsiveness.

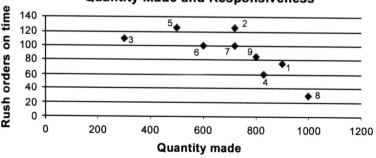

Figure 14.1

- Who is efficient? Team 8 in the lower right has great production—1000 units per week—but terrible responsiveness, with only about 30 rush-order deliveries being on time. On the other hand, both teams 2 and 5, toward the top, do great on responsiveness—they lead the pack at 125 rush-order deliveries on time—but they're not so great on quantity produced.

- If we are prepared to assign relative importance to responsiveness and production, then we might be happy ranking one of these teams above the others. But without that kind of a priori decision, both team 2 and team 8 could be considered efficient. No one else can beat them in one output without being beaten by them in the other.

- This is a common definition of efficiency in economics, although it's also called **Pareto optimality**. Roughly, a solution is Pareto optimal if there's no way to make one thing better than it already is without making something else that you care about worse.

- Let's take this idea a bit further. Look at team 9. Should we consider this team efficient? There are only three teams that make as many or more units than team 9—namely, teams 1, 4, and 8—and none of them has as high a responsiveness score.

- However, we can look at this another way. Suppose that we blended teams 1 and 2 to make a hybrid team. Imagine that we cut each of them in half to make mini-teams, using only half as much of each input and generating only half as much of each output. Put these two demi-teams together, and we're still consuming one team's worth of inputs—30 trained workers, 20 untrained workers, and 4 supervisors.

- But what about outputs? This half-and-half hybrid will make 810 units.

  Productivity of 1/2 team 1 and 1/2 team 2:
  $0.5(900) + 0.5(720) = 450 + 360 = 810$.

- Similarly, the hybrid has a responsiveness of 100.

  > Responsiveness of 1/2 team 1 and 1/2 team 2:
  > $0.5(75) + 0.5(125) = 100$.

- This imaginary hybrid team uses the same resources as team 9, but it makes 810 units per week, not 800, and delivers 100 rush orders on time, not 80. So, the hybrid uses the same resources as team 9 but outperforms it in both productivity and responsiveness. The conclusion is that team 9 isn't efficient when compared to this mix of teams 1 and 2.

- There's nothing that says that our hybrid team has to consist of a 50%-50% blend of teams 1 and 2. The hybrid that is really spectacular is made of about 66% of team 1 and 28% of team 2, because this hybrid outputs exactly what team 9 does, but it only uses about 95% of each of the input resources. So, team 9's efficiency isn't 100%; compared to this hybrid, it's only about 95%.

- For a problem like this one—identical inputs and only two outputs—there's an even easier, geometric way to think about what we're doing. Taking a weighted average of two teams and having the weights add to 1 corresponds graphically to identifying all of the points on the line segment that joins those two teams on the graph. Combining more than two teams allows you to reach any point inside the region of which those teams are the corners.

- What we're really looking for in a DEA problem is the frontier. What is the best that you can do by creating **virtual producers** that are weighted averages of the real producers? In our factory teams example, the frontier would look like **Figure 14.2** on the following page.

- Both axes are outputs, and more is better. So, higher is better, and farther to the right is better. The points on a straight line connecting two points are all hybrid, virtual producers made by blending the real-life producers that are the line's two endpoints.

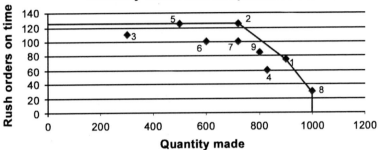

Figure 14.2

- How do we measure efficiency in this picture? If you are on the border, your efficiency is 100%, so teams 1 and 2 are both efficient. If you are not on the frontier, we draw a line from the origin, through your point, and onto the frontier curve. Compare the length of the line from the origin to the producer with the length of the line from the origin to the frontier. For team 9, for example, this would look like the following.

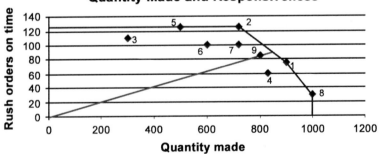

Figure 14.3

- Team 9 is 95% of the way from the origin to the production horizon, shown in black. That's what makes it 95% efficient, as we computed. But when you look at it, there's something a little wrong with this system. Anything on the frontier gets an efficiency of 1—of 100%—which means that team 5 is efficient. But compare team 5 to team 2. Both use the same resources—all of the teams do. Both have 125 rush-order deliveries on time. But team 2 makes 200 more units per shift.

- It's because of this that team 5 has what is called **weak efficiency**. It's not Pareto optimal, because team 2 shows that you can do better on one output without doing worse on the other. On the other hand, you can't do what team 5 does with less of each input, and that's why it gets a 100% efficiency rating.

Comparing Teams with Different Inputs
- So, teams 1, 2, and 8 are considered efficient, team 5 is weakly efficient, and the remaining teams are inefficient to various degrees, depending on their distance from the frontier. However, so far we've been assuming that everyone has the same inputs—and, of course, that's often not the case. How can we compare them, then? We can use essentially the same idea on a slightly larger scale.

- We're going to evaluate the efficiency of a specified real-world producer—the target producer. We do this by comparing it to a virtual producer, a hybrid blend of the real producers. In the previous example, we created a hybrid of just two teams, but you can use as many as you want. In order for this hybrid to have anything to say about the efficiency of the target producer, it has to meet two conditions.
  - Condition 1: The hybrid has to equal or exceed the performance of the target producer on every output.

  - Condition 2: The hybrid has to use up no more of each input than the target producer does.

- Suppose that a hybrid meets these conditions. In fact, suppose that for any input, the consumption of the hybrid is never greater than x% of the consumption of that input by the target producer. Then, we can conclude that the target producer is at most x% efficient.

- The best hybrid for the comparison is the one that meets our two conditions and uses as little of the inputs as possible. That is, we want the hybrid that meets the conditions and makes x%, the input fraction, as small as possible.

- In other words, to make a hybrid, we decide how much of each real-world producer goes into it. These are called the weights for each of the real-world producers. The weights can't be negative, but other than that, we can pick them as we like.

- Of course, this hybrid isn't relevant to our assessment of the target producer unless it satisfies our two conditions—our constraints. And our goal is to find the hybrid that meets these conditions and minimizes the input fraction for this hybrid.

- This is an optimization problem. And because a hybrid is a linear combination of the real-world producers, it's a linear program. Let's see what it looks like for the example of our worker teams, when we're trying to evaluate team 6's efficiency when compared to all of the other teams.

Team #	1	2	3	4	5	6	7	8	9
Untrained	20	20	20	20	20	20	20	20	20
Trained	30	30	30	30	30	30	30	30	30
Supervisor	4	4	4	4	4	4	4	4	4
Quantity	900	720	300	830	500	600	720	1000	800
Rush on time	75	125	110	60	125	100	100	30	85

Figure 14.4

- We start with a table listing the amount of each input that a real-world team uses (shown in dark gray), as well as the amount of each output that it produces from those inputs (shown in light gray). To specify a hybrid, we have to give the weights for each of these real-world producers.

- The numbers shown within borders of dotted lines are decision variables. We get to pick them. We're looking at a mix that is 90% of a team 2 combined with 5% of a team 5. But it's going to be the spreadsheet's job to find their best values.

Team #	1	2	3	4	5	6	7	8	9
Weights	0	0.9	0	0	0.05	0	0	0	0

Figure 14.5

- Next, we have to figure out how this hybrid behaves. How much of each resource does it use, and how much of each output does it make? We saw how to do this when we were creating the 50%-50% blend of teams 1 and 2. The gray-boxed numbers in a given row are multiplied by the corresponding weights in the boxes with dotted lines, and then results are added to give the total for that row. That's just the SUMPRODUCT operation that we used for our linear programming formulations.

- We record this in a new column in the sheet, to the right of our earlier table. We can see, for example, that this hybrid uses only 19 untrained workers, and it produces 673 units per shift.

Team #	1	2	3	4	5	6	7	8	9	
Weights	0	0.9	0	0	0.05	0	0	0	0	
Team #	1	2	3	4	5	6	7	8	9	Hybrid input used/ output made
Untrained	20	20	20	20	20	20	20	20	20	19.000
Trained	30	30	30	30	30	30	30	30	30	28.500
Supervisor	4	4	4	4	4	4	4	4	4	3.800
Quantity	900	720	300	830	500	600	720	1000	800	673
Rush on time	75	125	110	60	125	100	100	30	85	118.75

Figure 14.6

- How does this hybrid do? Look at the last column. The top three entries show that it uses less of each input than team 6, our target team. The last two entries show that it does better than team 6 on both productivity and responsiveness. We've just shown that team 6 is inefficient.

- Now let's push our hybrid a little more, by cutting down its inputs to, for example, only 97% of the inputs of team 6. Could our hybrid still function? What would it have available for inputs? The three inputs for team 6 are 20, 30, and 4, and we multiply each of these by 97% to get 19.4, 29.1, and 3.88.

Hybrid input used		Inputs allowed		Target (team 6) values		Input fraction
19.000	≤	19.400	(=	20	×	0.97
28.500	≤	29.100	(=	30	×	0.97
3.800	≤	3.880	(=	4	×	0.97

Figure 14.7

- We see that our hybrid function can certainly function with these reduced inputs. Our hybrid's input needs are more than satisfied by 97% of team 6's inputs. In fact, it would be satisfied with even a smaller fraction of the inputs—95% is good enough for this hybrid.

- But we want to find the hybrid that uses the smallest-possible fraction of team 6's inputs and still meets our performance conditions on inputs and outputs. And because we now have all of the pieces in place to do this, we just stick all of our pieces together, as shown in **Figure 14.8**.

Team #	1	2	3	4	5	6	7	8	9				Minimize	0.970
Weights	0	0.9	0	0	0.05	0	0	0	0					
Team #	1	2	3	4	5	6	7	8	9	Hybrid input used/ output made	Inputs allowed/ outputs required		Target (team 6) values	Input fraction
Untrained	20	20	20	20	20	20	20	20	20	19.000	≤	19.400	(= 20 ×	0.97
Trained	30	30	30	30	30	30	30	30	30	28.500	≤	29.100	(= 30 ×	0.97
Supervisor	4	4	4	4	4	4	4	4	4	3.800	≤	3.880	(= 4 ×	0.97
Quantity	900	720	300	830	500	600	720	1000	800	673	≥	600	= 600	
Rush on time	75	125	110	60	125	100	100	30	85	118.75	≥	100	= 100	

Figure 14.8

- The "hybrid" column shows what our hybrid uses and produces. Our "inputs allowed/outputs required" column gives us the conditions that our hybrid must satisfy—use at most $x\%$ of the inputs of the target team and still equal or exceed that team's performance. Our goal is to minimize that input fraction, and we're going to find the hybrid that does that. And *that* is the linear program. When we let Solver solve it, it gives the following.

Team #	1	2	3	4	5	6	7	8	9						Minimize		0.821
Weights	0.051	0.769	0	0	0	0	0	0	0								
											Hybrid input used/ output made		Inputs allowed/ outputs required		Target (team 6) values		Input fraction
Team #	1	2	3	4	5	6	7	8	9								
Untrained	20	20	20	20	20	20	20	20	20		16.410	≤	16.410	(=	20	×	0.821
Trained	30	30	30	30	30	30	30	30	30		24.615	≤	24.615	(=	30	×	0.821
Supervisor	4	4	4	4	4	4	4	4	4		3.282	≤	3.282	(=	4	×	0.821
Quantity	900	720	300	830	500	600	720	1000	800		600	≥	600	=	600		
Rush on time	75	125	110	60	125	100	100	30	85		100	≥	100	=	100		

Figure 14.9

- This shows that a hybrid that is about 5% team 1 and 77% team 2 can make everything that team 6 makes and can do it with only about 82% of the resources. That is, team 6 is only 82% efficient.

- To evaluate a different team, just replace the team 6 gray cells on the right side of the sheet by the values for the team being evaluated.

Important Terms

data envelopment analysis (DEA): A technique for evaluating the relative efficiency of decision-making units by comparing them to virtual producers. See **efficient**.

decision-making unit (DMU): The usual term given to a producer in data envelopment analysis.

efficient: In data envelopment analysis, a producer is inefficient if a virtual producer could create the same outputs as the producer under scrutiny while using less of each input. A producer that is not inefficient is efficient. See **weakly efficient**.

Pareto optimal: A solution is Pareto optimal if one cannot do better at accomplishing one goal without doing worse on at least one other goal.

virtual producer: In data envelopment analysis, a hypothetical producer created by taking a linear combination of actual producers.

weak efficiency: In data envelopment analysis, a producer is weakly efficient if it is efficient but some virtual producer creates its same outputs while using no more of any input and using less of at least one input.

Suggested Reading

Cook and Zhu, *Data Envelopment Analysis*.

Winston and Albright, *Practical Management Science*.

Questions and Comments

1. Suppose that you could assign a monetary value to each unit of input consumed and each unit of output created. Then, DEA would not be the best approach to comparing efficiencies of producers. How would you proceed instead?

    Answer:

    Simply divide the total value of the outputs by the total value of the inputs. The highest ratio corresponds to the investment with the best return.

2. There are alternative ways to formulate DEA. One is based on shadow prices. In this approach, the producer currently under review gets to set a table of prices per unit for the inputs and outputs. There are only two rules for the table. First, nothing may be given a negative price. Second, with this price table, no producer can have its total outputs worth more than its total inputs. If a producer can create such a table for which its own outputs are worth as much as its own inputs, then it

is efficient. On the other hand, if the best it could do is to turn $100 of inputs (according to this price table) into $80 of outputs, it would be only 80% efficient. Although it is not obvious, this approach is equivalent to the one in the lecture.

Where Is the Efficiency Frontier?
Lecture 14—Transcript

I don't think I've ever heard of a company that wasn't interested in improving its efficiency. Being more efficient means doing more with what you've got, or, equivalently, continuing to do what you're doing now, but managing to do it with less. You might not be able to say, this is the best way of doing things, but you'd certainly like to be able to compare what your organization is doing to what others are doing in your field. If you find a weakness, you can go to school on the more efficient organizations, find out what they've figured out that you haven't. And it would be good to know, if you are behind the curve, is just a little bit, or a lot? It's even more important in the public and non-profit sectors, where there isn't a clear dollar bottom line to demonstrate success or inefficiency.

But, how can you demonstrate inefficiency? Well, the game is turning inputs into outputs. Suppose that you start out with less of every input than I have and end up making more of every output than I do. Then, relative to you, I am not efficient. That's a no-brainer. If there's only 1 input and 1 output in our production, things are also very easy. Suppose that you turn 100 units of input into 250 units of output; I turn 200 units of input into 400 units of output. Then simple division gives our output/input ratios. You turn 1 unit of input into 2.5 units of output; I turn 1 unit of input into only 2 units of output. I am less efficient than you. In fact, since my 2 is only 80% of your 2.5, we could say that I am only 80% as efficient as you; 80% is my relative efficiency.

You can use this approach if you can reasonably assign dollar values to all of the inputs and the outputs, since that effectively makes our problem a one input/one output problem, with money being the input and the output. But how about when you have multiple inputs, multiple outputs, and no obvious way of pricing them? This often happens in the public and non-profit sectors. For example, how do you compare the efficiencies of different universities? Yes, they all value undergraduate programs, liberal education, research, and so on, but different schools weigh the relative importance of these factors differently. Or, how do you compare police forces, when each has its own resources, and limits, and its own particular challenges?

One more example. Program Follow Through in the 1970s. It was a federal program to help disadvantaged kids in the U.S. public schools. And the government, what's now the U.S. Department of Education, had collected the results for study. They wanted to compare the approaches of different schools to find out what worked and what didn't. But when the usual mathematical tools for an analysis like this were applied, none of them seemed to be working very well.

Roughly, those approaches worked like this. You looked at the data from a lot of schools. Then using tools like multiple regression, you try to find relations that say something like, if a school has this set of inputs, then on average it gives this quality of output. When you find a school that does a lot better than this average, you figure that they must be doing something right.

Well, the first problem with that is that it talks about output, not outputs. We have multiple outputs in the Program Follow Through, things like cognitive development, academic performance, self-esteem, and so on. When you try to fit a problem with multiple outputs into a technique that can handle only one, then you have to combine those outputs in some way to shoehorn them into that single output slot. And that means, essentially, that you have to decide a priori, the relative importances of the multiple outputs.

Besides, looking at deviations from average isn't really the right thing to do, anyway. It really doesn't matter how many average organizations you have when you're looking for those few exceptional examples on the frontier, the ones that might give you some insights into what constitutes best practices. No, for Program Follow Through data, using the usual techniques were getting answers that were unsatisfactory at best and absurd at worst.

So they came up with a new approach, one that ended up being useful for all the problems I've mentioned so far. They started with some previous research on efficiency, and combined it with our old friend, linear programming. The result is what is now called data envelopment analysis, or DEA. It was first published in its current form in 1978, but it's just as useful in the 21st century. Not only on problems like those we've already mentioned, but for tasks like benchmarking hospitals, rating the performance of banks, evaluating maintenance activities on air force bases, comparing performances of different university departments. DEA was even used to help identify promising sites for the new capital for Japan, one far away from Tokyo. Tokyo is a lovely city, but it sits on the meeting point of three tectonic plates and has earthquakes almost every week. Bill McGuire, Professor of Geohazards at University College London, has christened it the City Waiting to Die.

So, let's discover DEA for ourselves. We'll start by comparing just two producers. In standard DEA terminology, they'd be called DMUs, or decision making units, but "producers" is more familiar and captures the essence. There are two different ways to think about their relative efficiency, one focusing on inputs and one focusing on outputs, and one focusing on outputs. The output perspective would say, you're more efficient than I if you can make more of every output than I do without using more of any input than I do. The input perspective would say, you're more efficient than I if you can use less of every input than I do, and still make as much of every output as I do. I want to take this latter input perspective: Greater efficiency means using less of everything to make at least as much of everything. If you can squeak by and make what I make using only 60% as much of each input, then my relative efficiency when compared to you is only 60%. That's a decent start, but it leaves a lot of ground uncovered. You might use less of everything than I, but make less of some things and more of others; and then, there are more than producers in the world than just you and me. How does that change things? Well, let's start by looking at a special case, one in which all of the producers have the same supply of inputs.

Let's say I have 9 teams of workers at a factory, and that each team is doing, essentially, the same job, making units of the product and shipping them out. Each team has the same inputs. Now, those inputs might include a lot of things, like raw materials and so on, but to keep it simple, I'm going to focus only on personnel. Each team has 54 people, 30 trained workers, 20 untrained workers, and 4 supervisors. We'll also assume that they all face the same demand.

How about outputs? Suppose we decide that there are two outputs that matter for a team, productivity and rush responsiveness. Productivity is the average number of units that the team produces per week, rush responsiveness, the average number of rush orders that are delivered on time per week. So, for each team, I collect data on these two outputs. Since there are only two outputs, we can show them on a scatter plot, where the horizontal position shows the productivity of the team, and the vertical axis, its responsiveness. Just like in the graphical method for linear programs, looking at this easy-to-draw case can teach us a lot more in a general context. Ah, there we go.

So, who's efficient? Team 8 in the lower right has great a production, 1000 units per week, but terrible responsiveness, with only about a 30 rush-order deliveries on time. On the other hand, team 2 and 5, both toward the top, are great on responsiveness; they lead the pack at 125 rush-order deliveries on time, but they're not so great on quantity produced. If we are prepared to assign relative importances to responsiveness and production, then we might be happy ranking one of these two above the others. But, without that kind of a priori decision, both team 2 and team 8 could be considered efficient. No one else can beat them in one output without being beaten by them in another. This is a common definition of efficiency in economics, although it's also called Pareto-optimality. Roughly, a solution is Pareto-optimal if there's no way to make one thing better than it already is without making something else that you care about worse.

But let's take this idea a bit further. Look at team 9. Should I consider this team to be efficient? After all, our notion of efficiency is kind of a best in class. There are only three teams that make as many or more units as team 9, namely teams 1, 4 and 8, and none of them has as high a responsiveness score. Well, true enough, but look at it this way. Suppose I took teams 1 and 2 and blended them to make a hybrid team.

If you like, imagine I cut each of them in half to make mini teams, using only half as much of each input and generating only half as much of each output. I take these two demi-teams and stick them together, and we're still now consuming one team's worth of inputs; 30 trained workers, 20 untrained ones and 4 supervisors. But how about outputs? Well, this half-and-half hybrid will make 810 units, and here's how we get that; $1/2$ of the productivity of team 1 and the productivity of $1/2$ team 2 is $0.5(900) + 0.5(720)$, and that adds up to 810. The productivity for team 1 is 900, so half of team 1 gives you 450 units. Similarly, team 2 has a productivity of 720, so half of team

2 makes 360. Add them together, and it's 810. In the same way, the hybrid responsiveness is 100. We get it the same way, but this time using the responsiveness scores of the two teams, 75 and 125, to get the answer. Note that the multipliers are $\frac{1}{2}$ for each, since our hybrid is made up of $\frac{1}{2}$ of each of the two teams used to create it. OK?

Let's recap. We've just created a virtual team, an imaginary blend of two real teams. This hybrid uses the same resources as team 9, but it makes 810 units per week, not 800, and it delivers 100 rush orders on time, not 80. So, the hybrid uses the same resources as team 9, but it outperforms it in both respects, productivity and responsiveness. Conclusion: team 9 isn't efficient when compared to this mix of teams 1 and 2. And there's nothing that says that my hybrid team has to consist of a 50-50 blend of teams 1 and 2. The hybrid that I really like is made from about 66% of team 1 and 28% of team 2. Why? Because if you work it out, you find that this hybrid's outputs are exactly what team 9's are, but it only uses about 95% of the input resources. So team 9's efficiency isn't 100%. Compared with this hybrid, it's only about 95%, and that's the best you can do. Make sense?

For a problem like this one (identical inputs and only two outputs), there's an even easier, geometric way to think about what we're doing. Taking a weighted average of two teams and having the weights add to 1 corresponds graphically to identifying all of the points on the line segment that joins those two teams on the graph. Combining more than two teams allows you to reach any point inside a region of which those teams are the corners. And what we're really looking for in a DEA problems is the frontier. What's the best you can do by creating virtual producers that are weighted averages of real producers? In our factory teams example, the frontier would look like this.

Both axes are outputs, OK? And more is better. So, higher is better, and further to the right is better. Points on the straight line connecting two points are all hybrids, virtual producers made by blending the real-life producers that are the line's two endpoints. You can start to see why this technique is called data envelopment analysis; we're building a convex envelope that forms the frontier of possible production.

How do we measure efficiency in this picture? Well, if you are on the border, your efficiency is 100%, so teams 1 and 2 are both efficient. If you are not on the frontier, we draw a line from the origin, through your point, and onto the frontier curve. Look at how long the line is from origin to the producer, and compare it to the line from the origin to the frontier. For team 9, for example, it looks like this. If you measure carefully, you'll find out that team 9 is 95% of the way from the origin to the production horizon, shown in black. That's what makes it 95% efficient, as we computed just a minute ago.

But when you look at it, there's something a little wrong with this system. I just said that anything on the frontier gets an efficiency of 1, of 100%, which means that team 5 is efficient. But compare team 5 and team 2; both use the same resources, all of the teams do. Both have 125 rush-order deliveries on time. But team 2 makes 200 more units per shift! It's because of this that team 5 is called a weakly efficient producer. It's not Pareto-optimal, since team 2 shows that you can do better on one output without doing worse on the other. On the other hand, you can't do what team 5 does with less of each input, and that's why it gets a 100% efficiency. We'll have more to say about such producers later.

So, teams 1, 2, and 8 are considered efficient, and team 5 is weakly efficient, and the remaining teams are inefficient to various degrees, depending on their distance from the frontier. You can probably imagine what the corresponding picture would look like in a problem with three outputs. The frontier would be a dome with flat-paneled faces. In higher dimensions, the math still works, even if the imagination fails.

But, so far we've been assuming that everyone has the same inputs, and of course, that's often not the case. Different police departments, for example, may have different funding, different staffing, different numbers of cars, and so on. How do we compare them, then? Well, it turns out that we can use essentially the same idea on a slightly larger scale. Let me sketch it out for you. We're going to evaluate the efficiency of a specified real-world producer, the target producer. We do this by comparing it to a virtual producer, a hybrid blend of real producers. In our earlier example, we created a hybrid of just two teams, but you can use as many as you want. In order for this hybrid to have anything to say about the efficiency of the target producer, it has to meet two conditions.

Condition 1: The hybrid has to equal or exceed the performance of the target producer on every single output. Condition 2: The hybrid has to use up no more of each input than the target producer does. Suppose that a hybrid meets these conditions. In fact, suppose that for any input, the consumption of the hybrid never is greater than $x\%$ of the consumption of that input by the target producer. Then we can conclude that the target producer is at most $x\%$ efficient. And the very best hybrid for comparison is the one that meets our two conditions and uses as little of the inputs as possible. That is, we want the hybrid that meets the conditions and makes that $x\%$ as small as we can. We'll call that $x\%$ the input fraction for now, just to give it a name.

OK, let me say this same thing in a slightly different way. To make a hybrid, we decide how much of each real-world producer goes into it. These are called the weights for each of the real-world producers. A hybrid might consist of ½ of a producer 1, $\frac{1}{3}$ of a producer 2, and 2.5 producer 3s. The weights can't be negative, but other than that, they can be whatever they want. Of course, this hybrid isn't relevant to our assessment of the target producer unless it satisfies our two conditions a moment ago; those are our constraints. And our goal is to find the hybrid that meets these conditions and minimizes that input fraction for that hybrid.

Goal, constraints, decision variables, we've got an optimization problem. And since a hybrid is a linear combination of real-world producers, it's a linear program. Let's see what it looks like for the example of our worker teams when we're trying to evaluate team 6's efficiency when compared to all of the other teams. We start with a table listing the amount of each input that a real-world producer uses, shown in dark gray, as well as the amount of each output that it produces from those inputs, that's shown in lighter gray. To specify a hybrid, we have to give the weights for each of these real-world producers. Like this. I've given them red borders because they're decision variables; we get to pick them. For demonstration purposes, I'm looking at a mix that's 90% team 2, combined with 5% of a team 5. But it's going to be the spreadsheet's job to find their best values, as usual.

OK. Next we have to figure out how this hybrid behaves. How much of each resource does it use, and how much of each output does it make? We saw how to do this when we were creating the 50-50 blend of teams 1 and 2 earlier in the lecture. The gray-boxed numbers in a given row are multiplied by the corresponding weights, which are in the red boxes, and then the results are added to give the total for that row. You might recall, that's just the SUMPRODUCT operation that we used in our linear programming formulations.

We record this value in a new column in the sheet, to the right of our earlier example. We can see, for example, that our hybrid uses only 19 untrained workers, and it produces 673 units per shift. To make sure that you've got it, that 673 was computed like this. Put the weights in: $(0 \times 900) + (0.9 \times 720) + (0 \times 300)$, and so on. Add them up, 673. OK. So, how does this hybrid do? Look at the last column. The top three entries show that it uses less of each input than team 6, our target team. And the last two entries show that it does better than team 6 on both productivity and responsiveness, the corresponding entries in the team-6 column. We've just shown that team 6 is inefficient. But now, let's push our hybrid a little more, cutting down its inputs to only 97% of team 6. Again, I've picked the number 97% out of the air, for now. Could our hybrid still function? Well, what would it have available for inputs? It's an easy computation; the three inputs for team 6 are 20, 30, and 4, so we multiply each of those by 97% to get 19.4, 29.1, and 3.88. Again, can our hybrid function with these reduced inputs? And we see that it certainly can. Our hybrid's needs are more than satisfied by 97% of team 6's inputs. In fact, it would be satisfied with even a smaller fraction of the inputs; 95% is good enough for this hybrid.

But we want to find the hybrid that uses the smallest possible fraction of team 6's inputs and still meets our performance conditions on inputs and outputs. And since now we have all the pieces in place to do this, we just stick them all together, like this. The hybrid column shows what our hybrid uses and produces. Our inputs allowed/outputs required column gives us the conditions that our hybrid must satisfy—use at most x% of the target's inputs and still equal or exceed the team's performance. Our goal is to minimize that input fraction, and we're going to find the hybrid that does that. And that, ladies and gentlemen, is the linear program. When we let Solver solve it, it gives this, which shows that a hybrid that's about 5% team 1 and 77% team 2 can make everything that team 6 makes and do it with only about 82% of the resources. That is, team 6 is only 82% efficient. To evaluate a different team, we just replace the team 6 gray cells on the right with the numbers from the team being evaluated.

Let's try it with team 5. Here's what we get. And this is the problem we were talking about earlier, weak efficiency. Team 5 is on the frontier, so you can see it gets a 100% efficiency score, but the linear program compares it to team 2, and you can see that one of the constraints has nonzero slack, specifically, the productivity constraint, where 720 is strictly bigger than 500. If the efficiency is 1, having any of your output constraints with a nonzero slack is the hallmark of a weakly efficient producer. Unfortunately, a weakly efficient producer can hide in the linear program. How? Well, another equally good solution to this linear program, an alternative optimum, is to make the weight for team 5 equal to 1, and all of the other weights equal to zero. That comes down to comparing team 5 to the benchmark of itself. And not surprisingly, then all of the constraints have 0 slack, because team 5 does exactly as well at everything as itself.

This is one of the fiddly bits of doing data envelopment analysis; when there's more than one optimal solution, you always choose the one that maximizes the sum of all of the slacks in the output constraints, otherwise, a weakly efficient producer can go undetected. Actually, a lot of the recent literature concerning DEA focuses on the fact that there are two aspects to inefficiency, the technical efficiency, which is what we've been measuring so far, and this second factor arising from weakly efficient producers, who have slack in output constraints, what's frequently referred to as the matter of nonzero slacks.

Well, our example started with the assumption that all the teams had the same resources, and this made it possible to create a frontier in an intuitive way, on a scatterplot. But the linear program that we just created doesn't require this assumption. We've just built a machine that, with the obvious adjustments, can handle any number of producers, any number of inputs or outputs, and doesn't require identical inputs for producers. Let's look at an example. Let's take a look at hospitals. For our inputs we'll take the number of beds, the number of doctors with admitting privileges, and annual number of dollars spent on capital equipment, averaged over the last five years. We'll also have three outputs, number of patients admitted, number of patients discharged, and average number of beds in use. Let's assume that we have five hospitals in the region of study, like this.

You know how to read this chart now; each column tells you about one hospital. So hospital 1 has 185 beds, 26 doctors, and a bit over $2 million in equipment funding. With this, it admits a bit over 9000 patients, discharges 7000, and has 130 beds in use. The current run has us evaluating hospital 5 as the target. The red cells shows us that hospital 5 has been found to be inefficient, an efficiency of about 81%. If you run the program to evaluate each hospital in turn, by changing the gray column on the right-hand side, here's what you get.

Hospitals 3 and 4 are strongly efficient. Hospital 1? Ehhh, efficient, but only weakly so. The virtual producer found in the chart, a blend of half of hospital 3 and 0.2 hospital 4, uses the same resources as hospital 1, but discharges 25% more patients, and uses 25% more beds. Still, there is no way to do better than hospital 1 in all three outputs using only its inputs, so its efficiency is still 100%. But there's obvious room for improvement.

Hospital 2 isn't far off of the efficiency mark, with 97% efficiency, but hospital 5, a fairly small but decently staffed hospital, and equipped, is less than 81% efficient. If these hospitals are being supported by tax money, hospital 5 may have some explaining to do. The other hospitals in the study could accomplish what it does with only about $^4/_5$ of each resource. In fact, we can see from the weights of hospital 5 that a hospital 3 itself, scaled down to about 28% of its size, would accomplish just about as much as hospital 5 all on its own, but use only about 81% as much of the resources.

Which raises an issue, actually: The DEA model we've been working with assumes constant returns to scale. That is, it assumes that if you've got an company that can turn 5 units of input into 10 units of output, then you could scale that up and down to a company that uses 50 units of input for 100 units of output, or, 1 unit of input into 2 units of output. This assumption isn't valid for a lot of markets, and for that reason, a lot of economists didn't like DEA. It turns out, though, that you can handle variable returns to scale by adding some constraints to the weights. So far, the only constraints that we have on the weights is that they can't be negative. If a producer is strongly efficient in that model, the weights add up to 1. If the weights add to less than 1, the producer is operating under increasing returns to scale, while if the weights add to more than 1, the producer is operating under decreasing returns to scale.

We can handle variable returns to scale directly by imposing an additional constraint on the linear program, namely, all the weights have to add up to 1. It's not easy to visualize what this means in general, but I can give you an idea of what it does by looking at an example with only a single input and a single output. In this graph, it's different; the horizontal axis is the amount of input the producer has, and the vertical axis is the amount of output the producer makes from it.

In this picture, the requirement that the weights have to add to 1 restricts the hybrids that you can use. Specifically, the only hybrids that can be used as benchmarks are those within the envelope that you get when you connect each point in this picture to every other point, like this. It's called the convex hull of the producers, and those ones inside it, real or hybrid, are the only ones that you can use in this model. We're looking for the one in there that gives the best performance for the given amount of output.

If you're on this red line, you're efficient. Below it, and you're not. Note that this is quite different from our original constant returns to scale model. There, A wouldn't have been efficient, since it turns 3 units of input into 3 units of output, a 1-to-1 ratio, while C turns 9 units of input into 36 units of output, an input-output ratio of 1 into 3. Much better! But variable returns to scale model says that that C is benefitting from economies of scale. Taking $\frac{1}{3}$ of C is not a fair comparison to A. For that reason, even E is efficient, since no other producer has the burden of having so much input. Maybe there's nothing to do once your supply exceeds demand, except pay for extra inventory in cost to store it.

In contrast to this, a constant returns to scale model would allow you to take the convex hull that we just found and zoom it in and out from the origin, like this. Anything inside that red cone is a valid hybrid to use as a benchmark, and again, the top border of the valid benchmark is the efficient frontier. You can see here that in this model, only C is efficient. Not surprising, since it had the highest conversion rate of inputs into outputs, as we said at the beginning of the lecture, with 1 input and 1 output, that's all she wrote. Whether the constant returns to scale or variable returns to scale approach is the right one, depends on the producers that you're analyzing.

DEA is an attractive technique, because it mathematically boils down into a linear program, and that means that we have all the tools that we need to analyze such programs. And that includes being able to solve them, but it also means we can look at sensitivity analysis. Let's finish up with one quick example of how that can be useful. For that, I'll go back to our analysis of hospital 5, the one with the dreadful efficiency of less than 81%. Here's the right-hand-side range report for the linear program. Let's see what it tells us about the DEA program.

Especially useful here are the shadow prices. They allow you, among other things, to compute the marginal rates of substitution for the inputs. For example, take the shadow price of doctors and divide it by the shadow price of equipment, and you get $-0.073/-0.00068$, or 106.4.

What does that mean? It means that, as far as efficiency goes, being able to reduce the staff by one doctor would improve hospital 5's efficiency as much as cutting equipment budgets by \$106,400. If the hospital is trying to increase efficiency by tightening its belt, this ratio could be used to help it decide what kind of change might have less impact on the hospital's outputs. Similarly, the shadow prices on the outputs are reporting how much improvement in efficiency would be obtained by managing just one more unit of that output. From the report, we can see that the hybrid significantly outperforms hospital 5 on check-ins and check-outs, so even if hospital 5 improves somewhat in these regards, it doesn't help its relative efficiency. Using on average one additional bed per week, though, would increase its efficiency rating by about 2%, since the shadow price of this constraint is 0.02.

DEA has a lot going for it, especially the way that it doesn't require you to prioritize the values of differing outputs. But it's got its own problems, too. One of the biggest, in my opinion, is that the results that you get can depend considerably on what you consider to be the inputs and the outputs. In our hospital example, we took patients admitted as an output, based on the notion that a hospital that admitted more people served a larger community, or served the community more thoroughly. We also had discharges as an output, since a patient who is discharged, presumably, got well enough to go home. I confess that if a hospital quadrupled its admissions while its discharges fell to 0, though, I'd be concerned. Sounds like a plague with no one getting well.

The worse your ability to select meaningful inputs and outputs, the more nonsensical the results may be. There's also a problem when the number of inputs and output quantities becomes very large. A virtual producer has to outperform the target producer in all respects in order to identify it as inefficient, and as the number of inputs and outputs becomes large, this is harder and harder to do. More and more producers are classed as efficient in some ways, and they may be, but only in a very narrow niche market. DEA works well with multiple inputs and outputs, but not myriad inputs or outputs.

DEA has allowed us to consider and compare efficiency in a more comprehensive way. But what if efficiency isn't your only goal? Remember the factory manager who was so proud of his efficient robots. How do you evaluate the quality of a solution based on more than a single objective?

Multiobjective problems are real and important, and not just when evaluating efficiency. So I'm going to take our next lecture to explore how we can analyze such situations. And not surprisingly, we're going to need to consider some things that, so far, we've been able to ignore—trade-offs, priorities, and when pretty good is good enough. I'll see you then.

Programs with Multiple Goals
Lecture 15

B ecause linear programs have only one objective, our approach when faced with more than one so far has been to choose the one we really care about and focus on that. However, very frequently, ignoring the other objectives just isn't a satisfactory solution. In this lecture, you will explore how to analyze multi-objective problems. Specifically, you will learn one of many approaches to solving multi-objective problems that involves combining multiple objectives into one with the use of soft constraints and penalties.

A Multi-Objective Problem

- Each year, the National Broadcasting Company (NBC) sees about $4 billion in revenues. From a revenue-generating point of view, the company is in the business of delivering viewers to advertisers. You could say that NBC has only one goal: maximizing revenue from the sale of its finite inventory of commercial slots. Those commercials are the most important programming, in both senses, for a commercial broadcaster's bottom line.

- But that's a bit simplified, of course. This isn't a one-shot deal. NBC wants repeat business. So, it also has the goal of giving an advertiser as much as it can of what the advertiser would prefer. And the advertisers have many preferences for many different factors, including season, day of the week, and program placement. In addition, they don't want ads for their competitors appearing next to their own. And what the network offers to the advertisers has to work for them as a package.

- For NBC planning, there was more. Not all contracts close at the same time. Last-minute changes—new clients or clients dropping out—are part of the business. Of course, NBC wanted to make as much money as possible and to satisfy the current advertisers as much as possible, but it also wanted to minimize the number of prime commercial slots that it had to use on any given advertiser. It was trying to do this with each of its clients, and it was trying to do so in a timely fashion.

- And if this wasn't enough, NBC also wanted to encourage and reward big spenders, loyal customers, and clients who would take some slots in the less-popular programs with deals.

- In May, NBC would tell potential advertisers what slots it would have available in the fall. Advertisers would respond on an NBC form saying what they wanted in terms of number of ads, days, seasons, shows, budget, and so on. Then, an NBC planner, working by hand and trying to keep all of the goals in mind, would work up a proposed schedule. The process usually took three to four hours.

- Unfortunately, the proposed schedule was often unacceptable to the advertiser. So, that required reworking the schedule, often multiple times. And this had to be done with every advertiser. During the crunch period during which most of the ads for the year were booked, planners were usually working 12- to 16-hour days. There were too many goals, and too many of them were in direct conflict with the others. The result was a not-very-efficient booking system and a massive headache for everyone.

- How do you handle a situation like this, with multiple goals? The first thing you can do is combine them into a single goal by taking a weighted average of them. For example, suppose that a firm had two goals: maximize market share and maximize profit. Management could create a single objective function to capture their desires if they were capable of stating their indifference point in the trade-off between the two.

- For example, suppose that they decided they'd be indifferent between gaining 1% of market share and gaining $100,000 in annual profit. Then, if SHARE is the percent of market share, we could create an objective like the following.

    Maximize *PROFIT* + 100,000*SHARE*

- It's left unchanged if market share moves 1% in one direction and profit moves $100,000 in the other. And as long as management is satisfied that the "1% for $100,000" trade-off is good by them over a wide range of changes, this objective captures their feelings about the relative importance of these two goals, and we can use it to evaluate the combination of both. And there's nothing to limit us to two variables when using this approach.

- But before this approach could even be considered for the NBC problem, it needs some refinement. After all, we can talk about "maximizing advertiser satisfaction," but doing that really means hitting the targets specified by the advertisers on their request forms. If you can't nail their request exactly, you want to at least get close. The further you are from what they asked for, the less happy they're likely to be.

- You're not going to be able to exactly satisfy everyone's requests. For starters, many of the variables in this problem are integers, such as how many ads a company runs. NBC found that if you got it within about 3%, customers were satisfied.

- While this isn't a new idea for us, it is something quite new for us in terms of our programming work. We've been assuming so far in this course that a constraint is a rule you have to obey. These have-to-obey constraints can be referred to as hard constraints. For example, for NBC, the number of commercial slots it has available is a hard constraint.

- The money the advertiser is willing to spend is another constraint, although usually advertisers give a range with a hard upper limit. This is the idea of a **soft constraint**. The advertisers tell NBC what they want, but if they don't get it, they at least want NBC to get close. The further NBC is from the target, the bigger the problem it is. The size of the deviation is usually measured by a penalty, and this idea is extremely useful in handling either multiple goals or soft constraints.

- Suppose that we have a constraint that says NBC has to give the advertiser 30 commercial slots during a particular week. We have an integer variable specifying how many slots this advertiser is given on each show during the week. If we add all of those variables together, we want the total to be 30. Let's call the number of commercials we assign to this client TOT_1, for total in week 1. Then, we want the constraint to say the following.

  $TOT_1 = 30$

- As written, that's a hard constraint. NBC has to give them 30 ads. But suppose that NBC gives them too few—it was under the target. How much under would it be? Take how many they wanted—30—and subtract how many NBC gave them—TOT_1. That would be the amount NBC was under the desired number of commercials in week 1.

  $30 - TOT_1 = UNDER_1$, or $TOT_1 + UNDER_1 = 30$

- On the other hand, NBC might give them too many commercials this week, perhaps to make up for the fact that NBC squeezed them out of some other week. Then, NBC would be over the desired number. And to find out how much NBC was over, we'd take the number NBC gave the advertisers, TOT_1, and subtract how many commercials the advertisers wanted, 30.

  $TOT_1 - 30 = OVER_1$, or $TOT_1 - OVER_1 = 30$

- We don't know before we find a solution whether we're going to be over, under, or just right. But we can combine these two formulas in a rather clever way that covers all of the bases, as follows.

$$TOT_1 + UNDER_1 - OVER_1 = 30$$

- This is the general way that you handle soft constraints. At first blush, it looks like it won't work. After all, the math allows, for example, that $TOT_1 = 30$ and $UNDER_1$ and $OVER_1$ both equal 5. Certainly, $30 + 5 - 5$ equals 30, as required. The math works fine, but it doesn't make sense to say that we're both under and over 5, when in fact we're just right.

- But remember that being under or over is going to be a bad thing. We're going to penalize it. So, the program is not going to make either of these numbers, $UNDER_1$ or $OVER_1$, any bigger than it has to. It can always make one of them zero, and the other one will then be the correct value of how much under or over the target we actually are. And this is a linear relationship, so it works wonderfully in a linear program.

- When we want a soft constraint in a linear or integer programming problem, we're going to see these "under" and "over" terms in them. It always goes *value + under − over = target*. Sometimes, there's a penalty associated with deviations in either direction—values that are too high or too low—and that's the case with the advertisers. Sometimes, only one direction of deviation is penalized. For example, you may be unhappy if my car's trunk capacity is less than two suitcases, but I certainly don't mind it can hold more.

- And this is the approach that NBC took on their scheduling problem. Each commercial slot had a rating of how popular that slot would be to prospective advertisers. NBC's goal was to satisfy the current advertiser's needs using the least-highly-rated mix that would do the job. To this ratings total, NBC added penalties for missing targets on shows in a given week, shows in a given quarter, the budget for a given quarter, demographic targets, and so on. More serious violations had larger penalties.

Solving a Multi-Objective Problem

- NBC is running this optimization program for one advertiser at a time, coming up with a proposed schedule for them, getting it approved, and moving on. But think about how many integer variables are in this program—roughly one for each commercial length and each show aired during the year. Even if the advertiser is only interested in a band of programming that's five hours wide each day, that's still over 7000 integer variables. This problem would take a long time to solve, then, especially if it's being solved on small computers.

- But from what we know about integer variables, we have some ideas of how to get around this. Start with the LP relaxation of the program—that is, ignore the integer constraints. You can solve that quickly. And that's how the NBC operations research team began.

- Unless you're very lucky, your LP solution isn't going to be your final answer. It's going to have noninteger variables. The team proceeded by rounding all of these values down. But before doing so, they sorted all of the variable values on the basis of how big their fractional part was, in order of decreasing fractional parts. Then, they improved this solution by working down the list in order and rounding up the value of each variable—unless doing that broke a program constraint. If it did, they skipped it and moved on.

- Finally, they tweaked this schedule by adding a single commercial to it or deleting a single commercial from it to see if the result was improved. If it was, they took this as the new schedule and then tweaked this new solution in the same way; if not, they tried tweaking a different single commercial. After a specified number of tries, they stopped.

- This search procedure is a simple example of what's called a **tabu search**. You look at points in a particular area of the solution space, make small changes to them, and keep a list of the solutions you've recently tried. Those solutions are "taboo," so the procedure does not return to them.

- You wouldn't expect this procedure to find the optimal solution unless you did a very long tabu search, but that's okay. NBC was happy with a near-optimal solution. Their program generally ran in about 2 minutes, and some test runs suggested that the solutions were within a few percent of the optimal solution. It took about 10 minutes to put the data into the computer and then a little bit of time to look at the computer output and make any changes that the human scheduler felt were appropriate. The whole procedure took about 20 minutes per advertiser.

- Compare this to the three to five hours that the scheduling agents would spend per advertiser in creating the schedule by hand. And many of those handmade schedules had to be reworked because they didn't meet the advertiser's requirements or management's objectives. The computer-generated ones usually satisfied both groups on the first pass—and in 10% of the time required for a single try by hand.

- The operations research team went further. They created a system to estimate the demand for each show for the upcoming season, based on buying patterns from previous years. This allowed NBC to better set its original price list in May. The team also implemented a system so that one advertiser with different commercials would have two airings of the same commercial separated by as much air time as possible. All of these projects together are estimated to have increased NBC's revenues by at least $50 million per year.

soft constraint: A constraint that need not be completely satisfied. Generally, if it is not, a penalty is assessed, with a larger penalty applied to a larger violation of the constraint.

tabu search: Also called taboo search. A search procedure of a part of the feasible region in which a list is kept of recently examined solutions. The procedure avoids returning to solutions already found unsatisfactory.

Bollapragada, Cheng, Phillips, Garbiras, Scholes, Gibbs, and Humphreville, "NBC's Optimization Systems Increase Revenues and Productivity."

Ragsdale, *Spreadsheet Modeling & Decision Analysis*.

1. In the medical center problem in the video lecture, we computed driving distances from home to center along a rectangular grid. Given the geometry of the roads in the city, this makes logical sense. But if we decided to measure distances "as the crow flies," this problem would have been much more difficult to solve mathematically. Why?

    Answer:

    Distances would no longer be linear functions of the location of the center.

2. Suppose that you have the two goals of spending a lot of time with your family and making a lot of money at your job. Of the approaches discussed in this lecture, which would you favor in determining an optimal allocation of your time?

Answer:

Almost certainly the last approach, which allows you to explore trade-offs.

Programs with Multiple Goals
Lecture 15—Transcript

We've come to the point where we can use programming to address programming on commercial television. We looked last time at a technique for evaluating efficiency, called data envelopment analysis, or DEA. One thing that DEA has going for it, compared to many other analytics techniques, is the ability to assess quality based on multiple inputs and multiple outputs.

That makes a rather surprising application of linear programming, because linear programs always have a single objective function—a single goal. They have to, because in general, you can't insist that one solution be the best possible according to two different criteria. I brought this up in an earlier lecture when I talked about politicians promising the best possible health care at the lowest possible cost. Here, there are two goals, maximize quality of care and minimize cost of delivering it, and it doesn't take a genius to see that they're simply not both going to be accomplished by the same strategy.

Since linear programs have only one objective, our approach when faced with more than one has often been, well, pick the one that you really care about, and we'll focus on that. But we can do better. And a good thing, too, because, frequently, ignoring the other objectives just isn't a satisfactory solution.

Let me give you a lovely, real-world example. NBC, the National Broadcasting Company, formerly owned by General Electric, more recently bought by Comcast. Each year, NBC sees about $4 billion in revenues. From a revenue-generating point of view, the company is in the business of delivering viewers advertisers. You could say NBC has only one goal, maximizing the revenue from the sale of its finite inventory of commercial slots. Those commercials are the most important programming, in both senses of a commercial broadcaster's bottom line.

But, that's a bit simplified, of course. This isn't a one-shot deal. NBC wants repeat business. So they have the goal of giving an advertiser as much as it can of what the advertiser would prefer. And the advertisers have lots of preferences. Season: If your product makes a good gift, jewelry, toys, appliances, and so on, then November and December, the holidays, are much more attractive to you than January and February. Day of the week: Ads for new movies, for example. Since the films open on Friday, ads are best on Thursday night. Program placement: If you're selling cars, you probably want to advertise on the shows that appeal to the 29- to 45-year-old age demographic. And of course, you don't want the ads for your competitors appearing next to your own. And what the network offers to you has to work for you as a package.

For NBC planning, there's more. Not all contracts close at the same time. Last-minute changes, new clients, or clients dropping-out are part of the business. Yes, it wanted to make as much money as possible, and to satisfy the current advertisers as much as possible. But it also wanted to minimize the number of prime commercial slots that it had to use on any given advertiser. And it was trying to do this with each of its clients, and trying to do so in a timely fashion. And as if that wasn't enough, NBC also wanted to encourage and reward big-time spenders with deals, and loyal customers, and clients who would take some slots in the less-popular programs, and so it went.

In May, NBC would tell potential advertisers what slots it would have available in the Fall. Advertisers would respond on an NBC form saying what they wanted in terms of number of ads, days, seasons, shows, budget, and so on. Then an NBC planner, working by hand and trying to keep all these goals in mind, would work up a proposed schedule. The process usually took about three to four hours. Unfortunately, the proposed schedule was often unacceptable to the advertiser, so that required a reworking the schedule, often multiple times. And this had to be done with every advertiser. During the crunch period during which most of these ads for the year were booked, planners usually worked 12- to 16-hour days.

Too many goals, too many of them in direct conflict with the others. The result was a not-very-efficient booking system and a massive headache for everyone. So, how do you handle a situation like this, with multiple goals? Well, the first thing that you can do, and here's a solution that we touched on before, is to combine them into a single goal by taking a weighted average of them. For example, suppose the firm had two goals, maximize market share and maximize profit. Management could create a single objective function to capture their desires if they were capable of stating their indifference point in the trade-off between the two.

For example, suppose they decided that they'd be indifferent between gaining 1% market share and gaining $100,000 in annual profit. Then if share is the percent of market share, we could create an objective like, maximize profit + 100,000 share. Left unchanged, if the market share moves 1% in one direction and the profit moves $100,000 in the other, it stays the same. And as long as management is satisfied that the 1%-for-$100,000 tradeoff is good over a wide range of changes, this objective captures their feelings about the relative importance of these two goals, and we can use the combination to evaluate both. And of course, there's nothing to limit us to two variables using this approach.

But before this approach could even be considered for the NBC problem, it needs some refinement. After all, we can talk about maximizing advertiser satisfaction, but doing that really means hitting the targets specified by the advertisers on their request forms. They may want 10% of their commercials in the 10:00 to 11:00 pm slot, not 5% and not 20%. If you can't nail their request exactly, you want to at least get close. The further you are from what they asked for, the less happy they're likely to be.

And face it: You're not going to be able to exactly satisfy everyone's requests. For starters, a lot of variables in this problem are integers, like how many ads the company runs. If they run 23 ads and want 10% running between 10 and 11, well, you can't run 2.3 ads, now, can you? So you're not going to quite satisfy what the request was no matter what you do. NBC found that if you got it within 3% or so, customers were satisfied.

While this isn't a new idea for you or me, what we just said is something quite new for us in terms of our programming work. We've been assuming so far in this series that a constraint is a rule that you have to obey, period. In today's lecture, I'm going to refer to these have-to-obey constraints as hard constraints. For example, for NBC, the number of commercial slots that they have available is a hard constraint. So was the money the advertiser was willing to spend, although they usually gave you a range. The advertiser might say, "we've got somewhere between \$4.5, \$5 million to spend this year," but if they do, you can't go back to the advertiser and say, "Oh, and it came in at \$5.5 million." If they said \$5 million was the limit, it is.

On the other hand, we've just come across the idea of a soft constraint. Here's what I want, but if I don't get it, at least get close. And the further you are from the target, the bigger the problem it is. How big a problem the deviation is, is usually measured with a penalty, and this idea is extremely useful in handling either multiple goals or soft constraints. Let me show you how it's going to work for NBC.

Suppose we have a constraint that says, we have to give the advertiser 30 commercial slots during this week. Well, we have an integer variable specifying how many slots this advertiser is given on each show during this week, and we add all of those variables up, and we want the total to be 30. To make this easy, I'm going to call the total number of commercials that we give this client in week 1, TOT1, for total in week 1. Then we want the constraint to say, $TOT1 = 30$. As written, that's a hard constraint. You have to give them the 30 ads.

But suppose you give them too few—then you were under the target. How much under would you be? Well, take how many they wanted, 30, and subtract off how many you gave them, TOT1. And that would be the amount that you were under the desired number of commercials; $30 - TOT1$ would be UNDER1, or, $TOT1 + UNDER1 = 30$, same math. On the other hand, we might give them too many commercials this week, perhaps to make up for the fact that we squeezed them out some other week. Then we'd be over the desired number. And to find out how much we were over, of course, we'd take the number we gave them, TOT1, and subtract off how many they wanted, 30. So then, $TOT1 - 30$ would be OVER1, or, $TOT1 - OVER1 = 30$, again, just algebra.

Of course, we don't know before we find the solution whether we're going to be over, under, or just right. But here's the cool thing. We can combine these two formulas in a rather clever way that covers all the bases. Like this, $TOT1 + UNDER1 - OVER1 = 30$. This is the general way that you handle soft constraints, and it's worth looking at carefully. At first blush, it looks like it won't work. After all, the math allows, for example, that $TOT1 = 30$, and that $UNDER1$ and $OVER1$ are both 5; $30 + 5 - 5$ certainly is 30, as required. The math works fine, but it doesn't make sense to say that we're both under and over 5, when in fact, we were just right.

Ah, but remember that being over or under is going to be a bad thing. We're going to penalize it. So the program is not going to make either of these numbers, $UNDER1$ or $OVER1$, any bigger than it has to. And it can always make one of them zero, and the other one will be the correct value of how much under or over we really are. Even better, this is a linear relationship, so it works wonderfully in a linear program. I could do the same thing in a spreadsheet with a command like IF, but we'll be seeing in the next lecture, doing something like that, makes the problem nonlinear, and much, much, much harder to solve.

So, when we want a soft constraint in a linear or integer programming problem, we're going to see these under and over terms in them. It always goes, value + under − over = target. Sometimes, there's a penalty associated with deviations in either direction, values that are too high or too low, and that's the case with the advertisers. Sometimes, only one direction is penalized. For example, I may be unhappy if my car's trunk capacity is less than two suitcases, but I certainly won't mind it can hold more.

And this is the approach that NBC took on their scheduling problem. Each commercial slot had a rating of how popular that slot would be to prospective advertisers. NBC's goal was to satisfy the current advertiser's needs using the least highly-rated mix that it could do the job. To this ratings total, they added penalties for missing targets on shows in a given week, shows in a given quarter, budget in a given quarter, demographic targets, and so on. More serious violations had larger penalties.

You know, one thing I really like about this course is that the further we go, the more I can share with you. Because if you've been thinking about the stuff we just talked about so far, you can probably make some pretty shrewd guesses about solving the NBC problem. They're running this optimization program for one advertiser at a time, coming up with a proposed schedule for them, and then getting it approved, and moving on. But, think about how many integer variables will be in this program, roughly one for each commercial length and each show aired during the year. Even if the advertisers are only interested in a band of programming that's five hours wide each day, that's still over 7000 integer variables. You'd expect that a problem like this is going to take a long time to solve, then, especially if it's being solved on small computers. And you'd be right. But the stuff we talked about when we discussed integer variables, also can give you some ideas about how you might get around this. Start with the LP relaxation of the program; that is, ignore the fact that some variables have to integer. You can solve that quickly, because the great thing about LPs is that they solve quickly, and that's how NBC operations research team began with this problem.

Of course, unless you're very lucky, your LP solution isn't going to be your final answer. It's going to have non-integer variables, like, run 3.9 commercials during this show. The team proceeded by rounding all of these values down, so a 3.9 becomes 3. If we round up, we increase the number of slots, and we may not have room. But before we do so, they sorted all of the variables on the basis of how big their fractional part was, in order of decreasing fractional parts. So 3.9, with a fractional part of 0.9, would be higher on the list than 12.1, with a fractional part of 0.1. Then they improved this solution by working down the list in order and rounding up to the next value for each variable, unless doing so broke a program constraint. If it did, they skipped it and moved on. Finally, they tweaked this schedule by adding a single commercial to it or deleting a single commercial from it to see if the result was improved. If it was, they took this as the new schedule, and then they tweaked this new solution in the same way, if not, they tried tweaking a different single variable. After a specified number of tries, they stopped.

This search procedure is a simple example of what's called a tabu search. You look at points in a particular area of the solution space, make small changes to them, and keep a list of the solutions you've recently tried. Those solutions are "taboo," and so the procedure does not return to them. For example, if you walked one block north and then one block east and ended up in a bad neighborhood, then you wouldn't want to make another trip in which you walk one block east and then one block north. You've been there, and it wasn't good. In a tabu search, you'd never make that second set of tweaks.

Now, you wouldn't expect this procedure to find the optimal solution unless you did a very long tabu search, but that's okay. NBC was happy with a near-optimal solution. Their program generally ran in about two minutes, and some test runs suggested that the solutions were within a couple of percent of the optimal solution. It took about 10 minutes to put the data into the computer, and then, a little bit of time to look at the computer output and make any changes that a human scheduler felt were appropriate. Call the whole procedure 20 minutes per advertiser. And compare this to the three to five hours that the scheduling agents would spend per advertiser in creating the schedules by hand. And many of those hand-made schedules had to be reworked, because they didn't meet the advertiser's requirements or management's objectives. The computer-generated ones usually satisfied both groups on the first pass, and in 10% of the time required for a single try by hand.

The OR team went further. They created a system to estimate the demand for each show for the upcoming season, based on buying patterns from previous years. This allowed NBC to better set its original price list in May. And on a roll, the team implemented a system so that one advertiser with different commercials would have two airings of the same commercial separated by as much air time as possible. All of these projects together are estimated to have increased NBC's revenues by at least $50 million per year—happier customers, a less stressful work environment, and lots and lots of lovely money, a nice example of multi-objective programming.

Unfortunately, its scope is a bit too grand for me to show you the details in a half-hour lecture. But since I like this course to have a roll-up-your-sleeves component, let me take the ideas of soft constraints that we've discussed in the NBC example and use them to show you, in detail, a different way to handle multiple objectives in a linear programming framework. It's used when you can prioritize the goals. We can accomplish the most important goal as well as we can. Then, without sacrificing performance on that goal, we accomplish the second most important goal as well as we can, and so on.

The goals that we're talking about here could either be of the varieties that we've discussed so far. We can maximize or minimize some quantity of direct interest, as has been the case for most of our linear programming examples.

Alternatively, we can minimize the penalty from violating one or more soft constraints, penalties like in the NBC example. We'll do both, in a classic application of linear programming, facilities location. Check out the map.

Welcome to Gridville, an appropriate name, given its street plan; 25 equally-spaced, east-west streets, from 1st Avenue on the south border of town up to 25th Avenue at its northern border. There are also 25 north-south streets, all equally spaced. That's 1st Street over on the west-end side, all the way over to 25th Street on the eastern edge of town. As you can see, whatever creativity the city fathers lacked in coming up with a town name and a street layout, they also lacked in naming the streets. It's a good thing that it's easy to find your way around. The corner of 15th Street and 20th Avenue is just at $x = 15$, $y = 20$, where the red dot is. I'm guessing this place makes a great retirement community for aged mathematicians; actually, that's more or less what our problem is about.

See those six blue dots labeled A through F? Those are six retirement homes in Gridville. The area of the dot is proportional to the number of patients in that home. The elderly in these facilities sometimes require specialized care. For this reason, they're going to build a medical center in the city. The red dot is the proposed site for the center. But it's not really a very good site, as the supervisors of Home A pointed out.

Along the city's roads, it's a 21-block-trip from A to the proposed site. And as you can see, Home A is also the largest home. So, after some discussion, the supervisors of the homes and the city planners agreed on some goals and the order in which they should be prioritized. Number one, no home should be more than 12 blocks from the medical center; if you can't meet this condition, then the center should be located so that the total number of blocks exceeding this limit for all homes combined should be as small as possible. We'll call this the overage.

For example, two homes that are each at 13 blocks from the center generate an overage of two, one for each home. That's better than one home, which is 15 blocks from the center, because it, by itself gives an overage of 3.

Second, the average distance of a patient from the center should be as small as possible. We could have additional factors, too, notably, the price of the property and restrictions on what properties are available for sale, but doing so wouldn't change the character of our problem, so let's keep it simple for purposes of demonstration. Let's see if we can build a spreadsheet that will solve this problem for us. We'll take it piece by piece. By the way, taking it piece by piece is more than just good teaching, it's good analytics. Always be on the lookout for natural fault lines in your problem. You want pieces that can be handled relatively independently from the rest of what's going on.

So, let's start with the decision variable cells, where the center is eventually going to be built. And I'll temporarily put in the current proposal of 15th Street and 20th Avenue. Our first objective said that, ideally, no drive from home to center should be longer than 12 city blocks. So we need to figure out how to get from the center to each of the homes. How far east or west do we have to go? How far north or south? If you take a second and think about it, you'll realize that this is really just a soft constraint.

Our soft constraints looked like, value + under − over = target. And we can apply this to east-west location in our current problem. Value becomes our east-west position of our home; target is the east-west position of the center; and under and over are how far we must go east or west, respectively, to get from the home to the center. Like this.

For example, the first row computes where you end up if you start on 10th Street and go 5 blocks east and 0 blocks west; namely, 15th—perfect. The row gives the correct directions for getting to the center, since it is indeed on 15th Street, as you can see from the bottom. Now the numbers currently in the red cells for row C aren't doing such a great job just yet; 5 blocks east of 6th Street put you on 11th Street, and we want our trip to end us at the center, on 15th. But this isn't a problem. Remember that the numbers that I put in the decision variable cells are just placeholders. Eventually, Solver will find their appropriate values. I'll put constraints in Solver that say that each one of the six east-west ending points have to equal the street number of the center, the red cell on the bottom.

And of course, I have to do the same kind of thing with the north-south trips. The home's avenue + blocks north − blocks south gives the ending avenue, and this has to be the avenue of the center. As you can see, when I filled in the red cells on the right side of this chart, I got the right numbers for Home A on 4th Avenue, you have to go 16 blocks north to get to the center on 20th, but I didn't even bother with sensible numbers for the other homes. Again, Excel will find the correct values when it solves the program. I just add constraints that say that the north-south trips all have to equal the avenue that the center is on.

OK, that part looks good. In fact, you could run Solver with just the constraints that we did so far and an arbitrary objective function to verify that the spreadsheet is figuring the directions correctly. Now we'll use that to figure the value of our first objective. That, you'll recall, was the total overage for the six homes. A home contributes an overage if it's further than 12 blocks from the center. To figure the total overage, I have to begin with how far each home is from the proposed site for the center. But remember that we just found the directions from each home to the center, and that's all we're going to need. To figure the distance of a home from the center, take its east-west distance and add it to its north-south distance. So Home A is 5 + 16, or 21 blocks from the center. That was the long trip that started this discussion in the first place, remember?

If you want to be a bit more general, the distance of Home A to the proposed location is really 5 + 0 + 16 + 0 blocks. Distance east or west or north or south still count as distance. Home F, for example, is 5 + 0 + 0 + 3, or 8 blocks from the center; you have to go five blocks east and 3 blocks south to get there. Anyway, adding all of the east, west, north, and south displacements for a home gives us its distance to the center, and we can record these calculations as formulas in our sheet. The distance isn't a concern, unless it's more than 12 blocks. And here come soft constraints again! The target is 12, and distance + under − over = target, will, when Solver is done with it, tell us how much above and below that target each distance is. It looks like this.

This time, for clarity, I've put the correct values in the red cells, the ones that make home distance + under − over equal exactly 12. Again, Solver will find the correct values, even if you type random numbers in these cells. Your calculation might not give 12, but you'll add constraints to Solver saying that this calculation has to give exactly 12; then when Solver runs, the constraints will work as required. OK, in this part of the problem, being over the limit of 12 blocks contributes to overage, but being under, is fine. So we add together all of the over-limit values and find the total overage of the six homes. In the current work, that's 18 blocks. Quite a lot, really.

Well, now we have all that we need to address the highest priority goal in our problem. The location of the center must minimize the total overage. That's the objective. Put it in Solver, and run it. And here's what you get. The location is selected; it's in the center box at 14th Street and 12th Avenue. As you can see from our overage calculation, this puts Home F beyond the 12 block limit, but only by 3 blocks. Total overage is 3, and that's the best that can be done. If you look at the map, you can see, it's a definite improvement!

OK, let's take a crack at the second goal. The second goal said that the average distance of a patient from the center should be as small as possible. We find this by taking a weighted average of the distance of the homes from the center. The weights reflect the number of patients in each home. A home with twice as many patients counts twice as much in this calculation. So, multiply distance of each home from the center by the number of patients at that home, add them all up, divide by the total number of patients in all homes, and you'll see the answer in the yellow box.

The average distance works out to be a bit over 10.5 blocks for the current location. Can we do better? Well, change the objective to minimizing this average distance and rerun Solver, and here's what you get. The darker blue box says not to build at 14th and 12th, but to build at 10th Street and 15th, like this. This cuts the average distance from the center to about 9.25 blocks, that's about a 1.25-block improvement. But it does so at the cost of telling the people at Home D to go take a hike! They're now 19 blocks from the center! E is 13, resulting in a total overage of 8 blocks.

But minimizing overage was our primary goal. We don't want to cut down average distance if we can only do it at the cost of overage! So when we change to our second objective, we must also do one more thing. We just found a moment ago that the best possible value of overage was 3 blocks, so we have to add a constraint to the program that says, for any solution that we're going to even consider, it has to have an overage of 3 blocks. When we add that constraint, then we can change to our new objective. Like this. Add the constraint to Solver, and rerun it, with the objective of minimizing average distance.

And the optimal solution moves a bit to the southeast of our first run. As required, the new location doesn't increase overage beyond its optimal value of 3 blocks, but it reduces average distance from about 10.5 blocks to about 9.9. Except for one thing! The location is at 12.5th Street and 13.5th Avenue! As you can see, it's smack dab in the middle of a city block! And this is why I mentioned earlier that this was an integer linear program. If we made the street or avenue numbers integer, we'd force the center to be located on a corner. Alternatively, we could rerun the program with only one of them being an integer, to put us on a street or avenue.

No matter which option you pick, though, the answer comes out the same. Build on the corner of 13th Street and 13th Avenue. Doing so will give you the minimum possible overage of 3 blocks, and subject to that, it will give you the minimum average distance, of about 10.137 blocks. Then if we had more than two objectives, we'd just keep playing this game. For example, we might have a third goal of having the maximum distance from the center be as small as possible. We'd add a new constraint that said that the average distance from the center has to be 10.137 blocks, that's the value we just found, and optimize for the new objective. Turns out we can't do any better in satisfying this third goal without doing worse on one of the other two. So, our center gets built on the corner of 13th and 13th; even if that sounds unlucky, there's no other choice that can outperform it.

There can be kind of a buyer's remorse when you prioritize your goals. In the current problem, for example, the minimum overage was three blocks. But, if giving up a little on this goal could gain you a lot on your second goal, maybe you'd want to revisit your decision. Well, we can do such an investigation easily, now that we've written the program. Right now, our program requires that the overage must be its theoretical minimum value of 3. But suppose we replaced this with a constraint saying that the overage must be less than or equal to some amount, and then vary that amount. For each amount, we rerun Solver and see the best we can do on goal 2.

There's actually a free add-in for Solver called SolverTable that will do this for you in one command, and I heartily endorse it. But however you do it, here are the results. As you move to the right across the chart, you accept higher and higher levels of total overage. And you can see that this does, indeed, let us lower the average distance of a patient from the center. We can shorten the average by about 0.4 blocks if we're okay with 4 blocks of overage. Raise the overage limit to 6, and you can shave off another 0.4 blocks or so from the average distance. Any of the dots on this curve are possible results. Any solution to the lower left of this curve is impossible. This kind of trade-off curve can be very handy when deciding what combination of goals best suits the situation at hand. Here, for example, it's very unlikely you'd want to let total overage exceed 6 blocks.

So, we've seen three different approaches to multiple objectives, combining them into one by taking a weighted average; combining them into one with the use of soft constraints and penalties, like NBC problem did; and prioritizing them, as we did in the medical center problem. The prioritizing approach also lets us build the possibilities curve for a pair of objectives, allowing us to explicitly see the trade-off between them.

These aren't the only possibilities. For example, there's an approach called analytical hierarchy process, or AHP. If you can break the decision down into sub-issues, in as many levels as you like, and can identify the relative importances of those sub-issues, then AHP gives you a way to put all of those preferences together into a ranked list of choices, best to worst. It's an approach that's much easier to learn about after you know a bit about matrices, which we'll cover later in the course.

See? Even if you're not NBC, you still have to figure out how to optimize the schedule!

Optimization in a Nonlinear Landscape
Lecture 16

In this lecture, you are going to leave linear programming behind and start exploring a landscape that is much more wild and wooly—nonlinear programming. As you will learn, when you get rid of the requirement that everything is linear, a lot of things change. This lecture will give you an idea of how nonlinear techniques work, why we use multiple techniques, and what to watch out for as you move from a linear to a nonlinear world.

Linear versus Nonlinear Programming

- The landscape of a linear program is like a tilted pane of glass. The constraints run in straight lines on the surface of that glass, and you don't need to be able to see to the horizon to be able to reach the lowest point—you only need to see the ground beneath your feet. When you do, walk straight downhill. When you hit a fence, follow it, again going downhill. When you reach a corner, if the new fence keeps going downhill, follow it. When you can't go downhill any more, stop. You've reached a lowest point.

- How do these instructions change for a nonlinear program? We'll use two decision variables so that we can visualize our mathematical landscape as a landscape in the traditional sense. But now that landscape is much more interesting, including rolling hills and valleys. The topography is much more complex, and it is made even more so by the fact that the constraints can be nonlinear, too. That means that the fences that bound our zone of exploration—our feasible region—can be curvy, or zigzag in and out. But our goal is the same as it is with a linear program: to find the lowest point.

- How much of a difference do these new possibilities make? For a linear program, if the feasible region has a lowest point, it's going to be on a fence line. In fact, for such a region, you can always reach minimum altitude just by checking all of the corners and picking the lowest one.

- But for nonlinear programs, that's not true anymore. The fence could run along a ring of mountains surrounding a valley, and the lowest point would be out in the middle somewhere. As a real-world example, your store might have 10 checkout lanes so that you can handle your busiest times. Open them all on a typical day and you're paying a lot of employees to stand by the registers. Open only one and you're incurring costs from lost sales and customer dissatisfaction. Those are the two fences. The sweet spot is somewhere in the middle.

- Another difference in a nonlinear world is that something might look like the lowest point but not really be the lowest point. It might only be the lowest point in your area—a local minimum, not the global minimum.

- Calculus isn't as useful for some practical problems as you might guess. In contrast, nonlinear programming approaches can allow multiple variables, complicated boundary conditions, and the characterization of multivariate critical points.

Solving Nonlinear Programs

- How do we solve problems where the landscape is not smooth and continuous? There are a number of approaches, and which you should use depends, not surprisingly, on the topography of the landscape. Let's start with rolling hills and valleys, meaning that at any point, we can figure out what direction is downhill. In mathematical terms, we can find the **gradient** of the objective. The following is a strategy that can take advantage of computer power.

- Start at a random point in the landscape. Find out which direction is downhill at that point, just like you would for a linear program. Walk a little ways in that direction. Then, stop and repeat the process. Keep repeating it, taking a fix and walking downhill a bit, until you can't make any more progress. This is the basic idea behind a collection of techniques called **steepest descent**, or gradient, methods. Excel's GRG nonlinear method is an implementation of one of these. The variants differ in exactly what downhill direction you take and how far you go in that direction.

- Can such a technique fail to eventually find the lowest point in a rolling landscape? Sure. If it hits the bottom of a little dip valley, or follows a slope downhill to a point along a fence, then it's just going to stay there, even though there might be much lower points somewhere else—a local minimum. And that is a problem with any gradient descent method.

- One way to try to avoid this is to begin at many points, rather than just one. It's kind of like having a Boy Scout troop out with you, each Boy Scout starting at a different spot. When you're done, you compare notes to see who did the best. The more starting points you have, the better your chances are.

- You can imagine a landscape like a hole in a miniature golf course, where the hole is surrounded by a volcano-like mound. The bottom of the hole might be the deepest point on the course, but how likely is it that you're going to find it with a steepest descent approach? Every time you check your bearings when standing on the volcano, you're going to end up going in exactly the wrong direction.

- It's good that this kind of thing doesn't happen too often in the mathematical landscapes of real-life problems, because in general, there's no algorithm that is guaranteed to find the global optimum in any reasonable amount of time. But things often work well. And there are large classes of problems where gradient descent is guaranteed to work.

- A key issue is the idea of convexity. First, let's define the convexity of the feasible region. Imagine that the fences defining our feasible region are opaque walls. Two people are standing at two points inside the region. Then, the region is **convex** if, no matter where the people stand, they can always look one another in the eye. There are no walls in the way. So, a square region is convex, but a star-shaped region is not.

Figure 16.1

- There's no hiding from one another in a room with a square floor plan. But if one person was standing at one of the small dots in the star-shaped room and the other person was standing at the other, they couldn't see one another.

- We can also define the convexity of a function—in our intuitive terms, of a landscape. It's a closely related idea. Informally, the landscape is convex if no matter where two people stand on it, they can still make eye contact. So, if they are standing inside a bowl-like depression, they can see one another. If they are standing around a mountain, or in the vicinity of a mountain pass, then there are places where they could hide from one another's view. The function giving the landscape is not convex in those cases.

- Why does convexity matter in finding optimal solutions? Let's revisit the two people in the star-shaped room. Imagine that the goal is to move as far to the south (down) in this room as possible. If the two people, represented by the two dots, were following gradient descent, neither of them could move any farther south in the picture. They would both be at local minima—but only the person on the right would be at the actual global minimum.

- So, even if the objective is linear, a non-convex feasible region can cause problems. But what if we have a convex feasible region—and a convex objective, too? We're going to find the global minimum. In fact, we can usually do so quite quickly. And that's good news, because many important problems fall into this category.

- There are some simple rules for identifying some functions as convex without doing any work. One rule is that sums of squares of linear expressions in the decision variables are always convex. And that means that variance is convex. So, because risk is often assessed in terms of variance, the objective of minimizing risk is often relatively straightforward.

- But not everything behaves. We've been dealing with smooth landscapes, where derivatives always exist. But the absolute value function, for example, has a graph shaped like a V. At the point of the V, there's no derivative. And, even worse, what about a cliff or mesa? Mathematically, now we're dealing with a function that is discontinuous. It has a jump. Again, at the edge of the cliff, the derivative is not defined.

- There are ways to handle even these kinds of problems, and the good news is that many of them have been implemented in computer software that can run on a PC. One neat family of techniques is referred to as **genetic algorithms**, or **evolutionary algorithms**. They take their paradigm from evolution in biology, and they work quite well.

Important Terms

convex: A region is convex if a line segment drawn between two points of the region always stays within the region. A square is convex; a star is not. One can intuitively think of a function being convex if someone walking on the surface of the function has "line of sight" to every other point on the function's surface. Hence, a single valley is convex, but two valleys separated by a ridge are not.

evolutionary algorithm: See **genetic algorithm**.

genetic algorithm: An optimization technique involving a stochastic search of possible solutions in a way that somewhat mimics DNA recombination.

gradient: A vector quantity that indicates the direction of "straight uphill" for a function of one or more variables. It is the higher-dimensional generalization of the derivative of a one-variable function.

steepest descent: Also called gradient descent. A collection of techniques for zeroing in on the maximum or minimum of a function. One first determines the direction in which the objective function is changing most rapidly in the desirable direction and then jumps from the current solution to one (nearly) in that direction.

Suggested Reading

Boyles, "Notes on Nonlinear Programming."

Winston and Albright, *Practical Management Science.*

Questions and Comments

1. The illustration depicts a hiker near a triangular lake. He spots a campfire beyond the lake and wishes to reach it—that is, to minimize his distance between himself and the fire. Assume that the hiker doesn't swim. Describe the path he will follow and his final location if he adopts a gradient descent approach. Now repeat the exercise, assuming that the lake has frozen over. What is the key mathematical distinction between the two feasible regions?

Figure 16.2

Answer:

With steepest descent, the hiker would walk straight toward the fire, hit the edge of the lake, move right until reaching the closest point to the fire, and stop. If the lake were frozen, he would continue in a straight line to the fire. The key distinction is that the feasible region is convex if the lake is frozen but non-convex if it is not.

2. If you know calculus, consider the function $z = x^4 + y^4 - 4x^2y^2$.

    a) Set $x = 0$. Verify that $y = 0$ is a minimum of the resulting function. This means that $(0, 0)$ looks like a minimum in the $x = 0$ plane. Verify that if you set $y = 0$, the function looks like a minimum in the $y = 0$ plane.

    b) Now set $x = y$, so that z is a function of x only. Verify that $x = 0$ is a maximum for this function. That means that $(0, 0)$ looks like a maximum in the plane $x = y$. This is the example given in the video lecture, reproduced here.

Answers:

    a) If $x = 0$, $z = y^4$. $z' = 4y^3$, and at $y = 0$, this is 0, so z has a critical point at $y = 0$. The second derivative test fails to identify it as a minimum, but in this case, it is obvious because y^4 is nonnegative for all y. The same argument holds for the plane $y = 0$, by symmetry.

    b) In this plane, $x = y$, so $z = 2x^4 - 4x^4 = -2x^4$. This equals 0 at $x = 0$ and is negative everywhere else, so $x = 0$ is a maximum in this plane. Again, the second derivative test would fail to reveal the nature of the critical point.

Optimization in a Nonlinear Landscape
Lecture 16—Transcript

The past nine lectures have been built around the idea that one mathematical technique can be applied to a lot of different kinds of problems. Look at the variety of problems that we've approached by linear programming. The reason that we've lumped all these programs together is because they all yield, mathematically, to the same kind of analysis. In a sense, once the problem becomes a linear program, it doesn't matter what the original context of the problem was. At least, it doesn't until you find the answer and have to interpret it in practical terms. Sometimes the intermediate steps between a formulation and an answer are sufficiently abstruse that it's tough to keep track of any meaning along the way.

And that's bad, because almost anyone who's good at mathematics, whether it's applying existing results and techniques or discovering new ones, is guided by a powerful intuition. Figuring out the best path to take, the path that gets you from start to the finish most efficiently, is a creative act that relies heavily on intuitive insight.

And that observation, too, has guided the presentation of the materials in this course. We've been performing the nuts-and-bolts work needed to find an answer, yes, but we've also been building an intuition about what the math is doing. And the one thing that we've relied on again and again is how good our natural intuition is when looking at patterns of objects in 2- or 3-D space. Using that strength can help us bridge the gap between real-world problems and pure mathematical calculation. We can use our intuition about visual space to guide our work on a myriad of problems that, from a real-world context, seem completely different from one another.

Look, if you're solving an optimization problem with no auxiliary variables, that is, no equality constraints, then you're stuck with one dimension for each decision variable contributing to the problem. And that means that if you want to visualize it, you're going to be visualizing a problem with at most three variables, and that's pretty restrictive.

But a lot of aspects of what's going on generally can be seen in these dimensionally low-level examples. Take linear programs: one variable, one dimension. The problem lives in a line. Constraints are points, or everything on one side of the point. The feasible region, if it doesn't go on forever in some direction, is a line segment. The objective tries to go as far to the right or to the left as possible without leaving the feasible region. At an end of the feasible region, one constraint is binding.

Two variables, two dimensions, the problem lives in a plane. Constraints are lines, or everything on one side of a line. The feasible region, if it doesn't go on forever, is a convex polygon and its interior. In any case, all of the sides are straight lines. The objective tries to go as far as it can in some specified direction in the plane, like north-northeast, without leaving the feasible region. An extreme point of the feasible region has two constraints will meeting.

Three variables; three dimensions. The problem lives in our normal three-dimensional space. Constraints are planes, or everything on one side of them. The feasible region, if it doesn't go on forever, is a convex polyhedron with its interior, a crystal with flat sides. In any case, all of its sides are flat. The objective tries to go as far as possible in some specified direction, like, that way, without leaving the feasible region. An extreme point of the feasible region has three constraints that meet, like the corner of a room, where a flat floor meets two flat walls.

These patterns hold for n dimensions as well, although, very few people can see them. In linear programs, the constraints are always flat, resulting in a feasible region with flat sides and no dents, what's called a convex polytope. As always, flat means that there are no kinks or bends or jumps. The objective always tries to go as far in some specified direction as it can without leaving the feasible region. And at an extreme point, at a corner, n constraints will meet.

Reasoning by analogy can be dangerous, of course, but the analogy can guide an intuitive guess, and then you can use rigorous techniques to prove that the guess is correct. And once you do, it's there for your future use. For example, suppose that you have a linear program with six decision variables but only five constraints, all inequalities. Then it can't have an optimal solution. Any linear program with an optimal solution has an optimal solution at a corner; that was the Extreme Point Theorem of Linear Programming that we mentioned when we were solving these things graphically. And in six dimensions, any corner has six five-dimensional walls meeting. I can't see it, but I can count. Our program only has five constraints, five walls. So the feasible region doesn't have any corners, and so it doesn't have any optimal points.

When reasoning about dimensions, intuition can lead you astray, too. Imagine you lived in a plane, like a tabletop, and you're examining what a loop of string could look like. You can simulate this by taking a loop of string, setting it down on the table, and pushing it around, making sure that every point stays on the tabletop. But you'd pretty quickly figure out that every loop looks pretty much like every other one, like a deformed circle. But, take that string in three dimensions, and you can have the same circular loop, or a loop with a knot in it. In fact, there are an unlimited number of different knots that you could make. This is new behavior for three dimensions. Take that knot into four and higher dimensions, and any such knot that we tied can simply be shaken out. A loop of string again has only one configuration.

The point that's relevant to this course is two-fold. First, develop a reliable intuition whenever you can; it'll greatly aid your understanding and ability to solve problems. Second, be careful, because sometimes new possibilities open up in higher dimensions. And with that in mind, we're ready to leave linear programming and to start exploring a landscape that's much more wild and wooly: nonlinear programming. Because when you get rid of the requirement that everything is linear, a lot of stuff changes.

To see why, let's see what's going on when there are only two decision variables. The visual intuition for this is a landscape in the normal sense. You can move around the countryside as you please, north, south, east, west. We'll use the third dimension, height, to represent the value of the objective function for each point in the landscape. So, for a maximize problem, you'd want the highest point; for a minimize problem, you'd want the lowest. To fix ideas today, I'm going to assume that we're looking for the lowest point.

We've used this imagery before for linear programs. The landscape there is like a tilted plane of glass, and the constraints run in straight lines on the surface of that glass, and you don't need to be able to see to the horizon to be able to reach the lowest point, you only need to see the ground beneath your feet. When you do walk, you walk straight downhill. And when you hit a fence, you keep going downhill. And when you reach a corner, if the new fence keeps going downhill, follow it. When you can't follow it downhill anymore, stop. You've reached the lowest point. Instructions a simpleton could follow.

But how about a nonlinear program? How do things change? Well, again, let's keep two decision variables so that we can visualize our mathematical landscape as a landscape in the traditional sense. But now that landscape is much more interesting: rolling hills and valleys, maybe ravines, mesas, ridges, maybe even pits. The topography is much more complex, and made more so by the fact that the constraints can be nonlinear, too. That means that the fences that bound our zones of exploration, our feasible region, can be curvy, or they can zigzag in and out. But our goal is the same as it was before, to find the lowest point.

How much of a difference do these new possibilities really make? Well, for a linear program, if the feasible region has a lowest point, it's going to be on a fence line. In fact, in such a region, you can always reach minimum altitude just by checking all of the corners and the picking the lowest one. But for nonlinear programs, that's not true anymore. The fence could run along a ring of mountains surrounding a valley, and the lowest point would be in the middle somewhere.

As a real-world example, your store might have 10 checkout lanes so that you can handle your busiest times. Open them all on a typical day, and you're paying a lot of employees to stand at the registers. Open only one, and you're incurring costs from lost sales and customer dissatisfaction. Those are the two fences. And the sweet spot is somewhere in the middle.

Another difference in a nonlinear world is something that might look like a lowest point, but not be a lowest point. It might only be the lowest point in your area, a local minimum, not the global minimum. Want a real-world example? Let's look at the great HD DVD and Blu-Ray war, and how it might have turned out differently. Purely for argument's sake, let's imagine that HD DVD is actually twice as good as Blu-Ray. Some fraction of the customer consumer market buys a player for one format, the rest buy the other. Some fraction of the disks manufactured are the one type, the rest are the other.

Imagine we start off in a world where equal quantities of disks in each format are available, then the logical choice for each consumer would be to buy HD DVD. We said it was twice as good. And if everybody then ends up owning an HD DVD player, producers would want to make only HD DVD, which would be fine with the customers. So we'd up in an HD DVD world, which is pretty much what we've got.

But, imagine that, by chance, we started with in a world where the product was 90% Blu-Ray. The higher quality of HD DVD wouldn't compensate for the lousy selection of compatible disks, so intelligent consumers would buy Blu-Ray players. And if consumers had Blu-Ray players, the best choice for producers would be to abandon the HD DVD and make only Blu-Ray disks. We end up in a Blu-Ray world. And it would stay that way. Neither producers nor consumers would have reason to change, even though the resulting world is only half as good as the other alternative. That's a local optimum, but not a global one.

Think about the ridiculous arrangement of keys on our keyboards, or the fact that, in America, we still use inches and miles, rather than the easier and more logical metric system, and you'll see that this idea isn't just abstract mathematics. But this was a solution that was in a corner of the feasible region—all HD DVD players and all HD DVD disks. Let's focus on the other possibility, when the lowest point is in the middle of a rolling landscape, somewhere. How can we find it?

Well one answer is magic, or so it must have seemed in the 1600s when the approach was invented. It's calculus—particularly, the derivative. The mechanics of differential calculus belong in a different course, but here's the magic of it. If you have a function that gives you the contours of the landscape, then its derivative gives you the slope of that landscape at every point. That is, how tilted the land is, and in what direction. And we've already seen one way that that information is valuable, since it tells you what direction you should go for downhill from where you are. But there's a second way that it's useful, too, and one that you undoubtedly saw if you ever took a calc class; it lets you narrow down where the lowest point could possibly be.

With the intuition that comes from our visualization, this becomes really clear. Imagine that the landscape is rolling in nature, no ridges, cliffs, and the like. And again, no fences near where you are. Then, if you are at the lowest point, the ground beneath your feet is flat, of level. I don't mean flat; you probably standing at the bottom of a bowl, actually. But you could imagine that a flat plane is buried under the ground that would reach the surface at the one spot where you are standing, and never rise above the ground anywhere in your immediate vicinity. That's called the tangent plane, and where you're standing, it's dead level. In calculus, the derivative, or to use more proper language, the gradient, tells you how that plane is tilted, or not tilted. So when you're at the lowest point, that's going to be zero.

And that means that if you can calculate the derivative with calculus, and you can, you can then use algebra to find out when the slope is 0; at least that's the hope. Sometimes solving the derivative = 0 equations can be rather a bear. But, even if you do solve them, you're not out of the woods. Yes, the ground will be flat if you're at the lowest point and you're not up against the fence. But it will also be flat if you're only at a local minimum. Or, think about it, a local maximum, or at a terrace point, where the ground goes up, levels off, and then goes up some more, like terrace farming. While those are the only possibilities with one decision variable, which is what you generally learn about in your calculus course, in a higher dimension, you can live with stranger things, like this.

It's a saddle shape. The technical name is a hyperbolic paraboloid, and it happens to be one of my personal favorite shapes. In fact, this landscape is the landscape for our HD DVD/Blu-Ray problem. On the left, it's an all HD DVD world, which you can see is a local maximum. On the right it's an all Blu-Ray world, which is the global maximum. But if you look at the place between them, where the ground is level, that's the lowest point on the path between the two maxima.

But that point isn't a minimum, because if you traversed the landscape from front to back, instead of left to right, that same point would look like a maximum. The more you move toward the front or the back of this picture, the more the consumers and producers are out of synch with one another, in terms of their format choices. In this direction, the mix is the best that you can do. That point in the middle with a 0 gradient is called a saddle point, for obvious reasons.

So, if you take the derivative of the objective, that gives you the landscape, then even if you can solve the resulting equations to find the points that are level, you still have some work to do to figure out whether these points are minima or not, and whether they're global minima, or just local ones. In higher dimensions, the math to sort out the possibilities can get a bit hairy. Solve a family of simultaneous nonlinear equations to find the eigenvalues of the Hessian matrix of partial second derivatives, and hope that it's nonsingular, that sort of thing.

And that only handles the case of when the minimum is out in the middle somewhere. If it's along a fence line, then the ground at the minimum doesn't have to be level. In fact, it probably wouldn't be. You can see that with our HD DVD problem: It has two local maxima and two local minima, and the ground is sloped quite steeply at all four of them. And there's another problem. We've been assuming rolling hills. In mathematical terms, we've been assuming that the function is differentiable, that the derivative exists. Sometimes, we even need more than this, such as the function also having continuous second derivatives. If the landscape includes cliffs, ridges, mesas, and the like, you can see where things can go wrong. At a sharp ridge, like a roof ridge, the derivative isn't defined. It doesn't have a single tangent, so you won't find it by looking for where the derivative is 0. It's not defined at the edge of a cliff, either.

As a simple example, imagine that we're making a product with a fixed demand. Each unit we make and sell gives us the same profit, so we have a nice, linear relationship. Until we saturate the market. Thereafter, each additional unit reduces profit by a fixed amount, since we paid to have the unit made, but we can't sell it; that's a linear decrease in profits beyond that point. The optimal production is obviously where we just meet demand, but you'd never find it via calculus. That's a roof peak, and there's no derivative there.

So calculus isn't as useful for some practical problems as you might guess. In contrast, the nonlinear programming approaches that we'll be developing will allow multiple variables that'll allow complicated boundary conditions, which usually don't appear until a second calculus course, and nonlinear programming can allow the characterization of multivariate critical points. So, how do we do problems where the landscape is not smooth and continuous? Well, there are a number of approaches, and which you should use depends, not surprisingly, on your landscape.

So let's start again with rolling hills and valleys. That means that, at any point, we can figure out what direction is downhill. In mathematical terms, we can find the gradient of the objective. So here's a strategy that can take advantage of computer power. Start at a random point in the landscape. Find out what direction is downhill at that point, just like you would for a linear program. Then walk a little ways in that direction. Then stop and repeat the process. Keep repeating it, taking a fix on which way is downhill, and walking that way a bit, until you can't make any more progress.

This is the basic idea behind a collection of techniques called steepest descent, or gradient methods. Excel's GRG nonlinear method is an implementation of one of these. The variants differ in exactly in what downhill direction you pick and how far you go before you check again. For our example, imagine that we have a landscape that looks like this in an aerial view. Here, we're in the southwest corner of an oddly shaped field, and darker colors indicate lower altitude. Straight downhill from where we stand is to the east, but you can see that we couldn't go very far that way before hitting a wall.

So, an algorithm like [Excel's] might choose a direction which has more anticipated room for improvement, maybe something like this. Of course, if the ground goes back uphill on the other side of the field, then that might have not been a great choice. So there's some finesse in deciding what direction you should go and how far you should go before you recheck. But that's the basic idea of gradient descent.

Can such a technique ever fail to find the lowest point in a rolling landscape? Well sure, if it hits the bottom of a little dip valley, or follows the slope downhill to a point along a fence, then it's just going to stay there, even though there might be much lower points somewhere else—the old local-minimum problem. And that is a problem for any gradient-descent method. One way to try to avoid this is to begin at many points, rather than one. It's kind of like having a Boy Scout troop out with you, each Scout started in a different spot. When you're done, you all compare notes, and you see who did the best. The more starting points, the better your chances.

Of course, you can imagine a landscape like one of those annoying holes in miniature golf, where the hole is surrounded by a volcano. The deepest point on the hole may well be at the bottom of the hole itself, but how likely is it you're going to find it with a steepest descent approach? Every time you check your bearings when standing on the volcano, you're going to end up going in exactly the wrong direction. It's good that this kind of thing doesn't happen too often in mathematical landscapes of real-life problems, because in general, there's no general algorithm that is guaranteed to find the global optimum in any reasonable amount of time. But things often work well. And there are large classes of problems where the gradient descent approach is guaranteed to work.

A key issue is the idea of convexity. Now, this lecture is about developing an intuition on these problems, so, I'm not going to get bogged down with a formal definition of convexity. But here's the heart of it. First, convexity of the feasible region, imagine that the fences defining our feasible region are opaque walls. You and I are standing at points inside the region. Then the region is convex if, no matter where we stand, we can always look at each other in the eye. No walls in the way. So a square region is convex, but a star-shaped region is not. There's no hiding from one another in a room with a square floor plan. But see those two little red dots in my star-shaped room? If you were standing at one of them and I was standing at the other, we couldn't see each another unless we had X-ray vision.

We can also talk about the convexity of a function, in our intuitive terms, a landscape. It's a closely related idea. Again, informally, the landscape is convex if no matter where you and I stand on it, we can still make eye contact. So, if we're standing inside of a bowl-like depression, no problem. If we're standing around a mountain, or in the vicinity of a mountain pass, then there could be places where we could hide from each another's view. The function giving the landscape is not convex in those cases.

Why does convexity matter in finding optimal solutions? Well, go back to you and me in our star-shaped room for a minute, and imagine the goal was to move as far south in this room as possible. South is down in my picture. Again, look at the two red dots representing you and me. If we were following gradient descent, neither of us could move any further south, could we? We'd both be at local minima, but only the person on the right would be at the actual global minimum.

So, even if the objective is linear, a non-convex feasible region can cause problems. But how about if we have a convex feasible region, and a convex objective, too? Your intuition may be telling you that now you're going to find the global minimum, and you'd be right. Usually you do so quite quickly. And that's good news, because a lot of important problems fall into this category.

I'll give you some simple rules for identifying some functions as convex without doing any work in my next lecture, but let me mention just one today. Sums of squares of linear expressions in the decision variables, sums of squares, are always convex, and that means that variance is convex. And risk is often measured in terms of variance. So the objective of minimizing risk is usually a pretty straight-forward one to handle.

But, not everything behaves. We've been dealing with smooth landscapes, where derivatives always exist. But the absolute value function, which we used in computing MAD and MAPE, for example, has a graph shaped like a V. At the point of the V, there's no derivative. And even worse, how about a cliff, or a mesa? Mathematically, now we're talking about a function that is discontinuous; it has a jump. Again, at the edge of the cliff, the derivative is not defined. These come up a lot in real life, too.

For example, the post office charges the same amount of money for heavier and heavier packages, until a certain weight is reached, and then wham, they jump to a new price. Most if-conditions will also do this, as in the cost is $4.00 per hour, but if you use 10 or more hours, the cost is $3.60 per hour. Again, we have a jump discontinuity, a cliff; 10 hours costs less than 9 hours and 1 minute.

If you use your topographic intuition, you can see the troubles that you can have. If you're at the top of a flat mesa, the ground around you looks level, so you have no reason to go any given direction. If you decide to move to a new position near your current one, the ground stays flat. In fact, it may even stay flat if the new position is at the bottom of the cliff. So you went from one height to another and never went downhill. And remember, the more dimensions in your landscape, the more possible directions you could choose to go. Imagine for a moment finding a treasure in a series of rooms where each room has only two doors, not too bad. You go in one; you go out the other.

Now imagine that you had to do this with each room having 1000 doors. Well, there are ways to handle even this kinds of problem, and the good news for us is that many of them have been implemented in computer software that you can run on a PC. One family of techniques is referred to as genetic algorithms; they're really neat. They're also called evolutionary algorithms. They take their paradigm from evolution in biology. I'll be honest; I'm still rather amazed that they work, and work as well as they do.

Let's see if I can give you a feel for them, since some of the harder problems that we'll be looking at are going to need them. In fact, if you write a program in Excel that uses many of its built-in functions, like IF, or CHOOSE, or VLOOKUP, or the like, your program is probably going to need a genetic algorithm. To get started, we have an initial population of possible solutions, maybe 500 of them, that are usually randomly generated. Unlike other optimization techniques that we've been discussing, genetic algorithms are stochastic; they involve randomness. Among other things, that means that if you apply genetic algorithms to the same problem twice, you probably won't get the same answer.

As an example, suppose that my problem involves three variables, x, y, and z. Then each solution would have the values of x, y, and z, of course. But here's the first twist; we'd represent them as coded into binary, strings of ones and zeroes. This echoes a strand of DNA, which is built of a sequence of four possible nucleotides. We only have two, but that's enough. Let's say that each variable is represented by a 16-digit string of 1s and 0s. Now, we have the possibility of any pair of solutions breeding. They do this by a process similar to recombination of genes. The procedure is called crossover.

In the genetic algorithm, we cut each 16-digit sequence in a particular location, maybe we cut the x sequence after the 5th digit, the y sequences after the 12th digit, and the z sequences after the 10th digit. These cut points are chosen randomly for each pair. Now, we're going to make two new x values from the two original ones. Let's see what that would look like. With these two x values and a crossover point being after the 5th digit, first, cut the sequence into two, after the 5th digit. And make two new x values by doing a swap. Take the head of the first old x value and stitch it to the tail of the second old x value; that gives one new x. And make another by taking the head of the second old x value and stitching it onto the tail of the first old x value, like this.

We'll crossbreed the y's and z's in the same way, and so from the original two solutions, we've created a pair of children. We're going to evaluate the parent solutions and their children by seeing whether they satisfy all of the constraints and seeing how well they do in the objective. These factors are used to determine how fit each potential solution is. We're going to pare down these candidates into a new population of the same size as the original pool of solutions. The more fit a solution is, the more likely it is to breed and appear in this new pool.

One way this can be done is something like a spinner in an old children's game, one that you might use to decide how many squares to move. Each candidate gets a wedge on the spinner, but a candidate that's twice as fit gets twice as big a wedge. Here's what it might look like with four candidate solutions, where the sea-green candidate was the most fit and the dark-blue candidate was the least fit. To decide who gets to fill each available slot in the new generation, spin the spinner. Whichever solution it stops on gets that slot. So the new population might contain multiple copies of a really good solution.

One problem with doing this is the same as a problem seen in modern ecology—loss of biodiversity. In this culling process, solutions that would have been great with a little tweaking might get lost, and surviving solutions might fixate on a local minimum. To prevent this, each member in each generation has a small chance of mutation, that is, of one of its digits randomly changing from a 0 to a 1 or from a 1 to a 0. Mathematically, this can teleport a solution completely across the solution of the landscape to a new position. The vast majority of the time, the new solution will be less fit than the rest of the population and will die out. But every once in a while we'll stumble on a better solution—and quite different from the other ones being considered—and then the search can proceed in this new area as well.

Because of the random aspect of the solution procedure, the technique will, eventually, find the optimal solution—eventually. But the bad news is, is it could take a very, very, very long time to do so. I've had small but hard programs run for hours before their solutions were found, and large, hard problems could be very time consuming indeed. In theory, you could have problems that would take longer than the life of the universe, so far, to solve. Fingers crossed on that one.

And, of course, genetic algorithms create a new generation of solutions by comparing fitness to those solutions that it's already has found. There isn't really any sure way of telling when you should stop the thing, because there's nothing that tells you when the optimum really has been found. From a practical standpoint, one usually lets it run until either no appreciable improvement has been seen over a specified number of generations, or until you just can't stand it anymore. And the stopping point isn't the only thing you have to decide. How big a population do you want? What mutation rate do you decide on? And so on. Most software will use default values, but you might get better performance if you tweak them.

But if you need genetic algorithms, you're probably going to be willing to put up with these shortcomings, because they can handle functions that none of our other techniques can deal with very well. They pretty much don't care what the functions are—even something weird like, how many times does this digit appear in the answer? You can make a set of variables that are all integers from 1 to 10, insist that each number be used only once, and make a rank ordering. Evolutionary solvers don't care. But, you could believe from the way I described the fitness function that they hate any but the most simple constraints. They like to have an upper bound and a lower bound on each variable, like $x \leq 10$ and ≥ 0, but anything more complicated usually slows them dramatically. We avoid this unpleasantness by using soft constraint and penalties, like the ones we used in the last lecture. They're easier to write for an evolutionary solver, since you can use things like IF statements.

Well, that's an overview of the nonlinear wilderness, the lay of the land, so to speak. Now you have an idea of how nonlinear techniques work and why we use multiple techniques, and what to watch out for in nonlinear world. In the next lecture, we'll be getting more specific about exactly how we do all of this. We'll gather together our equipment, square our shoulders, and set out over a landscape defined by a variety of practical examples. And because it's operations research, we'll have a better reason for reaching that peak than the mountaineer's explanation—because it's there.

Nonlinear Models—Best Location, Best Pricing
Lecture 17

In this lecture, you will explore the outer limits of programming, beyond linear programming to nonlinearity. You will learn about a practical problem and why a particular nonlinear technique may or may not be used to solve it. Specifically, the problem that is presented deals with facilities placement—addressing the challenge of finding the best places from which to conduct business. The idea that is addressed is one that allowed Delta Airlines to move from a small regional company to a major competitor in the U.S. airline industry.

Facilities Placement

- If an airline serves more than a few airports, providing direct flights between every pair of airports that it serves is unworkable. It's simply a matter of the number of flights you'd need. Connecting three cities with direct flights takes three flights in a triangle—or six, given that you have to fly both ways. But connect 30 cities with direct flights, and you'll need 930 flights. And that number grows rapidly as the number of airports increases.

- The solution that seems so familiar now was new in 1955, when Delta Airlines first pioneered the hub-and-spoke model. Atlanta was the hub city. If you wanted to go from A to B, you went from A to Atlanta, switched planes, and then flew from Atlanta to B. Flights from everywhere routed through Atlanta, and many still do today. Adopting this model made Delta a real competitor with Eastern Airlines. Since that time, all of the major airlines have adopted a similar system.

- Since those early days, Delta has also expanded to have additional hubs in Cincinnati, New York, and Salt Lake City, leading to a multi-hub map that is much more complicated, while extending the hub-and-spoke approach.

- The decision of where to place hubs takes into account a number of factors, of course, but very important among them is the goal of minimizing the total number of miles in the hub-and-spoke system. A slightly different goal is the following: Given the airports that we are going to serve and the number of people flying to and from those airports, what location for the hub airport would minimize the average number of miles flown by a passenger in this system?

- There might not be an airport in this perfect location, but if we can figure out where we'd like to have our hub, we can look to see what existing airports might be close to that location.

- For this problem, distance we care about is straight-line distances from the regional airports to the hub. And the math that gives the straight-line distance between two points is basically just the Pythagorean theorem. For example, if the hub is at coordinates (x, y) and the regional airport is at $x = 10$, $y = 20$, then the distance from the hub to the regional airport is the following.

$$\sqrt{(x-10)^2 + (y-20)^2}$$

- Those squares and that square root make the expression decidedly nonlinear. And minimizing average distance is going to mean taking a weighted average of the distances from the hub to each regional airport—weighted by how many people are traveling from and to each regional airport. So, that's a nonlinear objective. The good news is that we have essentially no constraints—no limitations on where to build the hub.

- If possible, we'd like to anticipate whether this problem is going to be tractable or not. This hinges more on convexity than it does on linearity. In general, it can be a fair amount of work to figure out whether a function is convex or not, but the following are a few helpful rules that you can keep in mind.
  - Linear expressions are convex.

- o Any even power of a variable is convex, such as x^2, x^4, and so on. So is e^x.

- o If you replace a variable by a linear expression in a convex function, it's still convex. So, $(3x + 4y + 2)^2$ is convex, because x^2 is.

- o If you take a weighted sum of convex functions, it's still convex as long as none of the weights is negative. So, $0.3x^2 + 0.4(y - 6)^2$ is convex.

- o Euclidean distance—that is, straight-line distance—is convex.

- Our objective in this problem is to minimize a weighted average of the distances from the hub to each regional airport. Euclidean distance is convex, so the objective is a weighted sum of convex functions with each weight being nonnegative—namely, the fraction of travelers who hail from that city. By our rules for convexity, that means that our objective, while nonlinear, turns out to be convex. So, a gradient-based technique is guaranteed to find the global minimum for this. In Excel, for example, the standard nonlinear Solver will do the trick.

- We start out with the data on the cities served by our airline. Their location is specified by how far east and how far north they are from the arbitrary reference point of Houston, Texas. The cells with dotted lines for borders are our decision variables, specifying the location of the hub, again with Houston as the origin. From this, we can find how far each regional airport is from this proposed hub. It's just the Pythagorean theorem. We record these in the "distance to hub" column. (See **Figure 17.1**.)

- Finally, the weighted average of these distances is computed on the bottom of the sheet, using the populations of the cities in millions of people as proxies for the amount of air traffic in and out of each city.

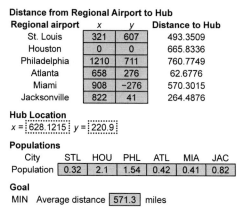

Distance from Regional Airport to Hub

Regional airport	x	y	Distance to Hub
St. Louis	321	607	493.3509
Houston	0	0	665.8336
Philadelphia	1210	711	760.7749
Atlanta	658	276	62.6776
Miami	908	−276	570.3015
Jacksonville	822	41	264.4876

Hub Location
$x = $ 628.1215 $y = $ 220.9

Populations

City	STL	HOU	PHL	ATL	MIA	JAC
Population	0.32	2.1	1.54	0.42	0.41	0.82

Goal
MIN   Average distance   571.3   miles

Figure 17.1

- This objective minimizes the total number of miles that customers have to fly, rather than the total miles in the hub-and-spoke system. The logic behind this is that a spoke with more traffic would need multiple planes per day. Alternatively, we could simply minimize the sum of the numbers in the "distance to hub" column; in this case, the resulting location is still in the same vicinity.

- We told Solver where the variables are and where the objective is, just as we did for linear programs. We didn't have to impose any constraints, because we didn't have any. The hub is allowed to be built anywhere. The only thing new was that we told Solver to use the GRG (generalized reduced gradient) nonlinear solving method rather than the Simplex LP method.

- The answer comes very quickly, as we expect with this convex problem. It says that the hub should be built 628 miles east and 221 miles north of Houston. That is, we'd like our hub to be about 63 miles southwest of Atlanta. Just like Delta Airlines, we like the idea of Atlanta as a hub.

Genetic Algorithms

- What happens when Delta grows and establishes new hubs? Imagine that the Atlanta hub was already established and that now we're looking for locations of new hubs. The distance calculations wouldn't be any more difficult than before; there'd just be more of them. But a new factor would appear—namely, now a regional airport would be assigned to some specified hub. Its distance from the others wouldn't matter. This would mean a 0/1 variable for each pair of regional airport and hub, answering the yes/no question of whether *that* airport goes with *that* hub. Because of the large number of resulting integer variables and constraints, we should size up the problem before we get too committed to one approach.

- Linear programming is out, because the objective is nonlinear. But because we have no constraints to begin with, we might want to consider a genetic algorithm. They hate all but the simplest constraints—upper and lower bounds on decision variables—but other than that, they allow almost unlimited flexibility in formulation.

- Let's expand the problem to include multiple hubs and approach it with a genetic algorithm. They can be slow, but that probably won't be a problem in this case. Let's say that we'll have up to four hubs and that each hub has to be at one of the cities that we serve. This restriction is desirable from a practical standpoint, and its addition will make the problem easier to solve. We don't need the coordinates of the hub now, just the city where it will be located.

- For this example, we have chosen 29 large U.S. cities for our airline to serve. Our system will have multiple hubs, with each regional airport being a spoke attached to one of the hubs. Our goal will be to minimize the total number of miles in this system. Additionally, we will make the reasonable assumption that each hub has to have direct flights connecting it to each other hub. If not, then passengers might have more than two transfers for a domestic flight. We're also going to take a different approach to distance, getting it from a table of distances (given in miles) from city to city.

City #		Atlanta	Austin	Baltimore	Boston	Chicago	Dallas	Denver
1	Atlanta	0	1315.28	927.35	1505.11	944.4	1157.42	1945.42
2	Austin	1315.25	0	2166	2724.01	1571.76	293.52	1240.77
3	Baltimore	927.35	2166	0	577.85	973.23	1947.28	2422.32
4	Boston	1505.11	2724.01	577.85	0	1366.63	2490.97	2838.62
5	Chicago	944.4	1571.76	973.23	1366.63	0	1290.15	1474.26
6	Dallas	1157.42	293.52	1947.28	2490.97	1290.15	0	1064.41
7	Denver	1945.42	1240.77	2422.32	2838.62	1474.26	1064.41	0

Figure 17.2

- This information is much easier to obtain than the mileage coordinates from Houston that we used before. This means that we have a way for the spreadsheet to find the distance from A to B from the table, but Excel actually has many functions for that sort of thing, such as INDEX or VLOOKUP, and because we're using a genetic algorithm, we can use any such functions we want, with a specification of our four hubs.

Hub Location				
Hub	1	2	3	4
City #	1	2	3	4
Name of Hub	Atlanta	Austin	Baltimore	Boston

Figure 17.3

- Each city is identified in our table with an integer, and we've temporarily put the hubs at cities 1 through 4: Atlanta, Austin, Baltimore, and Boston. To make it easier for a user to understand, we had the spreadsheet report the names of the hub cities as well. We used the INDEX function for this. We also have to have a variable for each regional airport, specifying to which hub it will be assigned. (See **Figure 17.4**.)

- The values in the cells that have dotted lines don't matter. We'll tell Solver that they are decision variables and have to be integers between 1 and 4—because there are four hubs, 1 through 4.

- Because we're using a genetic algorithm, though, we have another option. Once the hubs are determined, this problem is going to have its optimal solution if each city is assigned to its nearest hub. That

City #		1	2	3	4	5	6
Assigned to Hub		1	2	3	4	1	2
Distance to Hub		0	0	0	0	944.4	293.52
City #		Atlanta	Austin	Baltimore	Boston	Chicago	Dallas
1	Atlanta	0	1315.28	927.35	1505.11	944.4	1157.42
2	Austin	1315.25	0	2166	2724.01	1571.76	293.52
3	Baltimore	927.35	2166	0	577.85	973.23	1947.28
4	Boston	1505.11	2724.01	577.85	0	1366.63	2490.97
5	Chicago	944.4	1571.76	973.23	1366.63	0	1290.15
6	Dallas	1157.42	293.52	1947.28	2490.97	1290.15	0

Figure 17.4

is, we could tell the spreadsheet to compare the distances from this regional airport to each of the four hubs and then choose the hub that is closest. Ultimately, the Solver finds three hubs: Las Vegas, Memphis, and Baltimore.

Suggested Reading

Anderson, "Setting Prices as Priceline."

Questions and Comments

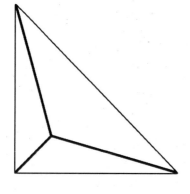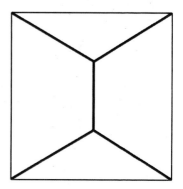

Figure 17.5

1. Like the Delta problem, a number of transportation problems involve minimizing some distance measure. One interesting one is to connect a set of n cities with roads so that the road system allows travel from any one city to any other, and the total length of all of the roads is minimal. Such a road system is called a Steiner tree.

It's an interesting fact that every intersection of roads in such a system always consists of three roads meeting at angles of 120 degrees. The solutions for the cases when the cities are on the corners of a right triangle and on the corners of a square are shown in the diagram. If each side of the square is 1 mile in length, joining all four corners to the center requires a total of about 2.828 miles of roadway, but the Steiner tree requires only 2.732 miles.

2. You are given the distances among four hubs: A, B, C, and D. Let AB be the distance from hub A to hub B, and define AC, AD, BC, BD, and CD in parallel fashion. It is easy to see if two hubs are identical, because the distance between them is zero. In our second problem, we needed to know the total inter-hub distance. For example, if A and B are the same and C and D are the same, then the total distance should be just AB.

Let $N_{AB} = 0$ if $AB = 0$ and $N_{AB} = 1$ if $AB > 0$. That is, N_{AB} is 1 if and only if A and B are different places. Define N_{ij} similarly for any pair of hubs i and j.

What expression would give the total distance in the system?

There are multiple answers, but here's one: $AB + N_{BC}AC + N_{BD}AD + N_{AB}N_{AC}BC + N_{AB}N_{AD}N_{CD}BD + N_{AC}N_{AD}N_{BC}N_{BD}CD$.

Nonlinear Models—Best Location, Best Pricing
Lecture 17—Transcript

There is nothing wrong with your television. Do not attempt to adjust the picture. We are now in control of the transmission. We control the horizontal, and the vertical. We can deluge you with 1000 channels, or expand one single image to crystal clarity, and beyond. We can shape your vision to anything our imagination can conceive. For the next 30 minutes, we will control all that you see and hear. We are exploring the outer limits, of programming, anyway, beyond linear programming; we're doing nonlinear programming.

In the last lecture, we developed our intuition about nonlinear optimization problems and how to go about solving them. We used the conceit of a physical landscape to guide our thinking. This lecture is the nuts-and-bolts side of nonlinear programming, examples of how you really do it, and how our work from last time can help us select a good approach. I've chosen two practical problem areas for today, locating facilities and retail pricing. We'll formulate a couple examples of each, and discuss why a particular nonlinear technique may or may not be the one for the problem. Our final pricing example will be looking at a system similar to those used by warehouse clubs, like Costco. In the process, we'll gain some insights into what makes Costco such a successful retailer.

But our first set of problems deal with facilities placement, addressing the challenge of finding the best places from which to conduct your business. And it's an idea that we'll start with with Delta Airlines, because it let them move from a small, regional company to a major competitor in the U.S. airline industry. You see, if an airline serves more than a few airports, providing direct flights between every pair of airports that it serves is unworkable. It's simply a matter of the numbers. Connecting three cities with direct flights takes three flights, in a triangle. Or, six, given that you have to fly both ways. No problem. But connect 30 cities with direct flights, and you'll find you need 930 flights. And that number grows rapidly as the number of airports increases.

The solution that seems so familiar now was a new one in 1955, when Delta Airlines first pioneered the hub-and-spoke model. Atlanta was the hub city. If you wanted to go from A to B, you went from A to Atlanta, switched planes, and then flew from Atlanta to B. Flights from California, New England, Texas, Florida, you name it, flights from everywhere routed through Atlanta, and many still do today. Adopting this model made Delta a real competitor with Eastern Airlines. Since that time, all of the major airlines have adopted a similar system.

Since those early days, Delta has also expanded to have additional hubs in Cincinnati, New York, and Salt Lake City, leading to a multi-hub map that's much more complicated, while extending the same hub-and-spoke approach. The decision [of] where to place hubs takes into account a number of factors, of course, but, important among them, very important, is the goal of minimizing the total number of miles in the hub-and-spoke system. A slightly different goal would be, given the airports that are going to serve and the number of people flying from those airports, what location of the hub airport would minimize the average number of miles flown by a passenger in this system? I grant you, there might not be an airport at this perfect location, but if we can figure out where we'd like to have our hub, we can look and see what existing airports might be close to that location.

If this problem sounds somewhat familiar, good for you. It has a lot in common with our retirement home example from two lectures ago. The problem there was, given the location of the retirement homes in the city, figure out where to build a treatment center convenient to them all. In that problem, distance between a home and the center was measured in taxicab distance, the shortest path from one place to the other using a combination of only horizontal and vertical segments. We did that because that's how the roads ran in this city. That made the math more messy, but at least our formulation was still linear, so our earlier techniques applied. But for the airport problem, things are different. Distance is as the crow flies, straight-line distance from the regional airports to the hub. And the math that gives the straight-line distance between two points is essentially just the Pythagorean Theorem. For example, if the hub is at coordinates x, y, and regional airport is at 10, 20, then the distance from the hub to the regional airport is $\sqrt{(x-10)^2 + (y-20)^2}$.

Those squares and that square root make the expression decidedly nonlinear. And minimizing average distance is going to mean taking a weighted average of the distances from the hub to each regional airport, weighted by how many people are traveling from and to each regional airport. So that's a nonlinear objective. The good news is that I have essentially no constraints here, no limitations on where to build the hub. If possible, we'd like to anticipate whether this problem is going to be tractable to a solution, or a pain in the butt. Well, we saw last time, that in a real sense, the line between sure-to-be-fairly-easy and maybe-not-so-easy doesn't really hinge on linearity, so much as on convexity. In general, it can be a fair amount of work to figure out whether a function is convex or not, but there are a few helpful rules that you can keep in mind. They act kind of like a toolkit for building convex functions, and [you'd] be surprised how often they apply.

First, linear functions, themselves, are convex. Second, an even power of a variable is convex, like x^2, x^4, and so on, and so is e^x. Third, if you replace a variable by a linear expression in a convex function, it's still convex. So, $(3x + 4y + 2)^2$ is convex, because x^2 is. Fourth, if you take a weighted sum of convex functions, it's still convex, as long as none of the weights are negative. So, $0.3x^2 + 0.4(y - 6)^2$ is convex. And finally, Euclidean distance, that is, straight line distance, is convex.

Our objective in the locate-the-hub problem is to minimize a weighted average of the distance of the hub to each regional airport. Euclidean distance, we just said, is convex, so the objective is a weighted sum of convex functions with each weight being nonnegative, namely, the fraction of the travelers who hail from that city. By our rules for convexity, that means our objective, while nonlinear, turns out to be convex. So we have a salad-bowl problem, and a gradient-based technique is guaranteed to find the global minimum for this. In Excel, for example, the standard nonlinear solver will do the trick. And the model is actually pretty simple. Let's take a look.

We start out with the data on the cities served by our airline. I'm giving their location by specifying how far east and how far north they each are from an arbitrary reference point. I'm using Houston, Texas as that point. The red cells below this are our decision variables, specifying the location of the hub, again, with Houston as origin.

From this, we can find out how far each regional airport is from this proposed hub; that's just the Pythagorean Theorem. We record these in the distance to hub column. Finally, the weighted average of these distances is computed at the bottom of the sheet, using the populations of the cities in millions as proxies for the amount of air traffic in and out of each city.

This objective minimizes the total number of miles that customers have to fly, rather than the total number of miles in the hub-and-spoke system. The logic behind this is that a spoke with more traffic would need multiple planes a day. Alternatively, we could instead simply minimize the sum of the numbers in the distance to hub column, the resulting location is still in the same vicinity, in particular this case.

I told Solver where the variables are and where the objective is, just as we did for linear programs. I didn't have to impose any constraints, since we didn't have any. The hub is allowed to be built anywhere. The only thing new was that I told Solver to use the GRG nonlinear solving method, rather than the Simplex LP method. This is a gradient technique like we talked about last time. In fact, GRG stands for generalized reduced gradient. The answer comes very quickly, as we expect with this convex problem. And here it is. It says that the hub should be built 628 miles east and 221 miles north of Houston. A graph makes it a little clearer. That is, we'd like our hub to be about 63 miles southwest of Atlanta. Just like Delta Airlines, we like the idea of Atlanta as a hub.

OK, suppose we wanted to take this problem to the next level. How would we do it? Well, clearly, the real-life problem involves many more airports, but that's just a matter of scaling. Our work ignores the curvature of the Earth. For a more accurate distance model, we'd need to use something like the haversine formula from spherical [trigonometry] to compute the distances from longitude and latitude. It's a messy formula, but conceptually, it's no harder than what we did. We might decide to minimize the cost of the number of miles flown per day by the airline, subject to meeting all of the demands; that could involve integer decision variables for how many flights were needed on each link.

But there's a more substantive change that would occur when Delta grew and established new hubs. The Atlanta hub, we could imagine, would already be established. But now we'd be looking for location of new hubs. The distance calculations wouldn't be any harder than before, but they should be more of them. But a new factor would appear, namely, now a regional airport would be assigned to some specific hub. Its distance from the other hubs wouldn't matter. This would mean a 0/1 variable for each pair of regional airport and hub, answering the yes-no question of whether that airport goes with that hub; 100 airports and four hubs would mean 400 integer variables and 100 constraints, one for each regional airport, saying that it has to get assigned to some hub. You can see why it can be smart to size up a problem before you get too committed to one approach.

Do we have an alternative in this case? Well, let's think about it. Linear programming is out, of course. The objective's nonlinear. But, since we have no constraints here to begin with, we might want to consider a genetic algorithm. They hate all but the simplest constraints, upper and lower bounds on decision variables, but other than that, they allow almost unlimited flexibility in formulation. For example, we'd only need one integer variable for each regional airport, giving the number of the hub to which it was assigned, like Hub #2.

So, let's expand the problem to include multiple hubs and approach it with a genetic algorithm. They can be slow, but I wouldn't expect that to be a problem here. Let's say that we'll have up to four hubs, and that each hub has to be at one of the cities that we serve. This restriction is desirable from a practical standpoint; its addition will actually make the problem easier to solve, too. We don't need the coordinates of the hub now, just the city where it will be located.

For this example, I've picked 29 large U.S. cities for our airline to serve. Our system will have multiple hubs, with each regional airport being a spoke attached to one of the hubs. Our goal will be to minimize the total number of miles in the system. And finally, I'm going to make what seems to me to be a reasonable additional assumption: that each hub has to have direct flights connecting it to each other hub. If not, then passengers might have more than two transfers on a domestic flight, and I don't want that.

I'm also going to take a different approach to distance, getting it from a table of distances from city to city. Here's the beginning of my table, with distance given in miles. Atlanta is a little over 1300 miles from Austin, and so on. The information is much easier to obtain than the compass point mileage coordinates from Houston that I used in the last example, and that troublesome curvature of the Earth is no longer an issue. It does mean, of course, that I'll have to find a way for the spreadsheet to find the distance from A to B from a table, but Excel actually has lots of functions for this sort of thing, like INDEX or VLOOKUP, and since I'm using a genetic algorithm, I can use any such functions I want.

OK, so, we'll begin like this, with a specification of our four hubs. Each city is identified in our table with an integer, and I've temporarily put the hubs at cities 1, 2, 3, and 4, from my table, Atlanta, Austin, Baltimore, and Boston. To make it easier for a user to understand, I also had the spreadsheet report the names of the hub cities as well. If you're interested, I used the INDEX function for this, a very useful function in picking out values from the sheet.

I also have to have a variable for each regional airport, specifying to which hub it will be assigned. Like this. So for example, look over there city number 5, Chicago. It's assigned to Hub 1, which for now, at least, is Atlanta. And so on for the rest of the 29 cities. And as always, the values that I put in the red cells don't matter. Later, I'm going to tell Solver that they're decision variables and that they have to be integers between 1 and 4, because there are four hubs, 1 through 4, and that's going to do it.

But because I'm using a genetic algorithm, I would have another choice here. Once the hubs are determined, the problem is going to have the optimal solution if each city is assigned to its nearest hub. That is, I could tell the spreadsheet to compare the distances from this regional airport to each of the four hubs, and then to choose the hub that's closest. Insisting on such a nearest-hub assignment would reduce the variables that one needs in this problem to only four, the hub locations. The nearest-hub calculation for a given regional airport would involve finding the MIN of four numbers in the spreadsheet, something that could cause us problems in a gradient descent approach, because that operation isn't differentiable, but once again, genetic algorithm doesn't care.

Nevertheless, I'm not going to require an assignment to the nearest hub. I like the idea that we could come back and modify this problem later, if the need arose. Perhaps, for example, we find that we're overloading one hub, and so we might want to occasionally assign a regional airport to a more distant hub. So, each regional airport has a variable specifying its assigned hub, and the sheet reports the distance between that airport and its hub. This distance is looked up with an INDEX command in the spreadsheet, much like we looked up the cities names.

OK. Nearly done. Excuse me, we have to compute our objective from this assignment, the total size of the flight system. The total distance from the regional airports to their hubs is trivial. We found each distance for each regional airport, and we'll just add them up. Surprisingly, a more annoying part to put in the spreadsheet is the total distance of the direct flights that connect each hub to all of the others.

With four hubs, there are six different pairs of hub cities, and adding up those distances is straightforward. But I want my spreadsheet to be a bit more flexible. After all, maybe it's better to use fewer hubs. If the sheet chooses the same airport for Hub 1 and Hub 2, then the distance from Hub 1 to Hub 4 is the same as the distance from Hub 2 to Hub 4, and that distance should only be counted once, not twice. It's actually a good problem to figure out how to efficiently decide the total inter-hub distance, but I'll leave that as an exercise for the interested viewer and include at least one solution in the guidebook.

So, the size of the hub-and-spoke system will look like this. Start with the total distance between regional airports and their hubs, the total spoke distance. Add the flights among the hubs, the inter-hub distance. That gives us the total distance in the network, the total system length. And that's what we're trying to minimize. The values shown are for the random values that I typed into the decision variable cells to begin with. All that's left are the Solver details. I told Solver that all of my variables were integers, that the hubs had to be cities numbered between 1 and 29, and that all regional airports had to choose a hub between 1 and 4. Remember that genetic algorithms want simple numerical upper and lower bounds on variables.

Those are the only constraints, which is good, because genetic algorithms don't like constraints. I don't expect this problem to be that hard for Solver, because of the fact that once the hubs are selected, there's very little left to do. But I want to make sure to get a good answer, so in Solver, I hit the options button and enter large values in the solving limits boxes, a population size of 500, up to 300 sub-problems, and 300 feasible solutions, and so on. And then, I turned Solver loose.

It came up with an answer in a minute or so, using only two hubs, Atlanta and Las Vegas. Everything from Denver to the west went to Las Vegas, and the eastern part of the country all went to Atlanta. Total distance: 26,864 miles. But, I let the evolutionary Solver keep running in case there'd be any further improvement, and after a couple of minutes, it did find a better answer. This one. It uses three hubs, Las Vegas, Memphis, and Baltimore. Each regional airport, of course, flies to the nearest hub, as suggested by the shading on the map. This new configuration shaved 2000 miles off the first one that the genetic algorithm suggested, that's an 8% improvement.

I can't swear there's not a better one, of course, one of the drawbacks of a genetic algorithm. But it's the best that I've found in several runs, and letting the algorithm run for about 10 minutes each time. I'll leave this example at this point, but you might want to think of other practical possibilities. The volume of traffic in a major city, for example, might make it worth flying from there to more than one hub. We could add this to our program, or use the program's answer as a starting point. Optimization problems don't replace a human decision maker, but they can help that decision maker to do a better job.

But, let's move on to retail pricing and see what's going on there. Start with the basics. Profit is revenue minus cost, and the revenue depends on two things, how much you charge for an item, and how much of it you sell at that price. How much you sell depends on your supply of that item and the demand for it. No problem so far. Take each price per unit, multiply by the number of units sold, add them up, and you've got revenue.

But in a pricing problem, your prices are decision variables. And the number sold of them almost certainly depends on its price. That means that to find the revenue for a given item, you have to multiply its price by a function that depends on that price. And that makes it a nonlinear expression. So linear programming isn't likely to be of use to us here.

How about gradient descent? Well, if your demand varies smoothly with the change in selling price, you should be OK. We talked about gradient descent in terms of solving a MIN problem, and this is a MAX, but you can use our logic from last lecture; you just turn the landscape upside down before you get started. MAXIMIZING revenue is the same as minimizing the negative of the revenue. If fact, if your demand is linear in price, so that each $1.00 increase in price loses you the same amount of demand, then the negative of the revenue ends up being convex. So the gradient method will find the global optimum with no problem at all. If the demand function is smooth but more complicated than that, then gradient descent might get stuck in a local minimum, the old concern, but it's not likely.

Let's take this problem to the next level, too, replacing the demand function by something that's certainly not smooth. Let's look at the pricing scheme for a warehouse club, like Costco or Sam's, and see why it works. And boy, does it! Costco Wholesale Corporation is the seventh largest retailer in the world, the third or fourth largest retailer in the U.S., and the largest retailer of wine in the world. This success is due to a lot of factors, but two of its pricing practices are important contributors to it. One is selling most items only in bulk. The other is the two-part tariff. Members pay an annual fee to be able to buy at Costco, after which they have access to the low per-item price in the store.

I'm going to explore these strategies with a simplified Costco. It'll be selling only one thing, cheese, in one-pound blocks. And its membership lasts only a day. It's essentially a permit to buy on that day. Ridiculous, of course. The difference between this Costco and the real Costco is actually only a matter of scale, though, and not of kind. If you can do this simpler problem, you can do the full-blown one. It's just a lot more of the same.

So, back to cheese. Like anything else, people are generally going to experience diminishing returns from buying additional cheese. That is, each additional pound of cheese is likely to be worth a bit less to the consumer than the one before it; 100 pounds of cheese is not worth 100 times as much to me as one pound is. But not all consumers of cheese value it equally. Not all of them have the same drop-off rates in the value of an additional pound cheese. To make things simple, I'm going to assume that no one wants to buy more than 10 pounds of cheese on their visit. I'm also going to assume for my example that I can break the market up into five segments. The people in each segment will have the same evaluation of the worth of n pounds of cheese. So, we'll need to know the marginal value of a pound of cheese to each of these five market segments.

Segment 1, for example, shows diminishing returns. The first pound, as you can see, is worth $9.00 to these customers, and the second one's worth almost as much, $8.50. Beyond that point, the value drops rapidly. If the customer already has 9 pounds of cheese, that 10th pound is worth only 20 cents to them. The other segments show diminishing returns too, except for segment 2. Perhaps these purchasers are themselves vendors, and so need a lot of cheese. They won't pay a lot, but every pound is worth $4.40 to them. In our problem, the limit on purchases is 10 pounds, just because that's the limit I set in this demonstration.

OK, let's begin by modeling this as a regular store, no membership fee, and customers can buy cheese in one-pound increments. What's the best price for us to set? I'll tell you that we buy cheese at our store for $1.10 per pound. Well, for the moment, imagine that we decided to charge $5.00 per pound. What happens? Let's see. The people in Segment 1 would buy one pound, since it's worth $9 to them, and then they'd buy another pound, since the second pound is worth $8.50. But they won't buy a third, since they'd be paying $5 for something that, to them, is only worth only $4.94. So, each of the 5000 people in the first segment buy 2 pounds of cheese, for a total sales of 10,000 pounds of cheese.

The people in Segment 2 buy nothing, since a pound is only worth only $4.40 to them, and we are charging $5.00. No sales there. We can follow a similar logic for the other three segments. It turns out that we sell 54,000 pounds of cheese at $5.00 per pound. Since we had to pay $1.10 per pound for it, our profit is $(5 - 1.10) \times 54{,}000 = \$210{,}600$. Easy calculation. But it's also highly nonlinear in the price that we set. The price determines the demand in each segment, and demand jumps as we pass that segment's price point for an additional pound of cheese. That's a discontinuous change, a cliff in our mathematical landscape.

So a gradient-based approach is out. We'll use a genetic algorithm again. Is that going to cause trouble? Well, we need to set an upper and lower bound on the variables, which means, here, what we charge for cheese. That's no problem. We'll set $1.00 for a minimum, since the cheese costs us $1.10, and $10.00 for the maximum, since no one will buy cheese for more than $9. Evolutionary solvers don't like a bunch of constraints, but we'll put the calculation of how much cheese each segment buys into the spreadsheet itself, so we won't need to include them as Solver constraints. Evolutionary solver is fine with that.

Figuring how much each segment buys is easy to do. With diminishing returns, the amount bought by a customer is just the number of times that a value larger than the selling price appears in their column. So at the $5.00 cost, the big spenders are in Segment 3, and they buy 4 pounds of cheese each. When we run the program, Solver finds that the best price to charge for a pound of cheese is $5.48. In this case, we sell exactly the same amount of cheese to each customer than we do at $5.00, but we make an additional 48 cents per pound, for a profit of about $236,500.

But can we do better? How about if we use the two-part tariff? Now a person who buys anything must first buy a membership fee. Well, in modeling this, we're going to have two decision variables, how much to charge for a pound of cheese, and how much to charge for membership. Given these, we have to figure out what each customer is going to do. So, first we compute the value of a possible purchase to a customer. Then we compute what it cost them to make that purchase. The difference in these two is called the surplus value of that purchase to that customer. And the customer wants the biggest surplus that he or she can get.

So let's see how this works. For now, imagine that the membership costs $4.00, and that we're selling cheese for $5.00 per pound. And let's focus on Segment 1. Let's look at each of their choices. First, of course, they could refuse to buy anything. That leaves them where they started, with a 0 surplus. But if they buy 1 pound of cheese, they're going to have to pay the $4.00 membership fee and also $5.00 for the pound of cheese. That's $9.00 for $9.00 worth of cheese. As far as that customer is concerned, they're no better off that they were when they bought nothing at all.

Ah, but how about buying two pounds of cheese? Since the first pound is worth $9.00 and the second is worth $8.50 to this customer, the two-pound purchase is worth a total of $17.50, and what do we charge them for it? A $4.00 membership and 2 pounds at $5.00 each, that's a total of $14. So the customer is spending $14.00 to get $17.50 worth of cheese. To them, that's a surplus of $3.50. And that's more attractive than either of the other options looked at so far.

And of course, we just continue from there. For this customer, 3 pounds is worth $9.00 + $8.50 + $4.94 = $22.44, but it costs $4.00 + $3.00 \times $5.00 = $19.00. That last pound added $5.00 to the cost, but returned only $4.94 in value. The surplus drops to $3.44. That's good, but it's not as good as the $3.50 surplus from buying only 2 pounds. As I'm sure you can see, every pound after the second adds less value than the $5.00 that it costs, so the customer buys two pounds of cheese and maximizes his personal surplus.

We can calculate the surplus for each segment by adding up the total worth of n pounds of cheese for that segment and subtracting off what the purchase would cost. I recorded those values in a new table, like this. At the bottom of the table, I put this new information to work. Scan the column to find the maximum surplus for that segment. Record how many pounds of cheese gave this maximum surplus. For those interested in the spreadsheet particulars, I used the MAX function, and the MATCH function, two more very handy functions.

I now know how many pounds each person in a market segment buys, and I know the size of the segment. Taking the sum product of these two quantities, pounds times segment size, added over all segments, tells me the total purchase of cheese. Multiply this by the profit per pound of cheese. Voila!

With the current pricing scheme, you can see it's a bit over $278,000. But you have to remember, that's just with my arbitrarily-chosen $4.00 membership and $5.00 price per pound of cheese. We'll let the spreadsheet's evolutionary solver find the best values. I specified that the cost per pound was between $1.00 and $10.00 in the constraints, and that membership fee was between $0.00 and $100.00. Those were the only constraints. Then, I let it run for a while.

The best answer found is to charge $8.57 for membership fees and to sell the cheese at $3.54 a pound. With these prices, we sell 74,000 pounds of cheese, which is rather staggering. That's a 270% increase in cheese sales over our no-membership-fee model! Our profit is almost $335,000, up 42% from the original model. The introduction of the membership fee, coupled with a reduced price per pound for cheese, drastically increased sales and profits.

So, we've seen the value of one part of Costco's pricing strategy, a membership fee. Now let's look at the other piece, selling only in bulk. I'm going to add one more decision variable to our Costco formulation, how large a package we sell. Specifically, we're going to sell packages of n pounds of cheese only, where n is a whole number. We'll have the spreadsheet compute how big n should be. Now, I could rewrite my program entirely in light of this new rule, but there's an easier way. My idea is equivalent to Costco saying to a customer, "Oh, we're selling cheese in 4-pound blocks, we are. You don' like 'at? You wanna buy a different amount? No problem, squire, but it's at a special rate for that. Hmmm, let's see, $100 more than you think it's worth that ought'a do it." That is, if the amount of cheese is not a multiple of our package size, we make the price so terrible, a surplus of negative $100 to the customer, that the customer never chooses that option, which of course, is exactly the condition we want to enforce.

So, the resulting and new spreadsheet looks like this. In the lower left, I have a new decision variable for the number of pounds of cheese in a bulk package. Here, it's 4 pounds. I modified my surplus table to say that, unless the pounds were a multiple of the package size, the surplus is −$100. Again, for spreadsheet fans, the easiest way to do this is to use the IF and MOD functions, two more things you wouldn't see in a gradient descent program.

Anyway, as you can see, if you keep our earlier pricing plan of $8.57 for membership and $3.54 per pound of cheese, then this bundling idea looks like a big mistake. Segments 2 and 4 now buy nothing, and segments 1, 3, and 5 buy 4 pounds each. Profit has dropped to about $238,000. But we'll run the program one last time, telling the spreadsheet that it can choose any integer from 1 to 10 as the package size, and see what happens.

Here's the result. You know that a downside of genetic algorithms is that they can be slow. This program ran for about five minutes on my fairly fast desktop computer before reporting back its answer. The solution is right there on your screen. Sell cheese in three-pound packages, setting the membership fee to $11.76, and the cost per pound as $2.48. That results in selling 99,000 pounds of cheese for a profit of over $348,000. That is, the three-pound packages increased the profits by an additional 4% and increased sales by 15,000 pounds of cheese!

So there you have it. The approach of a two-part tariff system and bulk sales, given the right market, can substantially increase both sales and profits. How? Memberships brought down the cost of an additional unit of product, encouraging customers to buy more. Bulk sales took advantage of the fact that the value of additional units of product can drop quickly. It forces the customer to buy a lot or not at all. Making it work, though, requires getting the right membership cost and the right price and package size for the products. The problem is complex. The answer is not obvious, but the improvements from implementing the best answer are enormous. And whether the problem is commercial, military, personal, whether the program is linear, integer, or nonlinear, improvement is what optimization is all about.

Randomness, Probability, and Expectation
Lecture 18

This lecture focuses on randomness, or uncertainty. We often have a pretty good intuition about random events that we deal with all the time, but when it comes to things that are unfamiliar or that occur only rarely, people are often terrible at estimating the probabilities of events or what average results they might expect if facing the situation many times. In this lecture, you will learn some essential tools used in probability, including calculations of the probabilities of compound events and expected value.

Probability

- **Probability** is a measure of the chance that an **event** occurs. Events that never occur get a probability of zero; events that always occur get a probability of one, or 100%; and everything else fits within that range, with more likely events getting higher probabilities.

- Consider the simplest example: flipping a fair coin. There is a 50% chance that the coin comes up heads; the probability of getting a head is 0.5. But that doesn't really mean that it comes up heads half the time. If you flip a coin twice, it's not uncommon for both to come up the same: two heads or two tails. In fact, half of the time, in two flips, you will get two of the same.

- So, where does this probability of 0.5 for a head come from? It's the classical probability of the event. With the coin, there are two equally likely outcomes: heads and tails. There is a probability of 1 in 2, or 1/2. In classical probability, if all outcomes are equally likely, then divide the number of successful outcomes by the total number of possible outcomes and you get the probability of success.

- However, in many real-world situations, this definition is practically useless. We often can't break down all possible outcomes into a collection of equally likely possibilities. And when we can't, we usually adopt a frequentist approach, finding the empirical probability based on historical data and relative frequencies.

- Empirical probabilities might not always consider all of the factors involved, but they are often the best we have, and they're often accurate enough to help us make good decisions. Unfortunately, sometimes we can't even get empirical probabilities.

- For example, what is the probability that AIDS will be cured in the next 10 years? Obviously, we can't use either classical or empirical probability for a question like this. We can't look at how often AIDS was cured in previous decades. And it may not really be sensible to use data for the cure rates of other diseases in predicting the eradication of AIDS. In this case, the best we have is a subjective probability—a (hopefully educated) guess. How reliable such a probability is depends on how much you can trust the person making the estimate.

- Regardless of where our probabilities come from, there are three things we can do with them once we have them. First, we can characterize the variation of the important quantities in a problem. How much on average will this risky investment pay? What's the worst that can happen? What's the chance of it losing money?

- Second, we can use probabilities to simulate situations. This lets us get a grip on how the random factors in the situation influence the final outcome of events. For example, what happens to traffic flow when we open a new lane on a particular highway?

- Underlying both of these uses is the third use: Probabilities of fundamental events can be used to figure out the probabilities of more complicated compound events. For example, how likely is it that Bank A or Bank B or both approve our loan? If we get the loan approval, how likely is it that our bid is accepted and that we can pay off the loan within five years? There are rules in probability that allow us to figure out the probabilities of combinations like this.

- The most important one is **joint probability**, which is the probability that one thing happens and another thing happens. More generally, we'd like to know how to find the probability that a whole collection of events occur.

- The answer is actually pretty simple. The word "and" in probability always goes with the operation of multiplication. The formula involves multiplying a string of probabilities together. The probability of event A is written as $P(A)$, and the probability of event B is written as $P(B)$.

- The idea of "the probability of A being true given that B is true" is important in many computations, so it's given a special name: **conditional probability**—in this case, the probability of A given B. The word "given" is represented in symbols as a vertical bar, as follows.

    $P(A \mid B)$ = probability of A, given B = the probability that A is true, given that B is true.

- The following is the representation of this idea in symbols.

    $P(A$ and B and $C) = P(A) \times P(B \mid A) \times P(C \mid A$ and $B)$.

- This kind of formula always works. In fact, it gets even easier when the events happen to be **independent**. A collection of events is independent if knowing whether some of them are true doesn't change the probabilities that the others are true.

- For example, this is always the case when the events have no connection with one another. If two people each pick a person randomly, whether one chosen person is married is independent of whether the other chosen person is of legal age. That is, knowing one chosen person's marital status neither increases nor decreases the chance that another arbitrarily chosen person is a minor. On the other hand, if two people agreed to choose the same person, the marital status and whether the person is a minor would be dependent events. Minors are less likely to be married than people of legal age.

- Anytime you're looking at joint probability, you always want to ask yourself if the events are independent. If they are, it's easy to determine the probability that all of the events happen. Just find the probability of each event individually, and then multiply all of those probabilities together.

  For independent events, $P(A \text{ and } B \text{ and } C) = P(A) \times P(B) \times P(C)$.

Errors in Reasoning
- The overconfidence effect is a well-established psychological bias. People consistently rate their subjective confidence in their answers higher than their objective accuracy.

- In a spelling test, for example, respondents correctly spelled only 80% of the words of which they were "100% certain." And the pattern continues for all levels of confidence. When people claimed to be 80% certain, they were right far less than 80% of the time.

- A 1981 study showed that 93% of American drivers rated themselves as being better than the median. By definition, the median is the halfway point in the population, so only 50% are better than half of all other drivers.

- Given any "concrete" information, people often seem to disregard the relevant probabilities. It's called ignoring the base rate, and it leads people to believe that a short, slender English man is more likely to be a jockey than a welder, even though there are over 100 times as many welders in the population as jockeys. It's a tempting error.

- This error in reasoning is terrifically common. Of course, it's true that if you are a jockey, the chances of your being slender are very high. But that does not mean that, if you are slender, your chances of being a jockey are very high. Every great pianist has two hands. But that doesn't mean that a person with two hands is likely to be a great pianist.

- Interestingly, a number of studies have shown that it's common for people to completely ignore the provided information of how frequently the target group appears in the population and instead to base their guesses on details of the people that have almost no predictive power for the question at hand.

Expected Values

- Because we're so often dealing with chance, we need to figure out how to judge whether one strategy really is better than another—because a strategy that would work out brilliantly under one set of chance circumstances may be the worst choice possible in another.

- The following might be an important question: How well does the strategy do on average? A bank making loans is going to sometimes reject a person who would have paid up and will sometimes accept a person who defaults. But they're interested in their long-term average return.

- The average payoff is also called the mean payoff, but when dealing with matters involving chance, mathematicians often call it the **expected value**. This term is misleading, because it isn't the value that any one person would actually expect.

- If 10 raffle tickets for a $50 prize are sold to 10 people, each person expects to win nothing—but the average winning is $5. The facts are that 10 people played, and the total winnings were $50, so the average winnings per person was $50/10, and that's what we call the expected value, even though it's impossible to win $5.

- But once you get past possible confusions about the name, computing expected values—averages—in many cases is pretty straightforward. Suppose that you have a quantity that can only take on a finite number of values. You know these values and the fraction of the time that each occurs. That's the probability. Then, to find the expected value, you take each possible value, multiply it by the probability that it occurs, and then add all of these products together.

- But keep in mind that unless a situation is being faced many, many times, knowing the average is never enough. If we're doing a risk analysis for an event that happens only once or a few times, we need to watch out for variability in the data. Still, the average is an important place to start, and expected values show a lot of useful and important mathematical properties that will often allow us to find expected values easily.

- One such property is that they behave very well under addition. Take any two activities A and B, and the average result from doing both A and B is the same as the average from doing A plus the average from doing B—regardless of whether A and B are independent or intimately connected.

Important Terms

event: A statement about the outcome of an experiment that is either true or false for any given trial of the experiment.

expected value: The average (or mean) value of a numeric random variable.

independent: Two events are independent if knowing the outcome of one does not alter the probability of the other. In symbols, A and B are independent if and only if $P(A) = P(A \mid B)$. Two events that are not independent are dependent.

probability: A measure of the likeliness of the occurrence of uncertain events. Probabilities lie between 0 and 1, with more-likely events having higher probability. If all possible outcomes of an experiment are equally likely, the classical probability or a priori probability of an event can be defined as the fraction of those outcomes in which the event occurs. In other cases, probability may be empirically defined from the fraction of historical cases in which the event occurred.

probability, conditional: The probability that some event occurs, given that some other collection of events occurred. The probability that A occurs given that B occurs is written $P(A \mid B)$.

probability, joint: The probability that all events in some collection occur.

Suggested Reading

Grinstead and Snell, *Introduction to Probability*.

Piattelli-Palmarini, *Inevitable Illusions*.

Questions and Comments

1. Many people believe that "bad news comes in threes." If this is nothing but a fantasy, how can we explain this? To simplify the math, imagine that a day has a 10% chance of having "bad news" but that each day is independent of the rest.

   a) Show that if one receives bad news on a given day, there is an 81% chance of no bad news on either of the following two days. Because 10% of days have bad news, this is common and not seen as part of a pattern.

   b) Two bad days in a row are relatively rare: a 1% probability for any two particular consecutive days. This catches people's attention. People are now looking for the third bad day. Given that two bad days in a row have just occurred, show that there is a 19% chance that at least one of the following two days will have bad news.

Answers:

a) Each day is independent, so looking at two consecutive days, we're just asking *P*(*no bad on day 1* and *no bad on day 2*). Because a day has a 10% chance of being bad, it has a 90% chance of being good. Because bad news on one day is independent of what happens on other days, we can just multiply the probability of *no bad on day 1* with the probability of *no bad on day 2*. That's 0.9 × 0.9, or 0.81.

b) Again, days are independent, so we're really asking, for two consecutive days, "How likely is it that at least one is a bad day?" Note that, as stated, this is not an "and" problem—it's an "or" problem: *day 1 bad* or *day 2 bad* or both. But with a bit of cleverness, we can still get the solution.

Turn the question around. The opposite of *either of the next two days being bad* is *both of the next two days being good*. But in part a), we found the probability of two good days in a row as 0.81. So, the opposite of this, 1 − 0.81 = 0.19, is the chance that at least one of the two days is bad. Note that neither the answer to a) nor b) depended on knowing what kind of day today was. Each day is independent, by assumption.

2. In the "Cat or No Cat?" game from the video lecture, suppose that you knew that you were being shown the pictures of three people who owned cats and six who didn't. You want to maximize the average amount of money you make on your guesses. How should you guess now?

Answer:

Surprisingly, you should *still* guess that there are no cat owners, and your average payoff is still $600. This is a demonstration of the power implied by the statement that "expected values of payoffs always add." For any one picture, alone, guessing "cat" gets you on average $33.33, and guessing "no cat" gets you on average $66.67. So, the sum over all nine pictures still works the same as before, with you getting 9 × $66.67 = $600 on average. Indeed, in guessing that there are no cat owners, even though you know that there are three hidden in there somewhere, you are guaranteed $600, and any other guessing scheme will, on average, get you less.

So, should you never guess three cat owners and six non-owners? Well, there is one variant where you'd want to do that: if you were told that there were three cat owners and six non-owners but that you'll only get paid if you guess every one correctly. In this situation, clearly, you'll want to match the demographics. Your chance of success turns out to be 1 in 84. You can work this out yourself, with an approach like that used for the English footballers. Imagine having nine poker chips, three of which have the letter C on their bottom faces. Think about flipping three coins in a row and finding a C each time.

Randomness, Probability, and Expectation
Lecture 18—Transcript

Randomness, uncertainty, is a part of life, and in this lecture, randomness takes center stage, where it will remain with us to the end. So let's start by refining our intuitions about randomness, drawing on ideas from probability.

We often have a pretty good intuition about random events that we deal with all the time, but when it comes to things that are unfamiliar, or that occur only rarely, people often are terrible at estimating the probabilities of events, or what average results they might expect if facing the same situation many times. Take a roulette table, when a streak of five or six black numbers shows up, a number of people will scramble to bet on black, which they believe to be hot, counterbalanced by a second crowd of people that will scramble to bet on red, based on the fact that they believe it is overdue.

These opposite conclusions are sparked by a common belief, that the streak that just occurred is unnaturally long, too long to be random chance. Well, to a pretty good approximation, you can figure out how long a streak you'd expect in a certain number of spins by a literal rule of thumb. I'll show you.

First, specify how many spins you're talking about. If the wheel spins every two minutes over an eight-hour day, that's 240 spins. Now, put up your thumb and say, two, then double the number that you say every time you raise another finger, so, 4, 8, 16, 32, 64, 128, 256. Stop when you get close to your number of spins. The average length of the longest streak is quite close to the number of digits that you just had raised. So, in just 8 spins, I'd expect a run of about 3 in a row, 2, 4, 8. In the case of 240, we'd expect a streak of about 8 reds or 8 blacks sometime during the day, by simple random chance.

We'd expect many more shorter streaks, too, of course. But these streaks that are sending gamblers to the table aren't surprising at all. In fact, it would be surprising if they didn't occur. A similar analysis applies to an athlete on a hot streak. In almost every case, there's no evidence that the streak is anything more than random chance.

Suppose you put 36 strangers in a room. How likely is it that any two have the same birthday? You might figure, since there are 365 days in a year, the answer's about 1 in 10 for 36 people. It's actually over 83%. You can ignore the calculations, or work through it. But notice the graph. With only 57 people, the probability of a shared birthday rises to over 99%.

Here's another one. Take those same 36 people; take their keys; scramble them up, and then redistribute them randomly. How likely is it that someone got their own keys back? And if you played this game with more people, would it be more or less likely that at least one person gets his or her own keys?

Well, the answer does depend, not surprisingly, on the number of people. But that dependence quickly becomes incredibly weak. Once you get to about five people, the probability is almost exactly $1 - \frac{1}{e}$; that's the same e that we used in our transformation of variables work. That's a bit over 63%, and that figure is right if you're returning keys to 5 people, or 36 people, or 5000. We can check the answer by brute force for 5 people; 120 possible orders for key returns, and in 44 of them, shown in red, no one gets his or her own keys. In the remaining 76 cases out of 120, someone does; 76 out of 120 is 63.3%, as we said. These arrangements in which no number stays in its own place are called derangements, and the formula that gives their probability is very beautiful, and not obvious. The brute force verification for only five people was already quite enough. I had to look at 120 orders. For 36 people, it would be over 10,000 trillion trillion trillion.

Another example, suppose you pick one of these people and they tell you that they have two siblings. You ask if they have a brother, and they say yes. How likely is it that they have a sister, too? If you think the answer is one half, guess again. It's $\frac{2}{3}$. Here's the math for that one, using tools that we'll discuss today. You are more likely to be killed by a toaster than a shark. In fact, more people are killed by collapsing sand castles than by sharks. And sometimes an impressive-sounding statistic tells you almost nothing at all; 50% of doctors graduated in the bottom half of their class; 100% of their patients will die within six months of their birthdays. It can be surprising how often statements like that upset people.

Well, these weaknesses in probabilistic thinking aren't just so much cocktail conversation; they can undermine predictions and decision making of all sorts, and occasionally, people who understand probability can use that to their advantage. One way that auditors catch people who are cooking the books is by looking at the relative frequencies of the leading digits in the numbers that they report. People who make up the numbers usually think that each digit is equally likely to start a number. But real data spread over several orders of magnitude tends to follow something called Benford's Law. Amazingly, about 30% of those numbers tend to start with the single digit 1, while numbers starting with 8 or 9 appear only about 5% of the time, each. Smaller first digits are more likely than bigger first digits. When an auditors sees a big deviation from that law, they take a closer look.

And taking a closer look is just what we'd better do, too, if we don't want to be blindsided by randomness. In regression and time series and data mining, we dealt with some randomness, but it was confined to the characterization of error. Do a statistical analysis of the variation, and use it to construct confidence intervals on how much you can trust the prediction. But in many situations, random events play a much more central role in how things will turn out. World War II veterans, like my father, will never forget the date of June 6th, 1944, D-Day, the invasion of Normandy and the beginning of the liberation of France by the Allies. But D-Day was tentatively scheduled for June 5th. On June 4th, General Eisenhower pushed it back a day, due to uncertainties in weather about the 5th. The weather predicted for the 6th was better, but not good. But the next window for a desirable tide was about two weeks later, and during that window, history now shows that a major storm would have made landing impossible—world history, turning on the roll of a meteorological die. More prosaically, when banks issue loans, the success or failure can hinge on how accurately the banks can assess the risk of not being paid back, and how well they can make its decisions in light of that risk.

Most of our work in optimization has been deterministic, so randomness hasn't played a role beyond the controlled deviations that we explored with sensitivity analysis, but if we want to go further with optimization, to optimize in the face of uncertainty, then that's going have to change, too.

Well, these are the kinds of things that the next section of the course is about, analyzing and evaluating situations in which we have only partial control—public relations problems, research opportunities, waiting lines, market share, homeland security, bidding on jobs. And we can't do any of this until we get a handle on uncertainty, on randomness, on balancing risk and reward. And that, in turn, means getting a handle on three things, basic probability, worst-case to best-case scenarios, and what we could reasonably expect to have happen if we face the same situation many times. That's what I want to focus now, and along the way, I want to give you a clearer sense of why we need these tools and why intuition is not enough.

Probability—it's a measure of the chance that an event occurs. Events that never occur get a probability of zero; events that always occur, get a probability of 1, or 100%; and everything else fits in that range, with more likely events getting higher probabilities. Going further than that isn't as easy as you might think. The first people we know about to seriously think about probability were Pierre de Fermat and Blaise Pascal, in the 1600s. Both were prodigious mathematical thinkers, both were very interested in gambling, and in their letters to one another they talked about this, and that's generally taken as the beginning of the study of probability.

Probability theory today is a formal discipline of remarkable complexity and depth, but you can go pretty far with just a good grip on the basics. Let's consider the very simplest example: flipping a coin—50% chance of coming up heads, yeah? Yeah. The probability of getting a head is 0.5. But that doesn't really mean that it comes up heads half the time, does it? I mean, look, if I flip a coin twice, I won't be surprised if they both come up the same, two heads or two tails. In fact, half of the time, in two flips, you will get two of the same. So where does this probability of 0.5 head really come from?

It's the classical probability of the event. With the coin, there are two equally likely outcomes: heads and tails. Heads is one of those two. Probability of 1 in 2, or $1/2$. If you draw a card from a deck of 52 cards, and 13 of those 52 are spades, then the chance of drawing a spade is 13 out of 52, $13/52$, or $1/4$. That's the whole deal with so-called classical probability. If all the outcomes are equally likely, then divide the number of successful outcomes by the total number of possible outcomes, and you have your probability of success. Except that, in a lot of real world situations, this definition is practically useless. You often can't break down all of the possible outcomes into a collection of equally likely possibilities. And when you can't, we usually adopt a frequentist approach, finding the empirical probability based on historical data and relative frequencies.

It's the same idea as a batting average. If a player has been at bat 400 times and has gotten a hit in 100 of those 400 attempts, his batting average is 250, 0.250. Barring other information, we'd take the probability of this player getting a hit when he steps up next time as $^1/_4$. There can be problems with this approach, of course. If one can conduct identical experiments in a large number of times, fantastic! But in our baseball player example, how about the effect of professional improvement over time, or aging? Still, empirical probabilities are often the best that we have, and they're often accurate enough to help us make good decisions. Unfortunately, sometimes we can't even get empirical probabilities.

For example, what's the probability that AIDS will be cured in the next 10 years? Obviously, we can't use either classical or empirical probability for a question like this. We can't look at how many times AIDS was cured in previous decades. And it may not really be sensible to use data for the cure rates of other diseases in the prediction and eradication of AIDS. Here, the best we have is a subjective probability, a (hopefully educated) guess. How reliable such a probability is depends on how much you can trust the person making the estimate. But, whatever our probabilities come from, wherever they are, there are three things we can do with them, once we have them.

First, we can characterize the variation of important quantities in the problem. How much, on average, will this risky investment pay? What's the worst that can happen? What's the chance of it losing money? Second, we can use probabilities to simulate situations. This lets us get a grip on the random factors in the situation and how they influence the final outcome of events. For example, what happens to traffic flow when we open a this new lane on this highway?

And underlying both of these uses is the third use. Probabilities of fundamental events can be used to figure out probabilities of more complicated, compound events. For example, how likely is it that Bank A, or Bank B, or both approve our loan? If we get the loan amount, how likely is it that our bid is accepted and that we can pay off the loan within five years? There are rules in probability that let us figure out the probabilities of combinations like this, and we're going to cover just a few, the ones that we need in later lectures.

The most important one is joint probability, joint in the sense of joint checking account. Joint probability is the probability that this happens and that happens. More generally, we'd like to get an idea of how the probability of a whole collection of events occur. The answer's actually pretty simple. The word "and" in probability always goes with the operation of multiplication. To get the formula, I always think of a person trying to make it through an obstacle course; every obstacle is an event from our collection, and making it through the course means clearing all of the obstacles. So, line up the events in any order you like, and run the course.

In order to complete it, first you have to clear obstacle A. The probability of doing that is just $P(A)$; that's how we write the probability of A. No surprises. Now you go on to event B, and it's natural to think that the chance of clearing this second obstacle is just $P(B)$, the probability of B. But that's overlooking an important piece of information. Because by the time that you get to this point in the course, we know something about you: We know that you cleared obstacle A. So what we really want at this point isn't just the probability that a random someone can clear obstacle B. It's the probability that someone who cleared A will also be able to handle B. For my little runner, after A, B looks like pretty easy sailing. And so we go on this way. The probability that we need for the third obstacle is the chance of clearing C, given that you already cleared A and B. Get it?

Once more, the word "and" in probability calculations always goes with the operation of multiplication, so if we want the probability that all of the events occur, we just take this whole string of probabilities and multiply them together. That's it. That's joint probability. Let me give you an example, and given my obstacle course metaphor, I'll choose one from sports. In 2007, three English teams made it to the quarterfinals in the European Champions League in soccer. Three English teams, five non-English teams. The pairing for teams for the quarterfinals was going to be decided by random draw, but there was a rumor flying around that the draw was going to be rigged so that no English team played another one in the quarterfinals. If that happened, it would obviously be good for England. And sure enough, when the pairings were made, no English team was pitted against another. Does this suggest some sort of hanky panky? Let's work it out.

The English teams were Chelsea, Liverpool, and Manchester United, which I'm showing as C, L, and M. Merely for convenience, I'll mark the five non-English teams as E, for European, even though the English teams are clearly Europeans in this competition, too. I'd mark them as C for Continental, but Chelsea already has dibs on that. So, to say that no English team plays against another means that Chelsea gets a European opponent, and Liverpool gets a European opponent, and Manchester gets a European opponent. And the word "and," as we said, always means "times" in probability. So, let's run our obstacle course.

First, Chelsea has to get a European opponent. Well, Chelsea could get any of the 7 opponents in the first round, and five of those 7 opponents are European. So the chance of Chelsea avoiding an English opponent is 5 out of 7, or $5/_7$. That was easy. Now, stage 2. We've cleared the first hurdle, and Chelsea is playing a European team. Cross out Chelsea and its opponent. Our picture now looks like this, six teams, four of which are European.

OK, now the second hurdle. Liverpool has to get a European opponent. Well, there are five remaining teams that Liverpool could play, and four of them are European. So the chance that Liverpool gets a European opponent, given that Chelsea got one, is 4 our 5, or $4/_5$. The chance that Chelsea and Liverpool both got European opponents is the product of these: $5/_7 \times 4/_5 = 4/_7$.

OK. Last hurdle. Manchester United. Given all that we've already said, Chelsea and Liverpool have European opponents. Cross them out. I really don't really know which European team to cross out, but it doesn't matter for this argument, so I'll just pick one. There are only three teams left that could play Manchester United, and all of them are European. So, given that Chelsea and Liverpool have continental opponents, it's guaranteed that Manchester does too. The probability is 1. So the chance of the whole sequence occurring, that is, no English team facing another one in the quarterfinals, is $^5/_7 \times {}^4/_5 \times 1 = {}^4/_7$, which is about 57%. So, in fact, the fact that no English team ended up facing another didn't imply foul play; it's actually the most likely outcome of a fair draw. The talk of a fix was fed by mistaken ideas about probabilities.

On the other hand, this idea of the probability of A being true given that B is true is terribly important in a lot of computations, so it's given a special name. It's called conditional probability—in this case, the probability of A given B. The word "given" is represented in symbols as a vertical bar, like this. So what we just said in symbols is [P(A and B and C) = P(A) × P(B | A) × P(C | A and B)]. And this kind of formula always works.

In fact, it gets even easier when the events happen to be independent. A collection of events is independent if knowing whether some of them are true doesn't change the probabilities that the others are true. For example, this is always the case when the events have no connection at all one another. If we each pick a person randomly, whether my person is married is independent of whether your person is of legal age. That is, knowing my choice's marital status neither increases nor decreases the chance that your arbitrarily chosen person is a minor. On the other hand, if we agreed to choose the same person, the marital status and whether or not a person is a minor would be dependent events. Minors are less likely to be married than those of legal age.

Any time you're looking at joint probability, you always want to ask yourself if the events are independent. If they are, it's a piece of cake to figure the probability that all of the events happen; just find the probability of each event individually, then multiply all of those together, like this. This is part of the idea behind building backups for vital systems, as in spacecraft. A vital component may only have a 1 in 100 chance of failing, not because of a design flaw, but just from mischance. That's small, but maybe you wouldn't like to risk your life on its continued operation. On the other hand, if we have two such units, then you are only in trouble if both of them fail. The failure of the one is independent of the failure of the other, so the chance that they both fail is $^1/_{100} \times {}^1/_{100}$, or 1 in 10,000. Add another backup, and you're down to 1 in a million, and so on.

Do we really need to understand this? I've been suggesting that if you ask a person, even a bright person, to make decisions in a situation in which they have only limited, probabilistic ideas of what may happen, that their proposals are often questionable and reflect fundamental gaps in their understanding of chance. Well, let me give you an example and see how you would handle it. I'm going to choose a topic that doesn't require any special expertise, to keep the playing field level. Take a look at my lineup of nine randomly chosen Americans. Now, according to a study of the American Pet Products Association, about 1 in 3 Americans have at least one cat. Let's take that figure as being correct. I'm asking you to divide these people into two groups, those that you think own a cat, and those that you think don't. Thumbs up or thumbs down on cat ownership for each person. It's my new game show—Cat or No Cat?

Each one of these people that you get right, you'll get $100; $0 if you guess them wrong. Get them all right, and you'll earn a big $900. So, make your guesses as to who's what. Your goal is to make as much money as you can. If you want, pause the video for a second if you need more time. The results should be educational.

OK, I'm going to show you the correct answers of which people in this randomly selected group of Americans are cat owners. You ready? Here they are. Give yourself $100 for each person you got right, a cat owner that you said owned a cat, or a non-owner who you said didn't. If you got six or more right, earning at least $600, you're in a very small minority. But if you played this game according to probability, you should have made even more than that. Let's walk it through.

Imagine, for starters, and I think that this is actually true, that people's faces don't really tell you much information about their cat ownership. We will revisit this point later, but let's assume that it's true for our original analysis. Whether you ignored faces or not, you may have picked three people as cat owners and six as non-cat owners, since cat owners make up about a third of the population. That's the most common choice, and in the past, I've had more than a few people challenge me indignantly on the fact that only two of our photos are of cat owners. They're $\frac{1}{3}$ of the population, so why aren't there three? Heh—funny question. Remember how obvious it was that with a coin, a 50% chance of flipping heads, might flip the same way twice in a row? Given the size of the population, whether any one of these nine people is a cat owner has virtually nothing to do with the cat ownership of any other. They're independent, just like coin flips.

So for whatever reason, suppose you chose three [as] cat-owners and six as non-owners. How likely is it you'd get them all right? That's a joint probability—and—and we just got done saying that the events are independent. That means you can just multiply together the probabilities of the nine individual events that you're looking for. Each time you said a person had a cat, you had a $\frac{1}{3}$ chance of being right. Each time you said that a person didn't have a cat, you had a $\frac{2}{3}$ chance of being right.

So, if you guessed three cat-owners and six non-cat owners and weren't able to sharpen your guesses by looking at the pictures, your chances of getting them all right is the product of three $\frac{1}{3}$ and six $\frac{2}{3}$, which comes out to be about $\frac{1}{3}$ of 1%, a bit worse than 1 in 300. On the other hand, if you guessed that there were no cat owners in the sample, your probability of getting them all correct is $(\frac{2}{3})^9$, which is 2.6%, about 1 in 38. Still long odds, but much better than 1 in 300. In fact, the more people that you guessed as cat owners, the less likely you were to get everyone right.

Of course, that's was assuming that you're not learning much from the pictures. Maybe you think that you can read a face pretty well. Obviously, if your inspection of a face results in a more than a 50% probability that the person is a cat owner, then guessing cat owner for that person makes sense. But how much can you trust your subjective probability? There's a well-established psychological bias, you see, the overconfidence effect. People consistently rate their subjective confidence in their answers higher than their objective accuracy. In a spelling test, for example, respondents correctly spelled only 80% of the words which they said they were 100% certain about. And the pattern continues for all levels of confidence. When people claimed to be 80% certain, they were right far less than 80% of the time. A 1981 study showed that 93% of American drivers rated themselves as being better than the median. By definition, the median is the halfway point in the population, so only 50% of the drivers are better than half of all other drivers.

Interestingly, a number of studies have shown that in a situation like our cat game, it's common for people to completely ignore the provided information on how frequently the target group appears in the population, and instead, to base their guesses on details of people that have almost no predictive power for the question at hand. Given any concrete information, people often seem to disregard the relevant probabilities. It's called ignoring the base rate, and it leads people to believe that a short, slender English man is more likely to be a jockey than a welder, even though there are over 100 times as many welders in the population as jockeys. It's a tempting error, isn't it?

We'll come back and look at this kind of thing more closely when we discuss Bayesian probability, because the error in reasoning is terrifically common. Yes, it's true that if you are a jockey, the chances of being slender are very high. But that does not mean that, if you are slender, your chances of being a jockey are very high. Every great pianist has two hands. But that doesn't mean that a person like me, with two hands, is likely to be a great pianist, does it?

Since we're so often dealing with chance, we'd better figure out how to judge whether one strategy really is better than another, because a strategy that would work out brilliantly under one set of circumstances may be the worst choice for another. In our cat game, I assessed the quality of a strategy in terms of how likely it was to correctly identify the cat ownership of all nine people. Miss one, and you might as well have missed them all. But in a lot of circumstances, that's not right. A more important question may be, how well does the strategy work on average? A bank making loans is going to sometimes reject a person who would have paid up, and will sometimes accept a person who defaults. But they're interested in their long term average return.

The average payoff is also called the mean payoff; when dealing with matters involving chance, the mathematicians usually call it the expected value. Rotten name for everyone else, really, since it isn't the value that anyone would actually expect. If 10 raffle tickets for a $50-prize are sold to 10 people, each person expects to win nothing, but the average winning is $5. Ten people played, and total winnings were $50, so the average winnings per person was $^{50}/_{10}$, and that's what we call the expected value—even though it's impossible to win $5.

But, once you get past the possible confusions about the name, computing expected values, averages, in many cases is pretty straightforward. Suppose you have a quantity that can only take on a finite number of values. You know these values and the fraction of the time that each occurs; that's the probability. Then to find the expected value, you take each possible value, multiply it by the probability that it occurs, and then add all these products up. That's it.

Let's look at an example. I can make a $50,000 year-long investment in a company doing genetic research for a particular disease. I've estimated, and obviously, this is a subjective probability, that there's a 10% chance that they'll succeed this year, in which case I'll paid $500,000. There's also a 50% chance that they'll go bankrupt, in which case I lose my $50,000 investment entirely. The rest of the time, I just break even, getting my $50,000 back. In terms of expected payoff, is this investment attractive? Well, let's find the expected value. First, list all possible payoffs, taking into account the $50,000 that I have to invest, then the probabilities of each of these. Multiply each payoff by its probability, and finally, add the results. Expected value = $20,000. On a $50,000 investment, 20,000 is a 40% return, not too shabby for one year.

What this means is that if you had the opportunity to make this investment again, and again, and again, then, on average, it would return 40% at the end of the year. But of course, there is also risk. Half of the time, you lose your $50,000 outright! So even though there's a reasonable 10% chance of exceptional gain, there's a much larger 50% chance of total loss. This variation in the outcome is the hallmark of a risky investment—great variability in the possible payoffs. Yes, you can make a bundle. You can also go lose your shirt.

If it were possible to repeat this investment many times, this might not matter. In the long run, the average should be trustworthy. It's a result from statistics called the law of large numbers. But if we're talking about a one shot, or an opportunity that occurs only very few times, then that variability is going to be a serious concern. Some of the techniques I'm going to be showing you in the next few lectures focus on expected value, the average payoffs. But even when we're looking at them, it's important to keep in mind that, unless a situation is being faced many, many times, knowing the average is never enough.

The average temperature on the surface of the Moon is about 5°F, decidedly nippy, but tolerable with winter clothes. But that 5° average comes about by averaging temperatures from +253° in the daytime with −243° at night. So we'll need to keep an eye on variability, too, and will do so increasingly as we move through the upcoming lectures. Still, the average is an important place to start, and happily, expected values show a lot of useful and important mathematical properties that will often let us find them easily.

One such property is that they behave very well under addition. Take any two activities, A and B, the average result from doing both A and B. That's going to be the same as the average from doing A, plus the average from doing B, regardless of whether A and B are independent or intimately connected.

This isn't nearly as obvious as it might at first sound. For example, suppose that a person is paid $10 for correctly calling a coin flip, but gets nothing for calling it incorrectly. Then the expected payoff for that person is $5. (0.5) $10 + (0.5) $0. So our new addition law says that if you and I both play this game, then our expected total winnings will be 5 + 5, or $10. That's true regardless of whether we share a flip and call the same result, or share a flip and call opposite results, or each flip our own coin.

So, how about the cat game? Earlier, we saw that if you wanted to maximize your chance of getting them all right, you should guess no cat owners at all. But we also said it made better sense to try to maximize your average winnings. What's your best choice now? Well, let's think about guessing one person only. Probabilistically, $^2/_3$ of the time, they're a non-cat owner; $^1/_3$ of the time, they're a cat owner. If a correct guess gets you $100, then guessing non-cat owner will get you $(^2/_3)100 + (^1/_3)(0)$, or about $67, on average.

Guessing cat owner only gets you $100 one-third of the time, so it's expected payoff is $(^1/_3)100 + (^2/_3)0$, or about $33, only half as much. But a moment ago, we said that expected values from different activities always add. That means that [in the] nine pictures, you make an average of $67 bucks every time you say no cat and an average of $33 every time you say cat. So your optimal strategy is still to guess that none of them have cats. On average, you'll make about $67 on each picture, or a total of $600 with this strategy. With the 9 people in this lecture, you actually make $700. Using their common sense, very few people fare this well.

So, we've put together some essential tools in probability, calculations of probabilities of compound events and expected value, and we'll need them. We also know to watch out for variability in the data, especially if we're doing a risk analysis for an event that happens only once or a few times.

OK, we're now ready to see how probability helps us to navigate the interplay of decisions and chance that characterizes an unfolding situation. That's our next stop, decisions in the face of uncertainty.

Decision Trees—Which Scenario Is Best?
Lecture 19

very decision that you make, whether for yourself or as part of an organization, has the potential to affect your overall reputation and prospects into the future. And, often, your control of that future is limited. Decision trees are an analytics tool for addressing just such scenarios—ones that unfold over time, interleaving the choices you make and the unpredictability of the rest of the world. In this lecture, you will learn how to solve a complex problem by using the relatively simple approach of decision trees.

Decision Trees

- With **decision trees**, sometimes you get to choose your path, and sometimes chance dictates it for you. A decision tree has many branches that, in the final analysis, will never come to pass. And, actually, that's the power of the technique—in generating all of the possible paths, and then navigating your way to the best decisions you can make in light of that collection of possibilities.

- For example, suppose that we run a mining company. We've had our eyes on two adjacent parcels of land that may have valuable deposits of minerals. Each deposit, if it's there, is worth $3.5 million, in fact.

- The owner is asking for $1.6 million for the pair, which is too expensive; after all, there's only a 30% chance that a parcel contains minerals. The owner agrees to sell us one of the parcels for $1 million. If we want, the owner will also sell us an option on the second parcel for $200,000. The option allows us to buy the other parcel for an additional $400,000.

- If we eventually want both parcels, this allows us to get them for the $1.6 million total. If the first parcel is a dud, we can walk away, losing the $1 million and the $200,000 option. But if the one parcel has minerals, the other one is 60% likely to have them, too. The owner is no fool, so without the option, we would have to pay more: $1.3 million to buy the second parcel.

- So, what should we do? We begin with three choices: Buy the first parcel and the option, buy the first parcel without the option, or forget the whole deal. Let's think about the top choice, where we buy the parcel and the option. We next find out whether the parcel contains minerals. That's a chance node with a 30% success rate.

- A chance node is represented by a circle on the decision tree; which way the situation evolves from that node is a matter of chance. Alternatively, squares represent decision nodes, where we're able to choose which option we prefer.

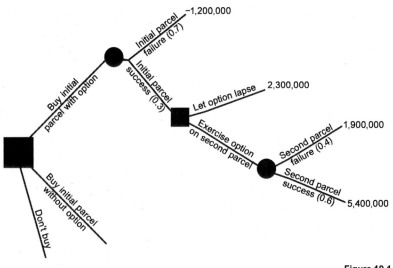

Figure 19.1

- We said that we'd walk away if the first property doesn't have minerals. There's nothing more to do in that case but lose the $1.2 million. But if the first parcel does contain minerals, we have another decision: whether to buy the second parcel.

- Because we bought the option, we can get the second parcel for only $400,000. Assuming that we buy it, we find out if the second parcel contains minerals as well. That's another chance node, with a 60% chance of success. Remember that if the first parcel had minerals, there's a 60% chance that the second one does, too.

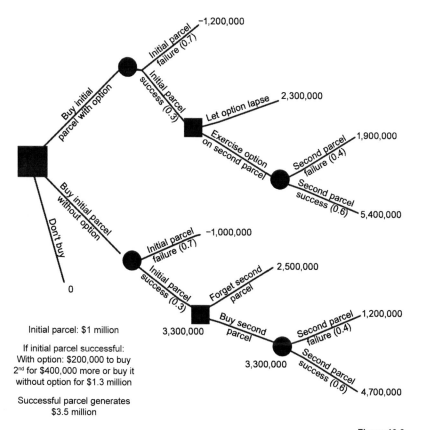

Initial parcel: $1 million

If initial parcel successful:
With option: $200,000 to buy
2nd for $400,000 more or buy it
without option for $1.3 million

Successful parcel generates
$3.5 million

Figure 19.2

- The payoff at the ends of the branches just adds up all of the losses and gains experienced as we move through the tree. Take, for example, the $1.9 million on the third branch down. What events occurred to get to that end? We bought the initial parcel and the option, had success with the first parcel, decided to pick up the option, but found nothing in the second parcel. So, we spent $1 million on the initial land, $200,000 on the option, and $400,000 for the second parcel for a total outlay of $1.6 million. But we made $3.5 million on the first parcel. The total profit is $3.5 - 1.6 = \$1.9$ million. All of the other payoffs work the same way.

- We can build the branch for buying the land without the option in exactly the same way. The whole tree is as shown in **Figure 19.2**.

- Let's see what we get when we **roll back** this tree, finding its optimal strategy. We work from the twigs on the right to the root at the left. For chance nodes, we find expected values. For decisions, we choose the most profitable option. We get what is shown in **Figure 19.3**.

- This tree says to buy the original parcel and the option on the second parcel, and if the initial parcel contains minerals, then exercise the option on the second parcel. On average, we'll profit by $360,000. This is not bad, given that our maximum outlay is $1.6 million— that's about a 22% return on investment. On the other hand, this enterprise is fairly risky. The most common outcome, which happens 70% of the time, is that you lose $1.2 million.

Utility Functions
- Decision trees, as we've been using them, have been assuming risk neutrality. That is, given two options, the one that gives the better average payoff is the more desirable. When the same situation is faced many times, this attitude makes sense.

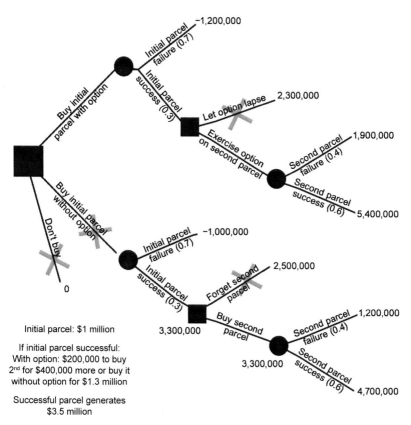

-1,200,000

Initial parcel failure (0.7)

Initial parcel success (0.3)

Buy initial parcel with option

Let option lapse 2,300,000

Exercise option on second parcel

Second parcel failure (0.4) 1,900,000

Second parcel success (0.6) 5,400,000

Buy initial parcel without option

Don't buy

0

Initial parcel failure (0.7) -1,000,000

Initial parcel success (0.3)

Forget second parcel 2,500,000

3,300,000

Buy second parcel

Second parcel failure (0.4) 1,200,000

3,300,000

Second parcel success (0.6) 4,700,000

Initial parcel: $1 million

If initial parcel successful:
With option: $200,000 to buy
2nd for $400,000 more or buy it
without option for $1.3 million

Successful parcel generates
$3.5 million

Figure 19.3

- The law of large numbers in statistics says that we're very likely to see overall results that are quite close to the expected value calculation when we repeat a situation many times. This is the principle that allows insurance companies to make money, even though they have less of an idea about your individual health or property than you do.

- Nevertheless, most people are not risk neutral, most of the time. Most people are risk averse. For example, suppose that someone offered you the following opportunity. The person flips a coin, and you call it in the air. If you're right, the coin flipper pays you $400,000. If you are wrong, you pay the coin flipper $200,000. Honestly, would you take the wager?

- Clearly, it favors you—on average, you'll win $100,000 on the flip. But half of the time, you'll be losing $200,000, and for most people, that downside risk is just too great. If you refuse the wager, you're risk averse in this situation. And the reason is actually fairly simple: The $400,000 you might gain isn't worth as much to you as the $200,000 that you might lose.

- To reflect these kinds of feelings in a decision tree, we need to replace each payoff in the tree with its corresponding **utility**. The idea is to define the utility of an outcome to a person in such a way that the expected value calculations we've been doing make sense, when applied to utilities.

- John Von Neumann and Oskar Morgenstern showed that, under four pretty weak assumptions, one could construct such a utility function for a decision maker. One assumption is transitivity—that if you like A better than B and B better than C, then you like A better than C.

- Another assumption is continuity. Suppose that you like A better than B, and B better than C. Someone offers you a choice. If you want, you can get B for sure, or you can enter a lottery. Win the lottery and you get A, but lose the lottery and you only get C. Continuity says that the person offering you the choice can make you indifferent between B for sure and the lottery if that person chooses the right probability, p, of winning the lottery.

- If the person offering the choice gave you a 100% chance of winning, you'd take the lottery, because you'd be sure to get your first choice, A. If the person gave you a 0% chance of winning the lottery, you'd take B, because the lottery is sure to give you your least-favorite outcome, C. And somewhere in between is a probability where you'd think B and the lottery were equally good.

- In the 1700s, Daniel Bernoulli suggested the following utility function. If you end up with a wealth of w, Bernoulli suggested using $\ln(w)$, the natural logarithm of w, as the measure of your utility. Don't worry about how to take natural logarithms by hand; the "ln" key on a calculator is happy to do this for you. The important thing about log functions is that they keep increasing as w increases, but they do so more and more slowly.

- In the wager example, half of the time you'd gain \$400,000 from someone, and the other half of the time you'd lose \$200,000. If you're risk neutral, you'd be a fool not to take this wager, because on average you win \$100,000.

- But let's say that your current wealth is \$400,000. Then, that person is offering you a chance to either double your current wealth or cut it in half. Using Bernoulli's log utility, your utility if you keep your \$400,000 would be $\ln(400{,}000)$, which comes out to be 12.9. If you win the wager, you'd end up with \$800,000, and $\ln(800{,}000)$ is 13.59. On the other hand, if you lose, you end up with \$200,000, and $\ln(200{,}000) = 12.21$.

- Looking at utilities, if half the time you get a utility of 13.59 and half the time you get a utility of 12.21, your expected utility from the wager is the average of these two, which is 12.9. And that was the utility that you started with. So, with Bernoulli's utility, you'd be indifferent between taking the wager and keeping your current money.

- If you had more money already on hand, the bet would look better to you. If you had a million dollars to start, the wager would leave you with either $800,000—utility ln(800,000) = 13.59—or with $1.4 million—utility ln(1.4 million) = 14.15. The average of these is 13.87, while keeping your original million gives you a utility of ln(1,000,000)= 13.81. Of course, 13.87 is better. In logical terms, with more money in the bank, the pain of a $200,000 loss is less severe.

- People have proposed other utility functions for risk aversion, but it's also possible to build a personal utility function to mirror any rational decision maker. Essentially, you take the lowest payoff on the tree, the highest payoff on the tree, and you offer the decision maker a lottery ticket that will pay either this lowest payoff or this highest payoff. What's that ticket worth? It depends on what the probability of getting the high payoff is (p). A high-p ticket is worth a lot more than a low-p one.

Important Terms

decision tree: A structure consisting of branching decision nodes and chance nodes used to analyze sequential decision making in the face of uncertainty.

rollback: The procedure of evaluating a decision tree by computing expected values at each chance node and choosing the most-attractive options at each decision node.

utility: In essence, "happiness points." Utilitarian models in economics posit that decision makers act so as to maximize their personal utility. A utility measure satisfies a number of sensible rules.

Suggested Reading

Dudley and Buckley, "How Gerber Used a Decision Tree in Strategic Decision-Making."

Ragsdale, *Spreadsheet Modeling & Decision Analysis*.

1. Some decision trees allow the possibility of doing research. This will always involve the addition of at least one additional decision node to the tree and at least one additional chance node.

    a) What decision would the decision node reflect?

    b) What chance event would the chance node reflect?

    c) If the research is conducted, the probabilities for events beyond that point in the tree may change. Why?

Answers:

    a) It will reflect whether you should do the research.

    b) What was the outcome of the research? This will always be in the form of "research says" or "research suggests" or "research predicts," never in the form of "research right" or "research wrong." That's because at the time that you get the research results, you don't yet know if they are correct!

    c) Because they are conditional probabilities. The purpose of research is to give you a better idea of how some future chance event will turn out. So, the research turning out a particular way will modify the chance that that future event occurs.

2. In decision theory, one sometimes computes the expected value of perfect information (EVPI), which tells you how much, on average, perfect information would be worth in the particular decision tree. Perfect information is knowing the outcome of every chance event in the problem before making any decisions.

Consider the following problem: A shoebox is equally likely to contain a $1 or a $10 bill. Someone gives you the choice of either taking the contents of the box or taking $5 from him or her.

a) How much do you make, on average, if you take the box?

b) Suppose that you knew what was in the box before choosing. This would be perfect information. Show that you would make $7.50 on average.

c) EVPI is how much better you would do with the perfect information than you could do on you own. Show that in this problem, the EVPI is $2.

Answers:

a) $5.50.

b) 50% of the time, you find that the box has the $10 bill, so you take it. 50% of the time, you find that the box has the $1 bill, so you take the $5 instead. Your average winnings are 0.5(10) + (0.5)5 = $7.50.

c) $7.50 − $5.50 = $2.

Decision Trees—Which Scenario Is Best?
Lecture 19—Transcript

Every decision that you make, whether for yourself or as part of an organization, has the potential to affect your overall reputation and prospects into the future. And often, your control of that future is limited. You'll exercise control over some aspects of the situation, but in many cases, some things are just out of your hands. Decision trees are an analytics tool for addressing just such scenarios, ones that unfold over time, interleaving the choices that you make with the unpredictability of the rest of the world.

Corporate image is incredibly important, and never more so than when you're marketing a product for infants. Take Gerber Products, Inc. They're probably most famous for the familiar Gerber Baby logo that appears on their jars of baby food, adorable. Still, this is deliberate, purposeful. Gerber has worked hard to cultivate its public image. But in 1986, change intervened, and they committed what many people consider to be the textbook example of a PR disaster. And 12 years later, as Christmas approached in 1998, Gerber was again on the hot seat. It was the focus of governmental review and media scrutiny, with the issue being the last thing that they wanted—baby safety. Here's what happened.

In 1986, the Food and Drug Administration received about 140 reports of glass shards in jars of Gerber baby food. The FDA looked into it and confirmed 21 cases of product tampering. These were probably the work of isolated individuals. But a week before, Beech-Nut, Gerber's competitor, had faced a similar situation, and Beech-Nut had promptly recalled its baby food from the store shelves. Gerber didn't. To many, Gerber seemed cavalier about the possible dangers. And when the state of Maryland responded by banning Gerber peaches in light of the glass shard scare, Gerber sued the state. From the point of view of reputation, it was not a pretty story. Gerber eventually recovered from the debacle, but it didn't forget it.

Then, in 1998, Gerber faced a second PR crisis. This one had its roots about 30 years earlier—polyvinyl chloride, or PVC, a remarkably useful plastic. You probably know the stuff; you've seen it around the house in plastic water pipes, food storage containers, toys, lots of places. It's also important in a lot of medical supplies. And, for some uses, it needs to be softened, to be made more pliable, and this is often done with a family of chemical plasticizers called phthalates.

Phthalates had been around and in use for about 30 years, and no one seemed overly concerned about their safety. But that was about to change. Enter the environmental group Greenpeace. Greenpeace had conducted research on phthalates and had concluded that they were carcinogenic in lab rats. It also said that phthalates would leach out of the plastic over time. Greenpeace was particularly concerned about products that children would chew or suck on. No stranger to the PR game, Greenpeace timed its press release to come out just before the Christmas season in 1998. The media picked up the story, 20/20 did a segment on it, and with Gerber having about 75 different products containing phthalates, the company was feeling the pressure to respond.

It got worse. Although phthalates had no known health issues, the U.S. Consumer Products Safety Commission, the CPSC, felt compelled to release a new statement expressing new doubts. As companies like Mattel and Disney began to pull back on products with phthalates, the spotlight focused on Gerber. About a month before Christmas, the CPSC told Gerber that were going to release a statement advising parents on the potential dangers of phthalates, and that Gerber would be one of the companies named in the statement. And that's why Gerber was on the hot seat. And in deciding how to handle it, one of the tools that they used was a relatively simple approach to a complex problem, a decision tree. In today's lecture, we're going to develop the same tool, apply it to this same problem, and see what it has to say.

To me, a decision tree always feels like a choose-your-own-adventure book. They're kinda fun. You start on the first page and the story advances a bit; then comes a decision point, and the book says something like, you're standing on the steps of the brownstone manor. If you want to ring the doorbell, turn to page 16. If you want to return home, turn to page 48. If you want to see who comes by and wait, turn to page 11. And then you turn to that page, and the story continues, based on the choice that you made. A low-tech hyperlink, I suppose. Anyway, decision trees work like that, except that in addition to the kinds of pages I just described, there'd be others where you don't get to choose. Chance does.

It's as if the book said, roll a die. On a 1 through 3, the weather stays fair; go to page 18. On a 4 or 5, it begins to rain; go to page 22. On a 6, it begins to snow; go to page 6. Obviously, in a book like this, there are a lot of pages that you'll never read, and that's like a decision tree, too. A decision tree has lots and lots of branches that, in the final analysis, will never come to pass. And actually, that's the power of the technique, in generating all of the possible stories, and then navigating your way through to the best ones to see the best decisions that you can make in light of that collection of possibilities.

Let's see how this works for Gerber. Given their situation, it has to decide whether to adopt a proactive or reactive stance to the situation. If proactive, it would aggressively address the issue by announcing that it would discontinue the use of phthalates in its products before the CPSC report was released. If reactive, it'd wait, see what the report said when it came out, and then decide how to respond. In either case, when the report was released, Gerber would know its contents. They thought that either the report would be a recall on phthalate-containing products, or else it would just be an advisory expressing concern. But whatever the report said, all of this would result in an impact on Gerber's profits, either for good or ill.

So how do we come to grips with this in a decision tree? Like this. Laid out like this, decision trees are surprisingly easy to understand. Read from left to right. The circles and squares that you see are nodes, but they're two different types. Squares represent decision nodes; that means we're able to choose which option we prefer. In our tree, Gerber begins by deciding on a proactive or reactive stance.

Circles, on the other hand, are chance nodes. They represent matters where we don't have control. In this particular tree, it's actually pretty simple, and once Gerber makes the original decision, things play out by chance, without any more choices on its part. The CSPC will do what it does, and then the consumer reaction will either be relatively good or relatively bad. The tree is also simple in that each node is a binary one, there are only two options leaving each node.

Still. Even though this tree is simple, we can use it to point out some properties that are true about decision trees in general. Trees start with a single node, the root. It may be a decision, like in this problem, or it may be a chance event, like looking out your window to see the weather before deciding how to dress for the day. Second, starting from the root and moving rightward, we are tracing out a possible sequence of events in the scenario. If you stop anywhere along the path, from the root on the left to the twig on the right, everything to the left of your current position is what you already know, and everything to the right is what you don't know, yet. If you like, the left is history; the right is mystery. This is an important thing to keep in mind if you're building a tree. For example, imagine that Gerber had a mole in the CPSC, so that it knew what the report was going to say before needing to make its decision about its stance. Then our tree would begin with the chance node for the CPSC report, followed by the Gerber decision of its stance, a better situation for Gerber, but not what happened.

OK, we've got the structure of the tree, but before we can do anything useful with it, we need two more kinds of information. First, we need to know how good or bad things end up being in any of the eight possible ways that this story can unfold. Second, all the chance nodes are not created equal. We need to know how likely each of the possible outcomes is for each of the chance nodes. Not surprisingly, Gerber had to estimate these. And here's what they came up with.

First possibility: Gerber announces removing phthalates from all of the products, and the CSPC report only expresses concern. In this case, Gerber figured that it would probably gain revenue from a positive public reaction to its proactive behavior, particularly when compared to slower-reacting competitors. So, say, a million dollars in increased revenue. Then again, there was some chance that the media attention could depress sales in spite of Gerber's actions. Again, let's say, a million dollar decrease in revenue. Gerber estimated an 80% chance of the good public reaction in this case. Let's put that on the tree too. Of course, it was also possible that the CPSC would order recalls. If they did, Gerber's proactive approach might let them keep their current sales about a 25% chance of that, but they thought it more likely they'd lose revenue to the tune of $1.25 million.

And if Gerber chose to respond reactively? The CPSC expressed only caution, Gerber thought it 25% likely they could maintain their current sales, but 75% likely that they'd lose $2 million. And that leaves the worst-case scenario, where Gerber responds reactively and the CSPC recalls phthalate-containing products. Surprisingly, Gerber thought there was still a 20% chance that they could actually gain a half a million dollars in this case, since they felt themselves more prepared for this nasty outcome than their competitors. Still, they thought it much more likely that they'd experience substantial losses, perhaps 5 million. Let's put all of this information on the tree.

This is the tree that Gerber used. But before going any further, I'd like to take a few minutes to address some concerns you may have about our work so far. First, perhaps it makes sense to say that Gerber is considering only two different stances. It may also be reasonable to limit the CPSC reaction to either recall or concern, since a recall may have serious impact on Gerber, while concern might only cause only minor effects. But surely, the last nodes in the tree aren't realistic. Surely there is a whole range of possible revenue, not just a pair for gain and loss numbers.

You're right, and in some work with decision trees, after a binary node with gain and loss, we might see another node, a fan with a whole wide range of values. Each spine in that fan would lead to a different payoff, and each would have its own probability. Saying this more formally, there'd be a probability distribution for all the possible different revenues.

But, the way that we're going to evaluate this decision tree is based on expected values; we're going to be looking for the best course to follow on average. So we could view our tree's upper payoff as saying, if we're proactive and the CPSC is only concerned, we think there's an 80% chance that things will go relatively well. If they go relatively well, on average, we'd expect a gain of about a million. That $1-million figure comes from evaluating the fan of all possible revenues. In fact, it may have come from a decision tree where many different favorable customer-reaction scenarios are considered.

OK, second concern. The probabilities on this tree are obviously subjective probabilities. We have to assume that the people at Gerber know their own business, but how much can we trust these figures? And at present, we don't even have Gerber's guess for how likely a negative CPSC report is. When we get one, how much can we trust that subjective probability? Well, that can be a difficult question to answer, but I can address a related question. How much do we need to be able to trust them? That is, how sensitive is our best decision to errors in these probabilities? This is a question we'll return to after we complete our original analysis. To get started, though, let's just take what we have, and further assume that Gerber thinks it's about equally likely that the CPSC report will be recall or concern. Then, here's our completed tree.

OK, we're ready to do the work to find our optimal strategy. At present, we're defining optimal as the strategy that results in the highest expected revenue for Gerber. Whichever option gives the highest revenue on average, that's the one we want. Such an evaluation is called risk neutral. The company is concerned only about the average result. It neither shies away from risk, nor is attracted to it. For a situation that's faced many times, this is actually a sensible attitude. But, there are lots of situations in which individuals or firms may face a situation only one time, and there they may demonstrate risk aversion, or more rarely, risk love. We'll talk about how to account for such attitudes in a few minutes. But for now, let's keep it simple and assume risk neutrality.

The key to analyzing any tree is to start the analysis in the right place. In their book *Thinking Strategically*, Avinash Dixit and Barry Nalebuff summed it up beautifully: "To look forward, you must reason backward." Start at the right-hand side of the tree, because at the rightmost nodes of the tree, have almost everything being their history. Once you figure out what to do there, you can roll back your analysis to the earlier nodes. Let's see how it works.

Here's the top node of the tree, where Gerber is proactive and the CPSC expresses only concern; 80% of the time, they make a million and 20% of the time, they lose a million. Our work from last lecture lets us easily figure out what the average revenue will be in this situation. Expected value, remember, is just probability times payoff, added up over all the cases. So here we get $0.8(1,000,000) + 0.2(-1,000,000)$, which comes out to be $600,000. So in this scenario, Gerber's expected revenue is $600,000. We can do exactly the same kind of calculation on all the other rightmost terminal nodes, too, and when we do, it looks like this.

As you can see, I've written the results of each expected payoff calculation over the corresponding node. For example, the bottom entry shows that if Gerber responds reactively and the CPSC orders recalls, Gerber's average losses will be $3.9 million. And now we just do the same thing again, this time for the CPSC nodes. For example, if Gerber responds proactively, then there's a 50% chance that the CPSC expresses concern, and a 50% chance of a recall. But we know that if they are concerned, Gerber expects to make $600,000 in additional revenue, while if it recalls, Gerber expects to lose $937,500.

Again, do an expected value calculation, probability times payoff, added up over all of the possibilities. $0.5(600,000) + 0.5(-937,500) = -168,750$. Half of the time you get the good payoff; half of the time you get a bad one. We just multiplied probability times payoff, added the results, and got our answer. We do the same kind of work on the lower node. Half the time they lose $1.5 million, and half the time they lose $3.9 million, so on average, they lose $2.7 million. We'll put these values on the tree too.

OK. We've worked our way back to the first node in the tree. Time for another averaging? If you think about it for a moment, obviously not. At chance nodes, the outcome is random. That's why we had to take expected values. But at decision nodes, obviously, you get to choose which path to take. Here, if Gerber is proactive, it loses about $170,000 on average. If it's reactive, its average losses are about 15 times as great. If what it cares about is expected revenues, its choice is pretty obvious; the company should be proactive. And that's was what Gerber did in real life. To finish the tree, we'll just barricade the rejected choices and put the payoff on the selected choice over the decision node, like this.

A more complicated tree is handled in just the same way. Begin at the end of the tree. For a chance node, compute an expected value. For a decision node, choose the best option and barricade the rest. Keep going this way until you reach the root node. When you're done, you'll have a selected option at each decision node. If and when you ever reach that node, follow the open edge. On average, you can't do better than this.

You can also use the resulting tree to get an idea of the risk that you face, in terms of possible outcomes and their probabilities. Every leaf you can reach from the root node along an open path represents a possible outcome to your situation, and you can find the probability of that outcome by multiplying together all of the probabilities along that path. The top path in our Gerber tree leads to the happy outcome of them gaining $1 million. The probability of this outcome is 0.5×0.8, or 40%.

The Gerber problem looks relatively simple, but the approach can be used to sort out much more tangled situations. For example, suppose I run a mining company. I have my eyes on two adjacent parcels of land that may have valuable deposits of minerals. Each deposit, if it's there, is actually worth $3.5 million, in fact. The owner is asking $1.6 million for the pair, which is too rich for my blood, after all, there's only a 30% chance that a parcel contains minerals. But, he agrees to sell me one of the parcels for $1 million. And if I want, he'll also sell me an option on the second parcel for $200,000. The option is to let me buy the other parcel for an additional $400,000.

So if I eventually want both parcels, this lets me get them for the $1.6 million total. If the first parcel is a dud, I can walk away losing the $1 million and the $200,000 for the option. But if the one parcel has minerals, the other one is 60% likely to have them too. The owner, no fool, says that, without the option, I'll have to pay $1.3 million more if I want to buy the second parcel. So what should I do?

This problem is considerably more complicated than Gerber's. Let's take a look at it in a tree. To begin with, we have three choices: buy the first parcel and the option, buy the first parcel without the option, or forget the whole thing. Let's think about the top choice, where we buy the parcel and the option. We next find out whether the parcel contains minerals. That's a chance node with a 30% success rate. We said we'd walk away if the first property doesn't have minerals. There's nothing more to do in that case but lose the 1.2 million. But if the first parcel does contain minerals, we have another decision, whether to buy the second parcel.

Since we bought the option, we can get the second parcel for only $400,000. Assuming that we buy it, we find out that the second parcel contains minerals as well. That's another chance node, with a 60% chance of success this time. Remember that if the first parcel had minerals, there's a 60% chance that the second one does, too. The payoff at the ends of the branches just add up above all of the losses and gains experienced as we move through the tree. For example, see that 1.9 million on the third branch down? What events occurred to get to that point? We bought the initial parcel and the option, had a success with the first parcel, decided to pick up the option, and found nothing in parcel 2. So, we spent $1 million on the initial land, $200,000 on the option, and $400,000 for the second parcel, for a total outlay of $1.6 million. But, we made $3.5 million on the first parcel. Total profit? $3.5 − 1.6 = $1.9 million. And all of the other payoffs work in the same way. We can build the branch for buying the land without the option in exactly the same way. In fact, when you're done, the whole tree looks like this.

Let's see what we get when we roll back the tree, finding its optimal strategy. It's straightforward. We work from the twigs on the right to the root at the left. For the chance nodes, we find the expected values. For the decisions, we choose the most profitable option. And here's what we get. What does it say? Buy the original parcel and the option on the second parcel. If the initial parcel contains minerals, then exercise the option on the second parcel. On average, you'll profit by $360,000. Not bad, given that your maximum outlay is $1.6 million, that's about a 22% return on investment. On the other hand, the enterprise is fairly risky. The most common outcome, which happens 70% of the time, is that you lose $1.2 million.

Well, we mentioned the idea of risk before. Decision trees, as we've been using them, have been assuming risk neutrality. That is, given two options, the one that gives the better payoff on average is more desirable. Now, when the same situation is faced many times, this attitude makes sense. The law of large numbers in statistics says that we're very likely to see overall results that are quite close to the expected value calculation when you repeat the situation many, many times. This is the principle that allows insurance companies to make money, even though they have less of an idea about your individual health than you do.

Nevertheless, most of us aren't risk neutral, most of the time. Most of us are risk averse. For example, suppose I offered you this opportunity. I flip a coin, and you call it in the air. If you're right, I pay you $400,000. If you are wrong, you pay me $200,000. Honestly, would you take this wager? Clearly, it favors you. On average, you'll win $100,000 on the flip. But half of the time, you'll be losing $200,000, and for most of us, that downside risk is too great. If you'd refuse the wager, you're risk averse in this situation. And the reason is actually fairly simple; the $400,000 that you might gain isn't worth as much to you as the $200,000 that you might lose.

To reflect these kinds of feelings in a decision tree, we need to replace each payoff in the tree with its corresponding utility. The idea is to define the utility of an outcome of a person in such a way that the expected value calculations we've been doing make sense, when applied to utilities.

John Von Neumann and Oskar Morgenstern showed that, under four pretty weak assumptions, one could construct such a utility function for a decision maker. The assumptions were things like transitivity, that if you like A better than B and B better than C, then you like A better than C. Another is continuity. Suppose that you like A better than B, and B better than C. I offer you a choice. If you want, you can get B for sure, or, you can enter a lottery. Win the lottery and you get A, but lose the lottery and you only get C.

Continuity says that I can make you indifferent between B for sure and the lottery if I choose the right probability, p, of winning the lottery. If I gave you a 100% chance of winning, you'd take the lottery, since you'd get your first choice, A. If I gave you a 0% chance of winning the lottery, you'd take B, since the lottery is sure to give you your least favorite choice, C. Somewhere in between is a probability where you'd think B and the lottery were equally good.

Let me demonstrate the idea of a utility by using the one originally suggested by Daniel Bernoulli back in the 1700s. If you end up with a wealth of w, Bernoulli suggested using $\ln(w)$, the natural logarithm of w, as the measure of your utility. There are some interesting theoretical arguments for this choice, but for now, we'll just see how it works. And let's not worry about taking natural logarithms by hand; the ln key on a calculator is happy to do this for you. The important thing about log functions is that they keep on increasing as w increases, but they do so more and more slowly. A moment ago, we looked at a wager where half of the time you'd gain $400,000 from me, and the other half of the time you'd lose $200,000. We said that if you're risk neutral, you'd be a fool not to take this wager, since on average, you make $100,000.

But let's say that right now you have $400,000 to your name. That's your current wealth. Then I am offering you a chance to either double your wealth or to cut it in half. Using Bernoulli's log utility, we'll see what happens. Your utility if you keep your $400,000 would be $\ln(400,000)$, which comes out to be about 12.9. If you win the wager, you'd end up with 800,000, and $\ln(800,000)$ is 13.59. On the other hand, if you lose, you end up with $200,000, and $\ln(200,000)$ is 12.21.

So, looking at utilities, half of the time you get a utility of 13.59, and half the time you get a utility of 12.21; your expected utility from the wager is the average of these two, which is 12.9. And that was the utility you started with. So, with Bernoulli's utility, you'd be indifferent between taking the wager and keeping your current money. In fact, if you had more money already on hand, the bet would look better to you. If you had a million dollars already on hand, the wager would leave you either with $800,000, utility $\ln(800,000) = 13.59$, or with 1.4 million, utility $\ln(1.4 \text{ million}) = 14.15$. The average of these is 13.87, while keeping your original million gives you a utility of $\ln(1,000,000) = 13.81$; 13.87 is better. In logical terms, the more money in the bank, the less the pain of the $200,000 loss is.

People have proposed other utility functions for risk aversion, but it's also possible to build a personal utility function to mirror any rational decision maker. Essentially, you take the lowest payoff on the tree, the highest payoff on the tree, and you offer the decision maker a lottery ticket that will either pay this lowest amount. Question? What's the ticket worth? Well, it obviously depends on the probability of getting the high payoff, call it p, a high-p ticket is worth a lot more than a low-p one. To see how we use this lottery-ticket idea to assign utilities, go back to the Gerber problem. The highest payoff was a million, the lowest was −$5 million. So, we'd ask Gerber how high p has to be before they'd be indifferent between a lottery ticket and a payoff on the tree. The first payoff on the tree was 1 million. Well, since the ticket either gives 1 million or a 5 million loss, Gerber would obviously only take the ticket if $p = 1$, if it always paid a million. Similarly, the ticket would always be better than the guaranteed $5 million loss, unless $p = 0$. Any higher p gives at least some chance of dodging the $5-million loss and making the million instead. So the utility for $1 million is 1, and the utility for −$5 million is 0.

But the other payoffs depend on Gerber's risk aversion. For example, if they are proactive, the CPSC only voices concern, and the public responds badly, the payoff is −$1 million. To turn this into a utility, we asked Gerber this; you either have to pay $1 million or take this lottery ticket. Sometimes the ticket makes you a million; sometimes it loses you $5 million. How big a chance of winning does that ticket have to have before you don't care whether you pay the sure 1 million debt or take your chances on the ticket?

This might take Gerber some time to decide. Maybe they'd decide that a ticket with an 80% chance of winning would be just good enough to tempt them; 1 time in 5, they pay $5 million, but 4 times in 5 they make a million. If so, the utility for this payoff is 0.8. Someone who was risk neutral would be willing to go for a p-value of $^2/_3$, but Gerber is more risk averse than that. We'd ask Gerber this same question for each payoff in the tree and replace each payoff with the p-value that they give us. After this is done, we could roll the tree back just as we did earlier today, and the answer would reflect how Gerber should behave in light of their own feelings about risk. Clever, eh?

Well, to wrap things up today, I want to return to the other issue we raised earlier, the matter of subjective probabilities, or even the payoffs, might be called into question. I said that we'd look at the question of how much our decision hinges on the accuracy of these values. Let's look at it with the Gerber tree, where we assumed a 50% chance that the CSPC would issue a recall. How sensitive is Gerber's decision to this number? There's a lot of software programs that will actually do such an analysis for you, but it's easy to do by hand. Let's look at the Gerber tree again.

Here, I've stopped the rollback right before we evaluate the CPSC reaction. If that 50% recall figure is wrong, let's replace it with a probability p. Now we roll back the respond proactively branch and also the respond reactively branch. We do this in a way that we always do, payoff time s probability, added up over all the cases. On the top branch, this becomes, $600,000(1 - p) + (-937,500)p$, and so on. For the bottom branch, you get this.

Responding proactively is going to be better if the top branch payoff is better than the lower branch one. Using the two payoffs that we just computed, this means that proactive is better if this inequality holds, $600,000 - 1,537,500p \geq -1,500,000 - 2,400,000p$. This is just a linear inequality, and it's easy to solve. Like this; $2,100,000 \geq -[862,500]p$. So we get that p has to be at least -2.43. Remember, the probabilities are always between 0 and 1. This -2.43 tells us that Gerber doesn't need to have a clue of the prediction of the CPSC, if the other figures are reliable. Even with no chance of recall, Gerber still does better, on average, with a proactive stance. This kind of analysis could be combined with a utility function model of Gerber's risk profile to get a more complete picture of the situation, but it's a decision for a proactive stance, and it's so far from a close call that it seems likely that a more complete analysis wouldn't change it.

Oh, the actual history? Gerber took a proactive stance, as I said. The CPSC issued a statement saying, in part,

> One pacifier and two models of feeding bottle nipples manufactured by the Gerber Products Company contained a related phthalate. Gerber has stopped manufacturing these products and is removing

phthalates from all future production. Gerber directed retailers to remove the phthalate-containing pacifier and nipples from store shelves.

Eventually the FDA, and the CPSC, and other organizations approved the continued use of phthalates. In the meantime, Gerber avoided another PR nightmare.

Decision theory lives in an interesting and important neighborhood of analytics. They take on stochastic events straight on, not just random errors, but identifiable events. At the same time, decision theory is an optimization technique; its job is to identify the best set of decisions. This is the first time that we've taken a close look at such a combination, a stochastic optimization technique, but it won't be the last.

You may be analyzing a situation that you face again and again from the perspective of risk neutrality. You may be finding your way in an uncommon problem, where your personal degree of risk aversion needs to be considered. Either way, decision trees offer a powerful way to reason forward by looking backward. Turning to decision trees is one of the best decisions you can make.

Bayesian Analysis of New Information
Lecture 20

T he chance that something is true changes as new and better information becomes available. This simple idea is the key to Bayesian analysis, named after mathematician and Presbyterian minister Thomas Bayes. In this lecture, you will learn that this type of analysis deserves to be used far more widely than it is. You will also learn that once you understand conditional probabilities, Bayesian probability is available to improve many kinds of decisions. In fact, when you have a conditional probability and need its reverse, then Bayesian probability is the answer.

Bayesian Analysis
- **Bayesian analysis** is named after the Englishman Thomas Bayes, an 18th century mathematician and Presbyterian minister who proposed a special case of what is now called Bayes' theorem. It's a powerful tool, but the idea behind Bayesian analysis is really pretty simple. The chance that something is true changes as new information becomes available. Bayes' theorem tells you how to compute the new probabilities.

- Let's say that your friend shuffles three playing cards—two red cards and the ace of spades—and deals them facedown in a row. Your card is the one on the left, and in a moment, your friend will reveal the card on the right. Do you have the ace of spades?

- The ace is equally likely to be in any of the three positions, with a 1/3 chance of each. This is called the **prior probability distribution**— it's the distribution before any new information comes in. So, your chance of having the ace is 1/3.

- But, now, your friend reveals the rightmost card. Whatever happens, you're going to learn something. Your friend turns over the ace, so that means that your card is not the ace. The new information just reduced your chances of having the ace from 1/3 to zero.

- And even if your friend turned over a red card, you'd still have new probabilities. The ace is now equally likely to be either of the two remaining cards, so your chance of having the ace is now 1/2.

- Alternatively, let's say that the card your friend turns over is *not* the ace, meaning that the ace is still out there, and it's equally likely to be either of the two remaining cards. So, your chance of having the ace went from 1/3 to 1/2. And if your friend had revealed the ace, of course, the change would be even more dramatic. Your chance of having the ace would have plummeted from 1/3 to 0.

- That's basically what Bayesian probability does. We begin with a prior probability distribution—a specification of the probability of each of the possible outcomes. Then, we get some additional information. Bayes's theorem then allows us to compute the updated probability for each possible outcome, the so-called posterior probabilities. Our original information is combined with the new information to tell us how our original probability figures should change.

Applying Bayesian Analysis

- Understanding Bayesian probability is important because failing to do so can lead to a lot of bad decision making, including racial prejudice, medical misdiagnosis, and bad public policy. The reason is that most people who are untrained in the mathematics of probability fail to distinguish between a conditional and its reverse.

- There's a medical condition called phenylketonuria (PKU) that results from an error in phenylalanine metabolism in infants. It results in mental retardation if not detected, but there is a screening that is 100% successful in spotting children with the defect, and there's a treatment for the condition. Screening is mandatory in all 50 states, so if your child has PKU, you're going to know.

- This test spots 100% of children with PKU, but what does it do with children who don't have PKU? After all, if we simply say that all babies have PKU, we have a test that spots 100% of the children that have the condition. We'll just be wrong about everyone else. Our test has no **specificity** at all. The missing information is that there are two common tests for PKU, and the better one is 98% likely to identify a PKU-free child as being PKU-free.

- How afraid should you be when your baby tests positive, with a test that is somewhere between 98% and 100% accurate? The question we want is not how likely a sick child is to test positive—that's 100%. The question is how likely a child who tests positive is to be sick. Let's use Bayesian analysis.

- Somewhere around 1 child in 15,000 is born with PKU. Imagine 15,000 children. On average, one of these children is going to be unlucky enough to have PKU—and our test is guaranteed to catch it. But there are 14,999 kids who don't have PKU, and the 98% reliable test, on average, will report that about 2% of them appear to have PKU. That's about 300 children. So, when this test that is over 98% reliable says that your child has PKU, the odds are 300 to 1 against your child having the condition. The rarer the disease, the more likely that a positive test result is a false positive.

- More people are killed by bee stings each year than by shark attacks. And an appropriate reaction is, so what? Because we don't want "probability of shark given death"; we want "probability of death given shark." If you are hip-deep in the ocean and a bee is approaching you from the shore as a shark closes in on you from the sea, run toward the bee.

- Most accidents occur within 15 miles of home. Of course they do—most driving occurs within 15 miles of home. The value that we need is $P(accident \mid near\ home)$, not $P(near\ home \mid accident)$. Numbers and probabilities have a cachet—they sound irrefutable. But even if the numbers are correct, you have to make sure that you're looking at the right probabilities.

- There has been congressional testimony that has examined whether marijuana is a gateway drug to cocaine and heroin. One argument presented was that the overwhelming majority of cocaine users used marijuana first. The figures you see vary, but many of them put the fraction of cocaine users who previously used marijuana at around 95%. For many people, this is a strong argument for keeping marijuana use illegal.

- But none of this makes the argument it purports to make. It's not just the logical problem of post hoc, ergo prompter hoc, nor even that an addictive personality would probably be exposed to marijuana before cocaine. It's that, if the argument were sound, an even stronger one could be made against milk, which more than 95% of cocaine addicts used, and usually from a very early age.

- What went wrong? Again, the problem is the reversed conditional. The question isn't what fraction of coke users started with marijuana. The question is what fraction of marijuana users go on to cocaine—which statistics have shown is about 25%. This is not a small number, but it's a far cry from 95%. What is amazing about the 95% statistic is how small it is; if correct, it means that 5% of coke users just vault over marijuana.

- A similar problem comes up with statistics about the fraction of highway fatalities that involve drunk drivers. The president of Mothers Against Drunk Driving quoted this statistic in her congressional testimony: According to the Department of Transportation, about 31% of traffic fatalities involve drunk drivers.

- But, again, knowing nothing else, this statistic tells us very little. After all, it means that over two-thirds of fatal accidents involve only sober people. If anything, that seems to imply that we should get liquored up before getting behind the wheel.

- This reasoning is, of course, nonsense. Again, we're looking at the reversed conditional. We don't need to know the probability that a fatal traffic accident involves a drunk. We need to know the probability that a drunk is involved in a fatal accident—as compared to a sober driver.

- Doing the Bayesian work leads to a conclusion that those above the legal limit for blood alcohol are at least 13 times more likely to be involved in a fatal traffic accident than a sober driver. And *that* is the statistic that's relevant to the drunk driving discussion.

- Many kinds of bigotry and prejudice are fueled, in effect, by confusing a conditional and its reverse. For example, when people think of terrorist attacks in the United States, 9/11 instantly springs to mind. And, overwhelmingly, the people involved in that attack were young, male, Arab, and Muslim. The hijackers themselves all fit the description and were probably all in their 20s and 30s. This means, of course, that if you know that you are looking at a 9/11 hijacker, you have a pretty good idea of what he looks like.

- But the problem, of course, is that most of the time, people aren't looking at a known 9/11 hijacker. They may be looking at someone who is young, male, Arab, and Muslim, though. And it's when you think about situations like these that you can realize how dangerous it is to confuse "*A* given *B*" and "*B* given *A*," because the vast majority of young male Arab Muslims are not terrorists, and thinking that they are is a cause for terror of a whole different kind.

Important Terms

Bayesian analysis: A technique for revising the probabilities of outcomes in light of new information.

prior probability distribution: Bayesian inference modifies the probabilities of events in light of new information. The probability distribution in effect before new information is provided is called the prior probability distribution. The new distribution, accounting for the new information, is called the posterior probability distribution.

specificity: The probability that an observation is classed as not having the trait of interest, given that it does not, in fact, possess that trait. Contrast to **sensitivity**.

Suggested Reading

Grinstead and Snell, *Introduction to Probability*.

Stone, "In Search of Air France Flight 447."

Questions and Comments

1. How does Bayesian probability relate to drug testing? What procedures do you think should be followed when someone tests positive for an illegal substance?

    Answer:

    It's similar to the PKU example in the lecture. Assuming that illegal drug use occurs in a relatively small fraction of the population, the chances of a false positive, even with a test of high accuracy, are considerable. If a person tests positive, following up with a second test is appropriate before other action is taken.

2. Think of several examples where people commonly make the error of reversing conditional probabilities—that is, they confound the chance of *A* when *B* is true and the chance of *B* when *A* is true.

Answer:

Examples are legion. Most follow the model of "because most A's are B, most B's are A" and can be made even worse if the A category is limited to one's own individual experience. For example, if a person is engaged in criminal activity, he or she will probably not want to submit to a polygraph. From this, we conclude (fallaciously) that someone who doesn't want to submit to a polygraph is probably engaged in criminal activity. If a proposed remedy to a problem works (medicine, treatment, prayer, etc.), then after the treatment, the problem will usually turn out okay. We conclude (again, fallaciously) that if such problems usually turn out okay after the remedy is employed, then the remedy probably works.

Bayesian Analysis of New Information
Lecture 20—Transcript

The chance that something is true changes as new and better information becomes available. Simple idea—but it's the key to Bayesian analysis. I want to start with an example, and I think that you'll see that the mathematics for it deserves to be known and used far more widely than it is.

On June 1, 2009, Air France Flight 447 left Rio de Janeiro, Brazil, on its way to Paris. About four hours later, the plane, an Airbus A330-203 was over the mid-Atlantic, having just crossed the equator. It was never heard from again. The plane went down, killing all 216 passengers and 12 flight crew, making it the deadliest accident in Air France history and the highest death toll on any A330-203 crash anywhere in the world. Although a search was quickly launched, it was clear that there was no hope of survivors, and indeed, it took five days before debris and bodies from the crash were located. Nine days after the crash, a total of 50 bodies were recovered in two distinct groups more than 50 miles apart.

But finding the impact site itself was a matter of considerable importance. Like other commercial airliners, the plane contained a flight data recorder and a cockpit voice recorder, the black boxes that had been ruggedized to survive a crash landing. Their contents would be invaluable in determining what happened to AF 447 between its last known location and the moment of the crash. Even though the water in the area was deep, and the seafloor mostly rugged, the recorders were equipped with locator beacons—pingers—that are designed to activate in contact with water. They were designed to operate for at least 30 days.

So, for about a month after the crash, two tugboats pulled passive sonar arrays through the rectangular area of about 50 × 60 nautical miles to the northeast of the last known location of Flight 447, more or less along its intended flight path. It was estimated that the arrays would have a 90% chance of detecting an active pinger, but no signal was heard. With the battery life of the pingers exhausted, a new phase of the search was initiated. This one used side-looking sonar in a region just south and east of the plane's last known position. It, too, was given a 90% probability of spotting the wreck if it were there; but it, too, found nothing.

So, phase three of the search began, with ships from the U.S. Navy and the Woods Hole Oceanographic Institute. They, too, used side-looking sonar and focused on the area suggested by a group that had researched drift in the area, roughly, a wide wedge extending to the northwest from the last known position of AF 447, out to about 40 nautical miles. Once again, the search was considered 90% likely to find the wreck if it were there. Once again, failure.

In July of 2010, more than a year after the crash, the French Bureau of Enquiry and Analysis for Civil Aviation Safety, BEA for short, decided a different approach. That approach was Bayesian analysis, using the information already available to create new maps giving probability distributions for AF 447's final resting place. They asked the U.S.-based consulting firm Metron to do the work. It was a natural choice; the chief scientist of Metron had previously helped to find the sunken nuclear submarine, U.S.S. *Scorpion*, back in 1968 using just this technique.

Bayesian analysis is named after the Englishman Thomas Bayes, an 18th-century mathematician and Presbyterian minister, who proposed a special case of what's now called Bayes' theorem. It's a powerful tool, but the idea behind Bayesian analysis is really pretty simple. The chance that something is true changes as new information becomes available. Bayes' theorem tells you how to compute the new probabilities, like this.

I have here three cards, two red ones and one black one, ace of spades, and, I shuffle them up and deal them out on the table. The one on the left is yours, and in a moment, I'm going to reveal the card on the right. The question of interest is, do you have the ace of spades?

Well, the ace of spades is equally likely to be in any of the three positions, with a $1/_3$ chance of each, that's called the prior probability distribution; it's the distribution before any new information comes in. So, your chance of having the ace of spades is $1/_3$. But now, I follow through with my promise to reveal the rightmost card. And whatever happens, you're going to learn something, like this. A red card, I don't have the ace, which means it's one of those two. The odds of you having the ace of spades has now gone from 1 in 3 to 1 in 2. And you got lucky.

That's basically what Bayesian probability does. We begin with a prior probability distribution, a specification of the probability of each of possible outcome. Then we get some new information. Bayes' Theorem then allows us to compute the updated probability for each possible outcome, the so-called posterior probabilities. Our original information is combined with the new information to tell us how our original probability figures should change.

So how does this apply to the Metron analysis of AF 447? Well, they needed a prior probability distribution, and that wasn't as simple as our-randomly three cards. They looked at how much time elapsed between the transmission of the plane's last known position and the time the next signal should have come in and determined the plane could not have flown more than 40 nautical miles in that time. So they started with a uniform distribution on a 40-nautical-mile circle. A uniform distribution gives the same probability to every point in the region, so Metron was assuming, for starters, the wreck was equally likely to be anywhere in that 40-nautical-mile circle.

But the accident certainly seemed to involve loss of control of the aircraft, and studies on similar crashes from similar heights showed that they generally occur within 20 nautical miles of where the emergency started. So Metron combined the uniform distribution with a normal curve, a haystack in probability, centered on the last known position. The further you got from the last known position, in other words, the less likely the plane's wreckage was to be found there.

They created a prior probability graph of concentric circles, like this one, with hot colors indicating higher probability, a heat map, like we discussed in our data mining visualization lecture. The outer circle of the map had a radius of 40 nautical miles. Anything else? Well, debris and bodies from the plane were found, and computer programs tried to model the possible drift paths they might have followed, but this is unbelievably hard. The simulations suggested that the bodies originated in a corridor through the last known position of the plane and running from southwest to northeast. Metron didn't weigh this heavily in creating its prior probability distribution, though, the analysis just wasn't that reliable.

So when they got done, their prior distribution had a patchy corridor of hotter colors where some of the probabilities were given slightly higher probability. Again, the central red region shows the area of highest probability, and the colors cool as the probability drops. This was Metron's best guess about where to look, ignoring the searches that had already been done. But, of course, ignoring those searches was exactly what Metron didn't want to do. They were like Sherlock Holmes in the story of The Silver Blaze. There, Holmes solved the crime by noting an odd clue, the dog that didn't bark. Here, Metron wanted to redraw the map on where to look for AF 447 by using the information of where it wasn't found.

This isn't quite a simple as it was in our card game. If I turned up my card and you saw that it wasn't the ace of spades, you could be sure, no doubt about it, that the ace was one of the other two cards, with equal likelihood. But the sonar scans for AF 447 weren't perfect, only about 90% likely to find the ship. This makes things trickier. In fact, if you imagine a completely general search, it could be worse still; there could be false positives, the equivalent in our card game of there being a chance that I'd tell you I had the ace of spades, when I really had a red card. To handle all of this, we need more machinery than we needed for our original card game. We need the full power of Bayes' Theorem.

To see how this works, I'm going to take the Air France search and simplify it just a bit. Metron was dealing with thousands of possible distinct wreck zones in their map, but we won't lose any of the concepts involved if we simplify this down to six, like this. [Stage] 1 looked for the pingers in the rectangular region including zones 2, 3, and 4; stage 2 of the search looked in zone 5; stage 3 searched the wedge including zones 1 and 2. Zone 6 was never searched.

Now we begin with a certain chance that the wreck is in each of these zones, our prior probability distribution. Zones 2, 3, and 4 are all on the hot part of our map, the areas of higher probability. This is especially true for zone

4, but it's also relatively small. Zone 6 is quite large, but most of it would require the plane to turn around from its last reported position, which is marked with an orange cross in the middle of the circle.

So let's start out with a prior probability distribution, just using our best initial guesses, like this. For example, zone 3 has the best chance, being large and including an area near the center of the search. So, let's say 25% chance of the wreck being there. The large zone 2 is along the flight path and is second with 20%, tied with the smaller but more central zone 4. We'll give the smaller but nearly central zone 5 the 15% chance, and zones 1 and 6 bring up the rear at 10% each. Although they're large areas, they require substantial flight path deviations to be the crash site.

What we're going to need to do is to modify our probabilities in terms of the new information that we get. The first piece of new information was from the passive sonars towed by the tug boats, a quite large sweep with a 90% chance of detecting the black box pingers. In our model, this sweep would cover zones 2, 3 and 4. This search revealed nothing. Assuming that the pingers were working, that means that either the wreck was somewhere else, or that the wreck was in zone 2, 3, [or] 4, but the sweep missed it.

Bayes' theorem computes how likely a particular result is in terms of the information you acquire. Here that means, given that you got this search result, how likely is it that the wreck is actually there. You'll probably recognize this as a conditional probability, the chance of one thing given another. Here, we'd like to say, given that the ship wasn't found in 2, 3 or 4, how likely is it that, nonetheless, the ship was in zone 2, or 3, or any other particular zone? In symbols, we'd write this. For the probability that the wreck is in zone 2, given that it was not found in zones 2, 3 or 4. Okay. How do we find this?

Well, that's a good question. If the search fails, how likely is it that the ship was actually there? You know, it's kind of a shame that's what we want, because if you flip that conditional around, it's really easy to answer. Flipping that around would give: If the ship is in the searched area, how likely is it that the search fails? And we were told that. It's 90% likely to find the wreck if it's there, so, if the wreck is there, the search is 10% likely to fail.

Unfortunately, probability that the wreck is there, given search failed, which is what we want, isn't at all the same as the reversed conditional, probability of search failed given the wreck is there. These two conditional probabilities are quite different. Look, given that two men are father and son, the chance they share a last name is quite high. But given that two men share the same last name, the odds that they are father and son [are] very small indeed. Happily, though, even when the conditionals are reversed and they're very different, they're related in a way that's going to let us crack the Air France puzzle. And the key is something we talked about a couple of lectures ago.

Remember this guy? We used him as a mnemonic for joint probability, probability of A and B. He has to complete the obstacle course by clearing both obstacles. And we said they'd do both. First he has to do one, and then, given that he succeeded in that, he has to do the other. The probability of finishing the course, of both A and B being true, is the product of these two probabilities, the probability of wreck in area) times probability of search fails, given wreck is in that area.

And we can do this calculation for any zone we want. For example, zone 2, the chance of the wreck being in the area is 20%, according to our prior probability distribution. That's the first obstacle. Now, given that the wreck is in that zone, how likely is for the search to fail? Well, the search includes zone 2, and if the wreck is there, it'll spot it 90% of the time. So, given that the wreck is in zone 2, there's a 10% chance of a failed search. So, a 20% chance of a zone-2 wreck times a 10% chance of missing it if it's there, gives 0.20×0.10, or 2%. That is, there's only a 2% probability of both of those statements being true, of the wreck actually being in zone 2, but our search not finding it.

Unfortunately, this isn't quite what we wanted. We wanted the reversed conditional. We know how likely it is that we'd fail if the wreck were in zone 2, but what we want is how likely is it that the wreck is in zone 2, given that we fail. Well, if you want the reversed conditional, reverse the obstacle course, like this. Now what does our runner face? To complete the course, to do both A and B, first he has to accomplish B, and given that he did, he has to accomplish A. That is, first, the search has to fail. Then, given that it fails, the wreck has to be in the area.

Now here's the key observation. Our little friend, regardless of his orientation, has to run the same course, as far as probability goes, at least. When he finishes the course, you get the probability that the wreck is in the area and the search fails, which is the same as the search failing and the wreck being in the area. Running left to right, we found this probability as 0.02. So running right to left has to give the same answer. And the second hurdle in running from right to left is exactly the probability that we want, the probability that the wreck is in zone 2, given that the search failed. But to find it, we have to find the probability of clearing the first hurdle. That is, we need to find the chance that the search fails.

Getting this is a little trickier than you might think. That's because there are actually six ways for the search to fail. The wreck, of course, is in one and only one of the six zones. That means that [if the] search fails, either it failed and the wreck was in zone 1, or it failed and the wreck was in zone 2, or it failed and the wreck was in zone 3, and so on. The total chance that the search failed is the chance that any one of these six combinations happened. Can we find the probability of each of these combinations? Sure! Easily.

Each combination is the chance that the search failed and that the wreck is in a particular zone. And we already did this calculation for zone 2, at least, running left to right through our course, and got an answer of 0.02, or 2%. And in an exactly analogous manner, we can do the similar combination for every other zone. Take a look. First, each term on the right is the prior probability that we originally assigned to each zone. Let's put them in. OK, now, how about the conditionals? Well, look at the first one. It says that, if we know that the wreck is in location 1, how likely is it that a search in locations 2, 3, and 4 will fail? Well, it's guaranteed, isn't it? Unless the equipment got false positives, thinking that it found a wreck where none existed, you can't find the ship if you're not looking where it is. And that's true for locations 5 and 6, by the same logic. For the other three locations, we use the logic that we used for zone 2. If the wreck is in zone 2, 3, or 4, the search is going to find it with a probability 0.9, so it has a 10% chance of failing. Like this.

So, let's come up for air, and see what we've got. If the search fails, then exactly 1 of 6 different things had to happen. The wreck could be in zone 1 and the search fails, or it could be in zone 2 and the search fails, and so on.

We've found the probabilities for each of these possibilities. So if the search failed, exactly one of them had to happen, and that means you can add up those six probabilities to find the chance of a failed search in zones 2, 3, and 4. Just add them up.

There was a 41.5% chance that the initial sweep in locations 2, 3, and 4 would fail. So, we can finally calculate the posterior probability that started all of this. Knowing that the sweep did fail, what's the probability that the wreck was in zone 2? From our runner, we had this, probability of wreck in zone 2 and search fails is the probability that the search fails times probability of wreck in zone 2 given search fails. And we know everything in this equation but the last term, which is the one we want. Plugging in the numbers of the other two quantities and solving gives us this, the probability that the wreck in zone 2, given that the search fails, is 0.048. That is, the search gives a 41.5% chance of failing, and given that it did fail, the probability of the wreck being in zone 2 just dropped from 20% down to 4.8%.

And that's what Bayesian probability is all about. We can do exactly the same kind of calculation for each other location to get the new probability of the wreck being in each of the six locations. In fact, we can just take the six joint probabilities that we found and divide each by the probability of a failed search in zones 2, 3 and 4, which was 0.415. Here's what the result would look like.

As you can see in red, factoring in the fact that the first search was a failure, the most likely resting place for the wreck is now in zone 5, with about a 36% chance of holding the wreck. In real life, the second search was conducted in zone 5, using side-looking sonar. This search failed as well. So, what we do is the same kind of work that we just completed, using Bayes' theorem. We use the red distribution as our prior and compute the posterior probabilities for each zone given that we had a failed search in zone 5. Here's what we get.

Check out the green bars. The failure of the second search makes zones 1 and 6 the most likely candidates for the wreck, each with about a 36% chance of containing the wreck. Zone 6 is huge, though. In real life, the third search was in zones 1 and 2. Its failure allows us to apply Bayes' theorem one more time, using the green distribution as the prior and resulting in this.

At this point, the wreck is about 83% likely to have already been found. Since it hasn't, the best bet is the only zone not yet searched, zone 6, with about a 58% probability. Metron's analysis was actually very much like our own, but with thousands of zones, rather than six, and with a more careful creation of the prior distribution. The map that they got after doing their analysis looked distinctly different from the one that they started with. Areas in which a failed search had been conducted showed dramatic cooling of their colors; the probability of the wreck being in these regions was greatly reduced. At the same time, the colors in the unsearched areas grew notably hotter. Their posterior probabilities, in light of the new information, were higher than the original ones. Just like our work, where zone 6 is now our best bet.

Except, except that in our analysis, the odds were quite good that the plane wreck would have been spotted by now, about 83%, since we're 90% likely to see a wreck if it's there. And we haven't seen it. Maybe we were just unlucky, and Air France 447 went down in the huge, if fairly unlikely, zone 6.

But Bayesian analysis showed that the odds were very good that the plane would have been found by now. If it really was in Zone 6, its size was going to make for a very slow search. But the people at Metron were beginning to think that perhaps that wasn't what happened after all. Before tackling zone 6, they explored another idea. What if, for some reason, the pingers malfunctioned? What if the search of zones 2, 3, and 4 was unsuccessful because there was no sonar ping to hear?

They redid their work under this assumption, more Bayesian analysis of the kind we just did, and the new heatmap showed a strong concentration of probability within about 10 miles of the plane's last reported position. Metron advised another search of this area, and after one week of additional searching, the wreckage was found in the location well within that high probability zone, about 10 miles north-northeast of the last known location. No one knows if the pingers failed to operate; the BEA says that they are highly reliable. The probability maps suggest that something either kept them from working or kept them from being detected. It's not certain but given the results of the Bayesian analysis, it seems a real possibility.

Understanding Bayesian probability is, I think, terrifically important, because failing to do so can lead to a lot of bad decision making, including racial prejudice, medical misdiagnosis, and bad public policy. The reason is that most untrained people fail to distinguish between a conditional and its reverse. By untrained, I mean don't mean bumpkins, only untrained in the language of mathematics and probability. Let me give you an example. There's a medical condition called phenylketonuria, or PKU, that results from an error in phenylalanine metabolism in infants. It results in mental retardation if not detected, but there is a screening that is 100% successful in spotting kids with the defect, and there's a treatment for the condition. Screening is mandatory in all 50 states, so if your kid has PKU, you're going to know it.

Ah, but did you notice, I didn't tell you everything. Yes, there is a test that spots 100% of kids with PKU. But what does it do with kids who don't have PKU? After all, if we simply say all babies have PKU, we have a test that spots 100% of the kids with the condition. We'll just be wrong about everybody else; that test has no specificity at all. So, here's the missing information. There are two common tests for PKU, and the better one is 98% likely to identify a PKU-free child as being PKU-free; 98% sounds pretty good. So, how afraid should you be when your baby tests positive? With a test that's somewhere between 98 and 100% accurate? Well, the question we want is not how likely a sick kid is to test positive, that's 100%. The question we want is how likely is it a kid who tests positive is to be sick.

So, let's use Bayes. Somewhere around 1 child in 15,000 is born with PKU. We can run the equations for Bayes' theorem as we did in the Air France example, but let me do it more intuitively this time. Imagine 15,000 children. On average, one of that 15,000 is going to be unlucky enough to have PKU, and our test is guaranteed to catch it. But there are 14,999 kids who don't have PKU, and with a 98%-reliable test, that means that, on average, it will get about 2% of them wrong, and tell us that they have PKU. That's about 300 kids. So when the test that's over 98% reliable says that your child has PKU, the odds are 300 to 1 against your child having the condition. And the rarer the disease, the more likely that the positive test result is a false positive.

When I said that most people untrained in probability have trouble with reversed conditionals, I meant it. Sixty medical students and staff at the Harvard Medical School were asked a question like our PKU scenario, except the test had a false positive rate of 5%, rather than 2%, and the condition occurred in one patient in 1000. So, what's the chance of disease in that case? Five percent false positives in 1000 patients means 50 wrong for every 1 that's correct; 1 in 50 is 2%, the chance of testing positive means that you actually have the disease. But only 18% of the Harvard staff and students got the correct answer. The most common answer was 95%; the average of the answers was 56%; the right answer, 2%. So even doctors need some help understanding reversed conditionals.

Here's another one. You've probably heard that more people get killed by bee stings each year than by shark attacks. And an appropriate reaction, really, is, so what? Because we don't want the probability of shark given death—we want probability of death given shark. Let me put it this way. If you are hip-deep in the ocean, and a bee is approaching you from the shore as a shark closes in to you from the sea, run toward the bee.

Most accidents occur within 15 miles of home. Of course they do! Most driving occurs within 15 miles of home! The value that we need is probability of accident given near home, not probability of near home given accident. It's important, because numbers and probabilities have a cachet; they sound irrefutable. But even when the numbers are correct, you have to make sure that you're looking at the right [probabilities]. I saw congressional testimony examining if marijuana is a gateway drug to cocaine and heroin. One argument presented was that the overwhelming majority of cocaine users started on marijuana first. The figures you see vary, but many of them put the fraction of cocaine users who previously used marijuana at about 95%. For many people, this is a strong argument for keeping marijuana use illegal. But none of this makes the argument it purports to make. It's not just the logical problem of post hoc ergo prompter hoc, nor even that an addictive personality would probably be exposed to marijuana before cocaine. It's that, if this argument were sound, then an even stronger one could be made against milk, which is more than 95% likely to be used by cocaine addicts, and usually they start it from a very early age.

What went wrong? Again, the problem is the reversed conditional. The question isn't what fraction of coke users started with marijuana—the question is, what fraction of marijuana users go on to use cocaine? For the statistics I've seen, that's about 25%, no small number, but a far cry from 95%. To tell you the truth, what amazes me about the 95% statistic is how small it is. If correct, it means that 5% of coke users just vault over marijuana. Want a joint? No thanks. I'm saving myself for cocaine.

A similar problem comes up with statistics about the fraction of highway fatalities that involve drunk drivers. The president of Mothers Against Drunk Driving quoted this statistic in her congressional testimony. According to the Department of Transportation, about 31% of traffic fatalities involve drunk drivers. Chilling. But again, knowing nothing else, this statistic tells us very little. After all, it means that over $^2/_3$ of all fatal highway accidents involve only sober people. If anything, that seems to imply we should get liquored up before we get behind the wheel. This reasoning is, of course, nonsense. Again, we're looking at the reversed conditional. We don't need to know the probability that a fatal traffic accident involves a drunk. We need to know the probability that a drunk is involved in a fatal accident, as compared to a sober driver.

Doing the Bayesian work leads to the conclusion that those above the legal limit for blood alcohol are at least 13 times more likely to be involved in a fatal accident than a sober driver. And that is the statistic that's relevant to the drunk-driver discussion. Many kinds of bigotry and prejudice are fueled, in effect, by confusing the conditional and its reverse. For example, when people think of terrorist attacks in the U.S., 9/11 instantly springs to mind. And overwhelmingly, the people involved in that attack were young, male, Arab, and Muslim. The hijackers themselves all fit this description and were probably all in their 20s and 30s. Which means, of course, that if you know that you are looking at a 9/11 hijacker, you've got a pretty darn, good idea of what he looks like.

But the problem, of course, is that most of the time, people aren't looking at known 9/11 hijackers. They may be looking at someone who is young, male, Arab, and Muslim, though. And when you think about situations like these, you can realize how dangerous it is to confuse A given B, and B given A. Because the vast majority of young, male, Arab, Muslims are not terrorists, and thinking that they are is cause for a terror of a whole different kind.

Once you understand conditional probabilities, Bayesian probability is available to improve all kinds of decisions. For example, in our last lecture, we were building a decision tree for whether to mine for minerals in two adjacent properties. What if we could test the property before making a purchase decision, and that our test was 75% likely to find minerals if they were there, and 5% likely to find them if they were present in less than economic quantities? Well, we said that there was a 30% base chance of worthwhile deposits of minerals being present. We can use Bayes' theorem to figure out how this changes our test for our deposits or not. It works out that there's about a 26% chance of a positive test, and 86% of the time, a positive test means that the minerals really are there. A negative test means that there's 90% chance of no good deposits. Add this research option to our decision tree, and it works out that doing the test before deciding what you'll buy will on average increase your profits by over $850,000, less the cost of the test. In the language of decision theory, the EVSI of the research is $850,000. EVSI is expected value of sample information. In more common language, the test is worth doing if you can do it for less than $850,000.

So, ladies and gentlemen, we can say thanks to mathematician and Presbyterian minister Thomas Bayes for giving us this lovely analytical trick. If you've got a conditional probability and you need its reverse, then Bayesian probability really is the answer to a prayer.

Markov Models—How a Random Walk Evolves
Lecture 21

In business and organizational decision making, Markov analysis, named after Russian mathematician Andrey Andreyevich Markov, is useful for many tasks, including anticipating bottlenecks on the factory floor or in a telecommunications network, predicting demands on hospital resources, characterizing web browsing behavior, designing maintenance schedules and predicting failure times, assessing the behavior of waiting lines, and evaluating whether admissions to a university program now will result in overloading the program in the future. It's no less useful in the sciences—for example, modeling diffusion across a membrane or drugs in a body. In this lecture, you will learn about the Markov process, including how to use Markov diagrams and transition matrices.

Markov Process

- The following is the mathematical skeleton of a Markov process. You have a system in which you are interested. At any given point in time, this system can exist in one of a variety of states. In many systems, the list of possible states is finite, but it doesn't have to be. As far as the management of a restaurant is concerned, for example, a table might have eight possible states, as follows.
  1. table reserved
  2. table available
  3. guests selecting orders
  4. guests ordering
  5. guests waiting for food
  6. guests eating
  7. guests waiting for check
  8. table unoccupied but needing cleaning

- A system generally stays in one state for some amount of time, and then transitions to another. Each such transition has a specific probability, and in a Markov process, it's one that depends only on the current state.

- So, for example, a table that is currently unoccupied but in need of cleaning may stay that way for the next minute, or may become a table that is available, or may become a table that is reserved. If we analyzed the system from the point of view of a diner, we'd have different states—and the diner would probably finally reach the state of pay bill and leave, although other possibilities, such as leave without eating, might exist, too.

- A Markov process addresses the following questions: What will be the long-term behavior of such systems? How will they evolve over time? And if the system ever does reach an unchanging final state—such as pay bill and leave—how long will it take to get there?

- The **Markov diagram**, in some ways, is an odd cousin of a decision tree. Each node is a state—it's a situation you could find yourself in. All of these nodes are like the chance nodes in a decision tree: There aren't any "decisions" for you to make, so each of those nodes has a certain probability of transitioning to another state. But, unlike decision trees, there may be more than one way of getting to the same state. In spite of these differences, though, a path through a Markov diagram, like a path through a decision tree, tells a story.

- Where things get really interesting with **Markov analysis** is when the Markov diagram bends back on itself, making a cycle. Conceptually, this means that it's possible to revisit a state that you may have been in before. For example, the restaurant table can return to being ready.

Applying the Markov Process
- Suppose that you run a direct marketing company. You send out catalogs once per year. You consider a new and promising customer to be in segment 1. If that customer doesn't buy in the first year, he or she slips to segment 2. If the customer doesn't buy in the second year either, he or she slips to segment 3, and you send him or her a reactivation package. If the customer still doesn't buy, you write him or her off as not worth it and stop sending him or her catalogs. But anytime that customer buys, he or she is back in segment 1.

- Of course, you don't know whether the customer is going to buy or not, but you can get probabilities for his or her actions from historical relative frequency data. For example, suppose that 30% of your segment 1 customers make an order when they receive a new annual catalog. Then, you could say that there's a 30% probability that a segment 1 customer reorders and remains in segment 1. Similarly, perhaps 1% of customers are lost per year to death or disconnection, so you'd take that as the probability of involuntarily losing a customer. We can put all of this into a Markov diagram that describes a customer's possible future relations with your company, as follows.

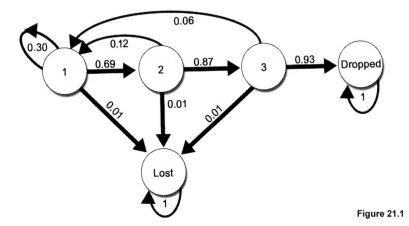

Figure 21.1

- We have two terminal nodes where things can end up: lost and dropped. In Markov analysis, we'd call these states **absorbing states**, because once you get into one of them, you never get out. But what goes on in the rest of the diagram is quite a bit more complicated.

- The segment 1, 2, and 3 nodes each have a chance of transitioning to segment 1—it happens when the customer makes a purchase. When the annual catalog is received, there is a 30% chance of a purchase from a segment 1 customer, a 12% chance for a segment 2 customer, and a 6% chance for a segment 3 customer. Each of these segments also has a 1% chance of transitioning to lost, because customers in any of these segments are equally likely to die or become disconnected. The remaining arrows capture the rest of the probability, because the total probability out of any node will equal 1 in a Markov diagram.

- It's pretty clear that, sooner or later, a customer is going to reach an absorbing state—that the customer will go three annual catalogs in a row without ordering, or become lost. But what's not clear is how long this is going to take.

- As a marketer, you're going to want to get a sense of how long you get to keep a person as a customer, as well as how many purchases you can expect the customer to make in that time. And with Markov analysis, we can do just this. First, we turn the Markov diagram into a Markov matrix (M), as follows.

$$\begin{array}{c} \\ \text{Seg1} \\ \text{Seg2} \\ \text{Seg3} \\ \text{Dropped} \\ \text{Lost} \end{array} \begin{array}{ccccc} \text{Seg1} & \text{Seg2} & \text{Seg3} & \text{Dropped} & \text{Lost} \\ \begin{bmatrix} 0.30 & 0.69 & 0 & 0 & 0.01 \\ 0.12 & 0 & 0.87 & 0 & 0.01 \\ 0.06 & 0 & 0 & 0.93 & 0.01 \\ 0 & 0 & 0 & 1 & 0 \\ 0 & 0 & 0 & 0 & 1 \end{bmatrix} \end{array}$$

- The row indicates the state you are transitioning from, and the column indicates the state you are transitioning to. Each row has its total probability adding to 1. So, the first row shows what happens to a customer in segment 1: a 30% chance of buying from the annual catalog and therefore staying in segment 1, a 69% chance of not buying and therefore sliding into segment 2, and a 1% chance of being lost.

- The last two rows show that once you're dropped, you're dropped, and that once you're lost, you're lost. That is, if before the transition, you're in the dropped state, then after the transition, you'll still be there, with 100% probability—and so on, forever. That's why it's called an absorbing state.

- To find out what happens in two years, we just multiply this **transition matrix** by itself: A year goes by and, then another year goes by. M times M, by the rules of matrix multiplication, results in the following.

	Seg1	Seg2	Seg3	Dropped	Lost
Seg1	0.1728	0.207	0.6003	0	0.0199
Seg2	0.0882	0.0828	0	0.8091	0.0199
Seg3	0.018	0.0414	0	0.93	0.0106
Dropped	0	0	0	1	0

- The top row says that a customer that is currently in segment 1 is 60% likely to be in segment 3 two years from now. And to find out about three years from now, we multiply this new M^2 matrix by M to get M^3, as follows.

	Seg1	Seg2	Seg3	Dropped	Lost
Seg1	0.112698	0.119232	0.18009	0.558279	0.029701
Seg2	0.036396	0.060858	0.072036	0.8091	0.02161
Seg3	0.010368	0.01242	0.036018	0.93	0.011194
Dropped	0	0	0	1	0
Lost	0	0	0	0	1

- That is, about 59% of the customers are gone by the third year, with about 56% being dropped for not buying catalogs and 3% lost. We could continue this way, year by year, to watch how the top row of this matrix evolves. To get the next year's matrix, we just multiply the previous year's matrix by M. **Figure 21.2** is a graphical representation of the evolution.

- As the graph shows, almost no customer lasts 10 years in this model. There is more than a 95% chance of being dropped from the mailing list for failing to make a purchase for three consecutive years. There's a 4% chance of being lost. And there's less than a 1% chance of still being active.

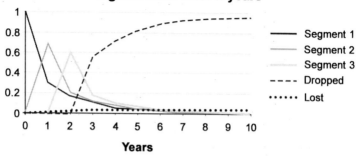

Figure 21.2

- There's another way of interpreting Markov transitions that is conceptually different but mathematically identical. We could imagine that we are speaking not of one individual but of a large population and that the transition probabilities indicate the fraction of the population currently in a given state that transitions to another state.

- For example, rather than saying that a segment 1 customer has a 30% chance of making a purchase, we could say that 30% of segment 1 customers make a purchase. From this perspective, our graph shows that in 10 years, we lose about 99% our customers, and 95% of our customers are dropped for not making an order in three consecutive years.

- One way to address this is to acquire new customers—for example, from mailing lists—to replace the customers who are dropped or lost. We can modify our Markov diagram to replace all lost and dropped customers by changing our lost and dropped states from absorbing states into ones that transition back to segment 1. That is, each person who is lost or dropped this year is replaced in the following year by a new segment 1 customer. The modified Markov diagram is as shown in **Figure 21.3**.

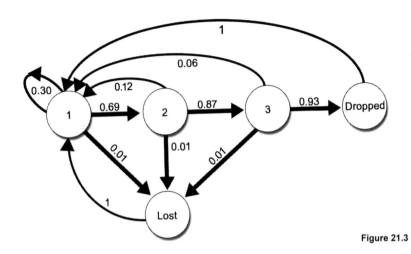

Figure 21.3

- The long-term behavior for this diagram is going to be very different from the previous one. In a system with one or more absorbing states, any non-absorbing state that can ever reach an absorbing state is eventually going to be empty. Such eventually empty states are called **transient states**.

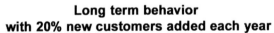

Long term behavior with 20% new customers added each year

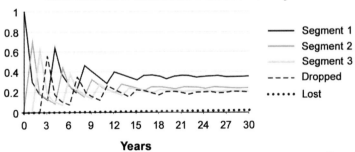

Figure 21.4

451

- Our basic approach to solving this problem is to use the transition probabilities to see how the system evolves through time, find what fraction of customers fall in each segment, and use this information and other cost and purchase information to determine how often mailings should be sent and to whom.

Important Terms

absorbing state: In a Markov system, a state that cannot be left once it is entered.

Markov analysis: A set of techniques for analyzing the behavior of systems that can be viewed as occupying one of a number of states, with a particular probability of transitioning from one state to another at various points in time.

Markov diagram: A graphical representation of Markov process. Each pure state of the system is represented by a node, and each possible transition from state to state is indicated with an arrow connecting those nodes.

transient state: In a Markov system, a state whose long-run probability approaches zero regardless of initial conditions.

transition matrix: In Markov analysis, a matrix whose entries p_{ij} give the probabilities of moving to a specified state j given that the system is currently in state i.

Suggested Reading

Elsner, Krafft, and Huchzermeier, "Optimizing Rhenania's Mail-Order Business through Dynamic Multilevel Modeling (DMLM)."

1. The same kind of work we did in this lecture can be applied more broadly. For example, imagine a species of creature that each year either dies or gives birth to a single offspring. That offspring itself will either give birth or die in the following year, and so on. We could represent this with the following transition matrix.

$$
\begin{array}{c}
\quad\quad\text{infant}\quad\text{adult}\quad\text{dead} \\
\begin{array}{c}
\text{infant} \\
\text{adult} \\
\text{dead}
\end{array}
\left[
\begin{array}{ccc}
0 & 0.2 & 0.8 \\
3.3 & 0.55 & 0.45 \\
0 & 0 & 1
\end{array}
\right]
\end{array}
$$

This is not properly a Markov matrix, because the middle row does not add to 1. Instead, the matrix says the following: 20% of infants survive to adulthood. Adults have a 45% annual mortality rate. Adults who survive have a litter of 6 infants ($3.3 = 0.55 \times 6$). The techniques of the lecture can still be applied to a matrix such as this. Because the questions are most easily done by looking at high powers of the transition matrix, you may want to do them on a spreadsheet. In Excel or Calc, the command to multiply two matrices is MMULT.

a) Show that in 50 years, an initial population of 2 adults grows to a population of about 700 adults and their offspring.

b) Show that if adult survival drops to 45% (for example, from overharvesting), the population dies out.

c) Show that if litter size drops to 4 (for example, from ecological damage), the population dies out.

Answers:

a) We can take an initial vector of [0 2 0] for 2 adults and then multiply it by the *M* matrix above 50 times. The result is [1952.8 670.3 14017.8], meaning that we have almost 2000 infants, 670 adults, and about 14,000 that have died.

b) Replacing the second row of the matrix with [2.7 0.45 0.55] and repeating the calculations in a) shows this. Note that $2.7 = 6 \times 0.45$.

c) Replace the second row with [2.2 0.55 0.45] and repeat part a).

Moral: Relatively small changes in transition probabilities can have huge long-term impacts.

2. One requirement for a Markov process is memorylessness. That is, where you go next in the Markov diagram depends only on chance and where you are now, not on how you got there. Think of processes for which this requirement is met, and ones in which it is not.

Answer:

Winnings on a slot machine is a memoryless process, as is radioactive decay. Many real-life events, such as arrivals at a grocery store, are close to memoryless. Human interaction with strangers is generally memoryless, but with the same individual, it is not. The history of previous interaction with an individual usually changes the probabilities of what happens next.

Markov Models—How a Random Walk Evolves
Lecture 21—Transcript

Sometimes, as the old saying goes, history is destiny. The trajectory of future events can be strongly influenced by what came before, in fact, sometimes that future seems almost fore-ordained. Is this a good thing? Well, it depends on what you mean by good. It can certainly be useful, if you're in the forecasting business. You'll remember from our discussion of time series that the whole analysis hinges on one crucial assumption, that what happened in the past is a good guide to what will happen in the future. And we were making that assumption in a very strong sense, that the observed trends and seasonality seen in the past could be expected to continue into the future.

But you might remember another topic discussed in that lecture, a random walk. We came across it in the context of an investment whose value each day changed by an amount and in a direction that was determined by the flip of a coin, by chance. I mentioned that some people think the short term variation in the Dow Jones index is, essentially, such a random walk. In a case like that, is the past a good model for the future? Well, yes and no. No, in that history is not destiny. The only thing that matters about your history is the very last part of it—your situation right now. How you got there doesn't count for beans. On the other hand, if you do know your situation now, then I can assign probabilities to what might happen next, probabilities that depend only on that current situation. And that allows a mathematical analysis of an entirely different kind.

That's what this current lecture is about, and it goes by the name of Markov analysis, named after the Russian mathematician Andrey Andreyevich Markov, who did a lot of the early work in the field. In business and organizational decision making, it's useful for tasks like anticipating bottlenecks, whether on the factory floor or in a telecommunications network, predicting demands on hospital resources, characterizing browsing behavior, designing maintenance schedules and predicting failure times, assessing the behavior of waiting lines, and evaluating whether the admissions to a university program today will result in overloading the program in years to come. It's no less useful in the sciences, for example, modeling diffusion across a membrane, or drugs in a body.

Here's the mathematical skeleton of a Markov process. You've got a system in which you are interested. At any given point in time, this system can exist in one of a variety of states. In a lot of systems, the list of possible states is finite, but it doesn't have to be. As far as the management of a restaurant is concerned, for example, a table might have eight possible states, table reserved, table available, guests selecting orders, guests ordering, guests waiting for food, guests eating, guests waiting for check, table unoccupied but needing cleaning.

A system generally stays in one state for some amount of time, then transitions to another. Each such transition has a specific probability, and in a Markov process, it's one that only depends only on the current state. So, for example, a table that's currently unoccupied, but needing cleaning may stay that way for the next minute, or may become a table that's available, or may become a table that's reserved. If we analyzed the system from the point of view of a diner, we'd have different states. The diner would probably finally reach the state of pay bill and leave, although other possibilities like, leave without eating, might exist, too.

What will be the long-term behavior of such systems? How will they evolve over time? And if a system ever does reach an unchanging final state, like pay bill and leave, how long will it take to get there? These are the kind of questions we're going to look at today. We're going to see how to represent these states and transition probabilities in a picture called a Markov diagram, a graphical, structural model of how the pieces fit together. We'll then convert that picture into a compact mathematical representation of the situation using what are called transition matrices. Finally, we'll use the insights of Markov analysis to extract a wealth of information from the result.

Let's introduce these new tools with a second look at a now-familiar problem, the downed plane example from last time. Markov analysis will give us a much easier way to envision some aspects of that problem. Then we'll apply our new Markov toolbox to something a bit more ambitious, and see how they helped a German direct-marketing firm avoid financial disaster.

So, first tool, the Markov diagram. In some ways, you could think of it as an odd cousin of a decision tree. Each node is a state; it's a situation you could find yourself in. All of these nodes are like the chance nodes in a decision tree; there aren't any decisions for you to make, and so each of these nodes has a certain probability of transitioning to another state, another node. But unlike the decision tree, there may be more than one way of getting to the same state. In spite of these differences, though, a path through a Markov diagram, like the path through a decision tree, tells a story. Here's the Markov diagram for our plane search.

What's going on here? Begin at the node labeled Start, at the top. From there, we follow an arrow, making a transition to a node specifying where the wreck is, zone 1 through zone 6. Each edge leading to a possible site of a wreck is numbered, giving the probability of following that edge. So 25% of the time, we arrive at the node for zone 3, which means that the wreck is in zone 3. Whatever zone node we reach, we then continue to the bottom of the diagram, ending at either at the node where the search found the wreck or where it didn't.

For example, if we follow this path, it means that the wreck was in zone 3, and then it was found. The first edge has a 25% chance of being followed, since originally we said there was a 25% chance of the wreck being in zone 3. The second edge has a 90% probability, since, if the wreck is in zone 3, then our search is 90% likely to find it. Our rule for joint probability thus tells us that the probability of this sequence of events is 0.25×0.90, or 0.225.

On the other hand, take the route from Start to zone 1 to found, this one. Never happens. There's a 10% chance that we originally transition to Node 1, that is, that the wreck is in zone 1, but if we get there, there's a 0% chance of making it to the found node. The search of zones 2, 3, and 4 didn't include zone 1, so the search is guaranteed to fail. And multiplying the probabilities on this path shows this $0.10 \times 0 = 0$.

This is how Markov diagrams work. If you want to know the probability of moving around the diagram following a certain path, just take all of the probabilities on that path and multiply them together. The method gives the correct answer, so long as the probability of ending up at a particular node after a transition depends only on what node you're leaving in that transition. That is, we've got a memoryless process, where how you got to a particular state doesn't matter. All that matters in terms of what happens next is where you are now. There are actually ways of introducing memory into a Markov process, but they're clumsy, and I'm not going to focus on them here.

Let's use our Markov diagrams to find out how likely it is that the search fails. That was an important probability in the last lecture. Well, there can be six different routes through the Markov diagram that end in this result, as shown by the six different colored paths here. The probability of any one path is obtained by multiplying the probabilities along it. Do it for each path and then add the results. That's the probability of reaching the failed node, that is, the chance that the search failed. It's actually the same calculation that we did in the last lecture.

Well, the picture is nice, but it would also be nice to have a more compact way to do this calculation itself. And there is one, one whose utility goes far beyond what we're using it for today. It's matrix arithmetic, or more specifically, matrix multiplication. So, let me give you a quick primer on that. If you already know it, it'll be review, and if not, it'll get you up to speed for what we need.

In mathematics, a matrix is nothing more than a rectangle full of numbers, so many rows, so many columns. If it's only got one row, it's also called a row vector. We usually put brackets around the matrix. The dimensions of a matrix are the number of rows and columns, so, here's a 6×2 matrix, six rows and two columns. You can add or subtract any two matrices that have the same dimensions, and you do it in the obvious way, just by adding or subtracting the corresponding locations. So, if matrix A has a 6 in the upper, left-hand corner and matrix B has a 2 there, then the upper, left-hand corner of A − B is just 6 − 2. You play the same game in each location in the matrices, which is why they have to have the same dimensions.

But the tricky operation with matrices is how to multiply one by another. A natural choice would be to play the same game you did with addition and subtraction, upper-left corner of A times upper-left corner of B, and so on. And you can do that operation, of course, if the matrices happened to have the same dimensions, but it turns out not to be very useful. There's a different way to define matrix multiplication that ends up being much more useful, and is exactly what we want for a Markov analysis.

In fact, I'm going to introduce matrix multiplication in general by thinking about the Markov diagrams in particular. I want you to imagine that our matrices have rows and columns labeled with states, and that the row of the matrix tells you where you're coming from, while the column of the matrix tells you where you're going to. In our airplane search Markov diagram, we begin at Start, but we can go to any of the six nodes labeled 1 to 6. I can represent this by a row vector.

One doesn't normally put labels on the rows and columns of a matrix, but including them here should make things a lot clearer in our current discussion. This one row matrix, this vector, lists the possible transitions from Start to a zone state, and it gives the probability of ending up at each zone after one transition. Those entries add up to 1, since there is a 100% chance that you end up somewhere after the transition.

It's not true that every row of an arbitrary matrix will add to 1, but every row of a Markov transition matrix will. This row vector can also be called the initial state vector; that's because it gives you the probabilities of the systems initially in one of the possible states. Here the probabilities that the wreck is in each possible location.

OK, how about the second transition, the transition from the zone nodes to the final result? That would look like this. Again, the numbers are just transition probabilities, with the probability the plane was found in each row of the left column, and the probability that it failed in the right column. For example, the 0.9 in the zone-2 row and found column means that, given that you start in the zone-2 node, there is a 90% chance you'll transition to the found node. That is, if the wreck is in zone 2, you are 90% likely to find it. So, each matrix is, if you like, a table of conditional probabilities. Given that you begin a transition in a particular row state, how likely is it that you'll transition to a particular column state?

Here's the thing to keep in mind. Matrix multiplication is so defined that it corresponds to the English phrase "and then" for Markov matrices. We want to say, from start, we transition to a zone, and then, from there, we transition to finding the results of the search. To write this mathematically, we just multiply these two matrices.

I've suppressed the labels here, but keep thinking about them. They're quite useful. Remember that, for each matrix, the row tells you where you're transitioning from and the column tells you where you're transitioning to. So again, the first matrix, the row vector, takes us from Start to a zone node, and the second matrix takes us from a zone node to a result node, either finding the wreck or failing to.

Now, I want you to see that in order for this to make sense, the number of columns in the first matrix, where the first matrix drops you off, has to be the same as the number of rows in the second matrix, where the second matrix picks you up. And a little more thought should convince you that the two together tell a whole story, which begins at Start and ends with either found or failed. That means that the answer to this computation should be a 1×2 matrix, 1 row, corresponding to Start, and two columns, corresponding to found or failed.

This is a general truth of matrix multiplication. When you multiply two or more matrices together, the final answer will have the same number of rows as the first matrix, and it will have the same number of columns as the last matrix. So a 2×3 matrix, times a 3×4, times a 4×5 will give you a 2×5 matrix. And each matrix in the product represents one transition, one step through the Markov diagram.

So, how do we actually do matrix multiplication? Like this. Take the row vector and stand it up on its end. Line it up with the first column of the second matrix. Now we multiply the two numbers in each row, and then add the six results, like this. Just go right down the column, $0.1 \times 0 = 0$; $0.2 \times 0.9 = 0.18$; $0.25 \times 0.9 = 0.225$, and so on. Add up all of six of these numbers and you'll get 0.585. Now, take our row vector, stand it up next to the second column of the second matrix, and do the same thing again. Multiply each entry in the row vector by the corresponding number in the second column, and then add the results. $0.1 \times 1 = 0.1$, $0.2 \times 0.1 = 0.02$; $0.25 \times 0.1 = 0.025$, and so on. The sum of all of these is 0.415. So the result of our matrix multiplication is this—which means that the search is 58.5% likely to result in finding the wreck and 41.5% likely to fail.

A couple important observations on what's been going on here. First thing, arithmetic. There was a lot of it, and multiplying matrices always involves a lot of it. Which is why as an undergraduate, I didn't give matrices the attention that they really deserve. But multiplying matrices by hand is no more necessary, or informative, than multiplying normal numbers by hand. A lot of scientific calculators can do matrix multiplication these days, and almost any spreadsheet can. Here's how you'd use Calc to do the math that we just did. The calculation is done by highlighting both the cells that I've filled here in green, and then typing in this command, which says to matrix multiply—MMULT—the A box and the B box. As always, if working in Excel, we use a use a comma instead of a semicolon as the separator.

But there is one freaky thing. Since this formula fills more than one cell, the green cells, in the spreadsheet, you don't just hit enter. When you enter it, use control-shift-enter. This tells the spreadsheet that you are entering what's called an array formula. Just be sure you highlight the right size rectangle before you start typing in the formula. For example, a $6 \times 5 \times 5 \times 7$ matrix is going to require you to highlight a 6×7 rectangle to hold the answer. But there's another observation that I don't want you to overlook, and it's this. The definition of matrix multiplication corresponds exactly to the words "and then" when looking at a Markov diagram. In other words, as soon as I could say the plane went down in a zone, and then, from that zone, it was either found or not, well, I knew I only had to multiply the what-zone vector by the found-or-not matrix to get the probability of the search being successful. I could do the needed calculations almost on autopilot or have a spreadsheet do them for me.

In spite of its odd definition, in most ways, matrix multiplication behaves like the ordinary multiplication you're used to. It's associative, it distributes over addition and subtraction, and so on. But there is one oddity that might surprise you. It's not commutative. A times B and B times A aren't the same. In fact, one of them may not even exist. If you think about the Markov interpretation, it's not hard to see why. If we reversed the order of our Markov matrices for the plane wreck, putting the second one first, like this, then the first matrix is taking us from one of the six zones to either found or fail. Logically, the subsequent "then" would have to start in found or fail, but it doesn't; it starts with only one entry, Start. In this order, the multiplication can't be done.

But let's kick our work up a notch. Where things really start getting interesting with Markov analysis is when the Markov diagram bends back on itself, making a cycle. Conceptually, this means that it's possible to revisit a state that you may have been in before. The restaurant table can return to ready. Being on hold this minute can transition to being on hold in the next minute, and it usually does.

A patient discharged from intensive care may wind up there again before leaving the hospital. A customer who bought something from a direct marketing catalog in the past may do so again, or may not. And it's this last situation that was getting the German direct marketing company, Rhenania, into hot water. It was losing customer base, losing market share, and losing profitability. And it was essentially Markov analysis that helped it recover, and in fact to become the second largest direct mailing firm in Germany.

Let's shrink the problem down to manageable size, without losing the essentials. We'll have you run the direct marketing company. You send out catalogs once per year. I'm a new and promising customer. You consider me to be in segment 1. If I don't buy in the first year, though, I slip to segment 2. If I don't buy during the second year either, I slip to segment 3, and you send me a reactivation package. If I still don't buy, you write me off as not worth it and stop sending me catalogs. But any time I buy, I'm back in segment 1.

Now, of course, you don't know whether I'm going to buy or not, but you can get probabilities for my actions from historical relative frequency data. For example, suppose that 30% of your segment-1 customers make an order when they receive a new annual catalog. Then you could say there's a 30% probability that a segment-1 customer reorders and remains segment 1. Similarly, perhaps 1% of your customers are lost to death or disconnection, so you'd probably take that as the probability of involuntarily losing a customer. But we can put all of this into a Markov diagram that describes my possible future relations with your company, like this.

This is quite a bit more interesting than the downed-plane chart. Again, we have two terminal nodes where things can wind up, this time, lost and dropped. In Markov analysis, we'd call these absorbing states, since once you get into one of them, you never get out. But what goes on with the rest of the diagram is quite a bit more complicated. The segment 1, 2, and 3 nodes each have a chance of transitioning back to segment 1; it happens whenever a customer makes a purchase. As you can see, when the annual catalog is received, there is a 30% chance of a purchase from segment 1, a 12% chance that a segment-2 customer will buy, and a 6% chance for a segment-3 one. Each also has a 1% chance of transitioning to lost, since customers in any of these segments are equally likely to die or become disconnected. The remaining arrows capture the rest of the probability, and since the total probability out of any one will add to equal 1, it's easy to figure out.

Well, it's pretty clear that, sooner or later, a customer is going to reach an absorbing state, that they'll go three annual catalogs in a row without ordering, or they'll become lost. But what's not clear is how long that's going to take. In the plane example, by the time you performed two transitions, you were guaranteed to have reached the end of the diagram, and the wreck was either found or it wasn't. Not so here. You could go round, and round, and round.

As a marketer, you're going to want to get some sense of how long you can expect to keep me as a customer, as well as how many purchases you can expect me to make in that time. And with Markov analysis, we can do just this. Let me show you how. First, turn the Markov diagram into a Markov matrix, like this. It's common to call the Markov matrix M. Remember, the row indicates the state from which you're transitioning from, and the column indicates the state that you are transitioning to. Each row has its total probability adding to 1. So the first row shows what happens to a customer in segment 1, a 30% chance of buying from the annual catalog, and so staying segment 1, a 69% chance of not buying and so sliding into segment 2, and a 1% chance of being lost.

Look at the last two rows, too. They show that once you're dropped, you're dropped, and that once you're lost, you're lost. That is, if before the transition, you're in the dropped state, then after the transition, you'll still be there, 100% probability, and so on, forever. That's why it's called an absorbing state.

To find out what happens in two years, we just multiply the transition matrix by itself. A year goes by and then another year goes by. M times M, by the rules of matrix multiplication. You get this. Again, look at the top row. It says that a customer that is currently in segment 1 is 60% likely to be in segment 3 two years from now. And to find out about three years from now, we multiply this new M^2 matrix by M, to get M^3, which looks like this.

That is, about 59% of the customers are gone by year 3, with about 56% being dropped for not buying from catalogs and 3% lost. We could continue this way, year by year, to watch how the top row of this matrix evolves. To get the next year's matrix, we just multiply the previous year's matrix by M. Here's a graphical representation of the evolution. As you can see, almost no customers last 10 years in this model. There's over a 95% chance of being dropped from the mailing list for failing to make a purchase in three consecutive years. There's a 4% chance of being lost. And there's less than a 1% chance of still being active.

There's another way of interpreting Markov transitions that's conceptually different but mathematically identical. We could imagine that we are speaking not of one individual, but of a large population, and the transition probabilities indicate the fraction of that population currently in a given state that transitions to another state. For example, rather than saying that a segment-1 customer has a 30% chance of making a purchase, I could say that 30% of segment-1 customers will make a purchase. From this perspective, our graph shows that in 10 years, we lose about 99% our customers, and 95% of our customers are dropped for not making an order in three consecutive years.

In the real world, Rhenania was seeing just this kind of drop off in its customer base. They followed the industry-accepted model of only mailing to people whose recent purchases more than covered the cost of selling to them. The result was a smaller and smaller volume of profitable customers. Of course, one way to address this is to acquire new customers, for example, from mailing lists, to replace the customers who were dropped or lost. To some extent, Rhenania did this. We can modify our Markov diagram to replace all lost and dropped customers by changing our lost and dropped states from absorbing states into ones that transition back to segment 1. That is, each person who's lost or dropped this year is replaced in the following year by a new segment-1 customer, like this.

The long-term behavior here is going to be very different from what we just saw. In a system with one or more absorbing states, any non-absorbing state that can ever reach an absorbing state is going to eventually be empty. Such eventually empty states are called transient states. Imagine a fountain with many basins draining into each another, maybe some pumps to send water back up to the top, but which includes a basin from which no water escapes. Then any basin that eventually feeds into this no-exit one will be empty in the long run. That was the case for dropped and lost in our original diagram, no exits. But on our new diagram, there are no absorbing states.

In fact, we can say more. Notice that, following the arrows, you can take a tour of this Markov diagram, that will start in 1, go through every single state in the system, and eventually return back to 1. For example, we could go 1, to 2, to 3, to dropped, then back to 1, then to lost, then back to 1 again. A Markov system with this property is called irreducible. In an irreducible system, no state ever runs completely dry.

The system also has a second important property. Because of that loop at state 1, we could take a trip from anywhere to anywhere in this diagram in exactly 20 steps. For example, suppose you wanted to go from lost to dropped in exactly 20 steps. Then start in lost, transition to state 1, hang in state 1 using its little loop, like you're in a waiting room, as long as necessary, and then at the appropriate time, leave it to make your way to from 1, to 2, to 3, to dropped, on the 20th move. There's nothing magic about the number 20 here. As long as you can always get from anywhere to anywhere in some fixed number of transitions, the system is called ergodic. Ergodic systems will eventually stabilize into an equilibrium. The equilibrium of an ergodic system is given by a string of numbers, a vector, called a steady state vector. And knowing what you already know can let you find it.

A state vector in this problem, steady or not, is going to look like this. It has a probability for each of the five states of 1, 2, 3, dropped, and lost. These have to add to one, as any row in any Markov matrix does. And in equilibrium, how things look after the transition is going to be exactly the same as how they look before the transition, steady state, so, $V \times M = V$. Start with V and then do a transition with the Markov matrix M. And you're right back at V where you started. Using the rules of matrix multiplication, this equation is really just a shorthand for five linear equations in the five unknowns, and that system can be solved to find the unique steady state vector. It takes less than a minute to set this up and solve it on a spreadsheet using Solver. Just enter those equations as the constraints of a linear program, pick any objective that you want, and solve it. When you do, we get this.

This is the long-run behavior that we'd expect to see. At the end of any year, we expect to see about 35% of ours customers in segment 1, 24% in segment 2, 21% in segment 3, and about 20% being lost or dropped. This 20% will have to be replaced by new customers for the upcoming year. We can verify this calculation by starting with everyone in segment 1, like this, and then watching the evolution of this system by multiplying this vector by our Markov matrix repeatedly. The results looks like this. And you can see that we're converging to the long-term, steady, state behavior that our equations predicted. Segment 1 has about 35%, and so on.

We can do a lot more with Markov matrices, especially those including absorbing states, but I'll just hint at some of the possibilities here. We can find the fundamental matrix, F, for our original one-customer model. It tells you the expected number of transitions that the customer spends in a column state before eventually being absorbed, given that they started in a row state. Here it is. So, for example, look at the top row, it says that a person who starts in segment 1 spends 1.72 years on average in segment 1. Since a customer starts in segment 1 and returns there only when making a purchase, we know that segment-1 customers on average make 0.72 purchases during their time with you. Similarly, a customer who's originally in segment 1 will spend an average of 1.19 years in segment 2, and 1.03 years in segment 3, for a total of 3.94 years as your customer.

So you send on average 3.94 catalogs and 1.03 reactivation packages, and I give you, on average, 0.72 purchases. If catalogs cost you $1 each and activation packages are $3 each, that means that you're spending about $7 on 0.72 sales, or a little under $10 per sale made. You'd better keep this in mind when you're working on your business model.

You can play fancier tricks with the F matrix to find what fraction of customers are eventually dropped, or eventually lost, or even the average time that a new customer who is eventually lost to death will spend as your customer. The existence of matrix algebra makes Markov analysis a powerful set of tools.

The real-life mail order firm that I mentioned earlier, Rhenania, learned this when they used these techniques, or ones similar to ours, to take a look at their mail order business. The transition diagram of their three-segment customer base looked very much like ours, and similar, although somewhat more sophisticated, version of our analysis was used to fine tune their operations. For example, they allowed the possibility of more than one mailing per year. If they mailed monthly, for example, a customer would have 12 chances to buy before being demoted to a segment-2 customer.

But their basic approach was the same as ours. Use the transition probabilities to see how the system evolves through time. Find what fraction of customers fall into each segment. And use this information and other cost and purchase information to determine how often mailings should be sent, and to whom.

Here's a little spreadsheet that I cooked up to do just this. You can see how as the number of catalogs per year increases, the size of the customer base grows. People who get more catalogs per year have more chances to buy, and so are retained longer. For a while, this translates into increased profits; the height of the green bar shows profits. But eventually, the additional sales from more catalogs is no longer enough to warrant more catalogs. And the profit, the green bar on the right of the graph, declines. In my model, using as many of the real-life figures for Rhenania as I could get my hands on, this happens with 26 catalogs a year. One every two weeks. That's just the answer that Rhenania found, too. They also found, by the way, that it was important to have the catalogs arrive on Saturday. German mail is reliable enough that they could make this happen.

The turnaround for Rhenania as these changes were implemented was dramatic. Before the implementation, Rhenania lost 30% of its customer base in four years. It had experienced declining market share and declining profits. Now, the base started to grow, the sales increased, they managed to acquire two of their competitors, and they moved from the fifth-largest direct marketing firm in Germany to the second largest.

Funny—the more you know about how chance works, the less you have to leave your fate to chance.

Queuing—Why Waiting Lines Work or Fail
Lecture 22

For something that, for the most part, consists of doing nothing, waiting is surprisingly unpleasant. When people have to wait in line, it's annoying and a waste of time; when products wait in line, it's costly. Either way, it's undesirable. But it turns out that the behavior of waiting lines—of queues—is subject to mathematical analysis. Queuing theory allows us to characterize the behavior of such queuing systems and provides us with some ideas about how to make them work better. This lecture will approach queuing theory from the perspective of Markov analysis, meaning that we will ignore everything about the past, except the current state of things.

Queuing Theory

- Suppose that you offer online computer services to customers, and doing so requires that you have two mainframe computers online at all times. When both machines are online, you're in business, and you generate a revenue of $6000 per day. Computers will go down from time to time, of course, but for the moment, let's say that you're taking that risk. Your data shows that, on average, a computer breaks about once every 6.5 months—once every 200 days.

- When one breaks, your income stream stops with it. In fact, it's worse than that. Your clients are making critical transactions, and if they can't rely on your service, they leave. You estimate a loss of $28,000 for every day that the system is down.

- For a big mainframe, it takes an average of 4 days to get a computer that fails back online. Online or not, a mainframe computer costs you $1000 per day, and if it's broken, the repairs cost $1000 per day on top of that. And the repair team can only work on one down computer at a time.

- So, your business is okay if it's not down too often. The following is a Markov diagram for your situation. It has only three states, depending on how many computers are up and running.

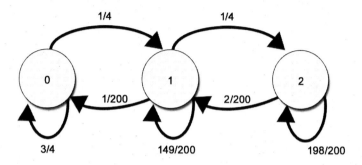

Figure 22.1

- In this diagram, each transition represents one day. The number in the middle of a circle indicates how many computers are up and running on a given day, and the arrows show the possible transitions from state to state. The arrows with the 1/4 probabilities reflect the fact that it takes an average of 4 days to repair a broken computer. If every day a broken computer has a 1/4 chance of being repaired, then it will on average take 4 days to repair it. Having the same chance of repair each day leads to what statisticians call a geometric distribution, and it's what you get out of this kind of discrete Markov process.

- In the same way, a working computer has a 1/200 chance of breaking per day, reflecting that it breaks, on average, once every 200 days. When two computers are working, the chance of a computer breaking doubles, to two in 200.

- Technically, there are some small errors in this model, because both computers could possibly break on the same day, or one computer could break on the same day that the other returns to service. The chance of these events occurring is very small, though.

- If nothing breaks or is repaired on a given day, the system loops back to the same state that it had at the beginning of the day. Sensibly enough, the total probability leaving each state is 1, which is how we found the probability for looping.

- This is indeed a queuing model, although it's probably not what you normally think of when you think about waiting lines. In this case, the "customers in line" are machines waiting to be repaired, and the entire population of potential customers consists of only two individuals.

- We can analyze this **queuing system** using Markov analysis. The system is **ergodic**—you can verify that you can move from any state in the picture to any other state in exactly 4 transitions—and that means that, in the long run, the system is characterized by its **steady state vector**. Using the techniques from Markov analysis, we find that the steady state vector is as follows.

$$\begin{array}{cccc} & 0 & 1 & 2 \\ \text{Start} & [0.00077 & 0.03843 & 0.96080] \end{array}$$

- This means that your system is up with both computers online about 96% of the time, and at least one is down about 4% of the time. We can then find the average profit per day by an expected value calculation: probability times payoff for each possibility, added together. In this case, 96% of the time, you make $4000 per day—that's $6000 revenue minus $2000 in operating expenses. The remaining 4% of the time you lose $30,000 per day, because you're losing customers, paying for machines, and making repairs. On average, that works out to about $2667 profit per day.

- This is not bad, but it's quite a bit less than the $4000 per day you'd be making if your machines never went down. The sizeable difference reflects the fact that you have a customer demand that is strongly affected by waiting; not having two computers online really costs you.

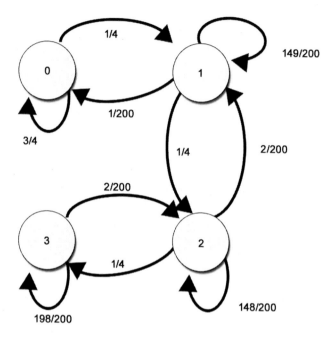

Figure 22.2

- So, is it worthwhile to have a backup computer? The bad part of the idea is that it's going to cost you an extra $1000 per day for another computer. The good part is that you could have one computer fail and still not hurt your business. Is the increased reliability worth the cost of providing it?

- Let's modify the previous Markov diagram and add the new machine. We add one more state for having three machines in operation, with one of them on standby.

- Either of the two active machines might fail when we're in this state, so there is a 2/200 chance of a failure on a given day. We're assuming that the machine that is only on standby can't fail. On the other hand, if a machine is broken, there is a 25% chance it'll be repaired during the day, so we have a 1/4 transition probability from two functioning machines to three. The steady state vector for this Markov system is as follows.

$$\begin{array}{cccc} 0 & 1 & 2 & 3 \\ \text{Start} \quad [0.00003 & 0.00154 & 0.03840 & 0.96003] \end{array}$$

- Now, like before, the system incurs customer dissatisfaction if fewer than two machines are in operation. But, now, that only happens 0.16% of the time. Even taking into account the cost of the backup machine, your average profits per day rise to $2947, as opposed to $2667 without the backup. That translates to an extra $100,000 per year for you and happier customers—a win-win situation. So, queuing theory recommends getting the backup.

- We have to address one weakness in our analysis. We've been assuming that each transition represents one day. That is, we assumed that each day, at most one thing could happen. But, of course, we could have two machines breaking on the same day or a repair and breakdown on the same day. Then again, every repair took a whole number of days—we never got the system back online in, for example, 16 hours. But in real life, we could.

- It's easy to address these issues by moving from a discrete Markov process, such as the one we've been using so far, to a continuous Markov process. In fact, most of queuing theory is done in this continuous process way. We just make a transition represent what happens during a very brief interval of time, of length dt. Our original problem without a backup, for example, would look like the diagram in **Figure 22.3**.

- There's not much difference between this picture and the earlier one, other than the addition of the dt terms. The diagram really just carries forward with the argument we made for daily transitions.

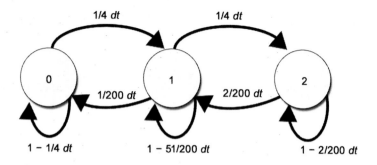

Figure 22.3

- We said that if the average time for a computer repair is 4 days, then on any given day, there is 1 chance in 4 that a broken computer will be repaired, which is where the 1/4s came from in our diagram. But the same logic that led to this 1/4 repair chance per day would suggest a 1/8 repair chance per half day, or a 1/16 repair chance for a quarter of a day, and so on. In general, during a tiny fraction dt of a day, the repair chance would be 1/4 times dt.

- In the limit as dt approaches 0, this picture then describes a continuous process, where transitions can occur at any time. The continuous model gives the same steady state as the discrete one does. And, again, the Markov model is assuming that the chance of transition out of a state during an interval doesn't depend on how long it's already been in that state. That is, the process is **memoryless**.

- Whether we use the discrete or continuous model, this system is going to approach the same steady state in the long run, but with the continuous model we no longer have the concern of two events occurring at the same time. With dt small enough, each transition is a very brief span of time. The chance of a repair or a breakdown occurring during a single transition is quite small, but the chance of more than one occurring during the same interval is so small as to be negligible.

The Human Factor

- When you think about a waiting line, you probably imagine people standing in a line, perhaps a very large number of them. And the size of the calling population—the people that could enter the **queue**—is much larger than the queue length that we generally observe.

- When the **calling units** are people, the solution isn't always found in mathematics. Psychological factors play an important role. Waiting feels longer when you're unoccupied. That's why there are often mirrors near elevators. That's also why the Disney parks always give you something new to look at as you wait for a ride.

- Understanding human psychology can make queues less painful to the people in them. And, even better, understanding the mathematics of queues can help us design systems to give us better traffic flow, better health care, faster emergency response, and more efficient production and service.

- Queuing theory isn't an optimization technique. It doesn't tell you the best thing to do. But it does allow you to consider the consequences of various possible choices, and then you can see what options look most attractive. Playing around with queuing models can reveal many nonintuitive solutions to practical problems.

Important Terms

calling units: General term applied to those individuals or objects that make use of a queuing system.

ergodic: A Markov system is ergodic if there exists some n such that, for any two pure states i and j, it is possible to move from state i to state j in exactly n transitions.

memoryless: A process where the probability of reaching a particular state after the passage of so much time depends only on the current state and not on the events that led to that state. Rolling a die is memoryless.

queue: Waiting line. The study of waiting lines is called queuing theory. Technically, the queue does not include any calling units that are currently in service, only those waiting for service to begin.

steady state vector: In Markov analysis, a state vector that is unchanged by the application of the transition matrix. A Markov system that reaches a steady state vector will never move from it.

system, queuing: All of the calling units either currently under service or in the queue are considered to be in the system.

Suggested Reading

Ragsdale, *Spreadsheet Modeling & Decision Analysis*.

Stevenson, "What You Hate Most about Waiting in Line."

Winston and Albright, *Practical Management Science*.

Questions and Comments

1. This lecture focused on the steady state of a queuing system, and the equations most often presented for queuing systems are for their steady states. Consider questions of traffic congestion. When might this model be the most appropriate? When would we be more concerned about the transient behavior of the system?

   Answer:

   It's an appropriate model when we are interested in the long-term behavior of the system when the conditions are not changing rapidly, such as mid-afternoon traffic on a beltway without an accident. But it's the transient behavior you care about when you want to know how traffic behaves with an accident or when the fans at a football game hit the parking lot.

2. Imagine two different single-server queuing systems that have identical input streams of calling units but with different mean service rates. You'd expect that, on average, customers with the faster server would spend less time in the system. But this may not be so. The inversion occurs when the slower server is steady, with a constant or near-constant service time, while the faster server is erratic, with wildly fluctuating service times. Provide a persuasive argument as to why this could be so.

Answer:

It's an extension of why we still see waits, even when average service rate exceeds average arrival rate. These figures are averages. An erratic worker sometimes will be much faster than average—sometimes much slower. When he or she is fast, the queue may empty, leaving the worker with nothing to do and wasting his or her faster rate. When he or she is slow, he or she creates a huge backlog, and the resulting waiting times for the people waiting to be served grow accordingly.

Queuing—Why Waiting Lines Work or Fail
Lecture 22—Transcript

Yes, thank you. My name is Scott Stevens. I only have a minute, but, there's a problem with my account, I was hoping—what? Yes. I'll hold. But I won't enjoy it, and I don't think I'm alone in this. For something that, for the most part, consists of doing nothing, waiting is surprisingly unpleasant. And we do a lot of it. Estimates suggest that the average American spends between 20 minutes and an hour a day waiting. Even at the low end, that's, 37 billion hours of American time lost to waiting lines each year—somewhere around nine months in the course of a lifetime.

The trouble is, of course, we often can't just decide not to wait. In a society where we rely on the services of others, waiting lines are almost omnipresent: stores, banks, restaurants, restrooms, traffic jams, amusement parks, airports—and some places where too long a wait can be critical, if not fatal, like hospitals and disaster response.

Then again, it's not just people who are doing the waiting. Every time that you go into a fast food restaurant, your food is literally waiting to be served. If it waits too long, it's no longer palatable. A longer wait on the food side results either in an unsavory meal for you or the expense of throwing it away for the restaurant. There's a cost when goods wait to be sold, even if it's just the cost of inventory, like cars sitting in a dealership. When people have to wait in line, it's annoying and a waste of time. When products wait in line, it's costly. Either way, it's undesirable.

But it turns out that the behavior of waiting lines, of queues, as they're called in the U.K., is subject to mathematical analysis. Queuing theory lets us characterize the behavior of such queuing systems and provides us with some ideas about how to make them work better. I think some of the results will surprise you. We're going to approach this field from the perspective of Markov analysis, meaning that we'll ignore everything about the past, except the current state of things.

Let's start with an example where the things in line aren't people. You offer online computer services to customers, and doing so requires that you have two mainframe computers online at all times. When both machines are online, you're literally in business, and you generate a revenue of $6000 a day. Computers will go down from time to time, of course, but, for the moment, let's say you're taking that risk. Your data shows that, on average, your computer breaks once every six and a half months or so, once every 200 days, on average.

When one does break, your income stream stops with it. In fact, it's worse than that. Your clients are making critical transactions, and if they can't rely on your service, they leave. You estimate a loss of $28,000 for every day that your system is down. For a big mainframe, it takes an average of four days to get a computer that fails back online. Online or not, a mainframe computer costs you $1000 a day, and if it's broken, the repairs cost $1000 per day on top of that. And the repair team can only work on one down computer at a time. So your business is OK if it's not down too often.

Let's draw a Markov diagram of your situation. It has only three states, depending on how many of your computers are up and running. In this diagram, each transition represents one day. The number in the middle of a circle indicates how many computers are up and running on a given day, and the arrows show the possible transitions from state to state. The arrow with the $\frac{1}{4}$ probabilities reflect the fact that it takes an average of 4 days to repair a broken computer. If every day, a broken computer has a $\frac{1}{4}$ chance of being repaired, then on average it takes 4 days to repair it. Having the same chance to repair each day leads statisticians to call the result the geometric distribution, and it's what you get out of a kind of discrete Markov process like this.

In the same way, a working computer has a $\frac{1}{200}$ chance of breaking per day, reflecting that it breaks, on average, once every 200 days. When two computers are working, the chance of a computer breaking doubles, to two in 200. Technically, here, there are some small errors in this model, since both computers could possibly break on the same day, or one computer could break on the same day that the other returns to service. The chances of these events are very small, though, so I'm going to ignore those things for now and come back to them a bit later.

If nothing breaks or is repaired on a given day, the system loops back to the same state it had at the beginning of that day. Sensibly enough, the total probability leaving each state is 1, which is how we found the looping probability.

This indeed is a queuing model, although it's probably not what you normally think of when you think about waiting lines. Here, the customers in line are machines waiting to be repaired, and the entire population of potential customers consists of only two individuals. We can analyze this queuing system using the work from our last lecture on Markov analysis. The system is ergodic, you can verify that you can move from any state in the picture to any other one in exactly 4 transitions, and that means that, in the long run, the system is characterized by its steady state vector. Using the techniques from last time, we get the steady state vector as this.

Meaning that your system is up with both computers online about 96% of the time, and at least one is down about 4% of the time. We can then find the average profit per day in an expected value calculation, probability times payoff, added up over all the possibilities. Here, 96% of the time, you make $4000 a day, that's the $6000 revenue minus the $2000 in operating expenses. The remaining 4% of the time you lose $30,000 a day, since you're losing customers, paying for machines, and making repairs. On average, that works out to be about $2667 a day. Not bad but it's quite a bit less than the $4000 a day you'd be making if your machines never went down. The sizeable difference reflects the fact that you have a customer demand that's strongly affected by waiting. Not having two computers online really costs you.

So is it worthwhile to have a backup computer? The bad part of the idea is that it's going to cost you an extra $1000 a day, but the good part is that you could have one computer fail and still not hurt your business. Is the increased reliability worth the cost of providing it? Well, let's modify that last Markov diagram and add the new machine. It's easy. We add one more state for having three machines in operation, with one of them on standby.

Either of the two active machines might fail when we're in this state, so there is a $^2/_{200}$ chance of a failure on a given day. I'm assuming that's on standby can't fail. On the other hand, if a machine is broken, there is a 25% chance it'll be repaired during the day, so we have those $^1/_4$ transition probabilities from two functioning machines, up to three. The steady state vector for this Markov system is this.

Now, like before, this customer incurs a dissatisfaction if fewer than two machines are in operation. But now, that only happens 0.16% of the time. Even taking into account the cost of the backup machine, your average profits per day rise to $2947, as opposed to $2667 without the backup. And that translates into an extra $100,000 per year for you and happier customers, a win-win situation. So queuing theory says, get the backup.

Before we go on, I do want to address one weakness in that analysis. We've been assuming that each transition represents one day. That is, we assumed that each day, at most one thing could happen. But of course, we could have two machines breaking on the same day, or a repair and breakdown on the same day, and so on. Then again, every repair took a whole number of days; we never got the system back online in, say, 16 hours. But in real life, we could. It's easy to address these issues by moving from a discrete Markov process, like we've been using so far, to a continuous Markov process. In fact, most of queuing theory is done in this continuous process way. We just make a transition represent what happens during a very brief interval of time of length, dt. Our original problem without a backup, for example, would look like this.

There's not really that much difference between this picture and the earlier one, other than the addition of those dt terms and what you see. The diagram really just carries forward the argument that we made for daily transitions. What we said was that if the average time for a computer repair is 4 days, then on a given day, there's a one chance in 4 that the machine will be repaired, which is where the ¼ came from. But by this same logic would have led to the idea that in a half day we'd have an $^1/_8$ of repair, or in a quarter day, $^1/_{16}$ of repair, and so on. In general, during any tiny fraction, dt, of a day, the repair chance would be ¼ times dt.

In the limit, as *dt* approaches 0, this picture then describes a continuous process, where transitions can occur at any time. The continuous model gives the same steady state as the discrete one does. And again, the Markov model is assuming that the chance of transition out of a state during an interval doesn't depend on how long it's already been in that state. That is, the process is memoryless. We'll come back to this idea in a few minutes.

But whether we use the discrete or continuous model, this system is going to approach the same steady state in the long run, but with the continuous model, we no longer have the concern of two events occurring in the same time. With *dt* small enough, each transition is a very brief span of time. The chance of a repair or a breakdown occurring during a single transition, quite small, but the chance of more than one occurring during the same interval is so small as to be negligible. Still maybe this situation doesn't feel like a waiting line to you. When you think about a waiting line, you probably imagine people standing in line, perhaps a very large number of them. And the size of the calling population, the people that could enter the queue, is much larger than the queue length that we generally observe. The number of people checking out of a grocery store might be sizable, but, it's far smaller quantity than the population of everyone who ever goes to that grocery store.

In such cases, mathematicians usually treat the population as if it were actually infinite. It makes the math easier, and the error introduced by doing so is generally quite small. Generally, if no more than 5% of the relevant population is usually at the queuing facility, you don't make much of an error by considering the population to be infinite. And if the population has unlimited size, then in general, the length of the queue could be anything, too. Sometimes, like with call waiting on a landline, the system can become full and not accept any more arrivals. But the simplest model for a queuing system is, if someone wants go get in line, let them get in line.

Let's look at this kind of queuing system, assuming that we only have one server. This would be like a convenience store with only one cashier. In keeping with our Markov idea, we're going to assume that there's an equal chance of a new arrival during any small time interval, but that if a customer is already being served, that customer has a fixed chance of completing service during each small time interval. This model is usually called the

M/M/1 model, which is short for Markov/Markov/1. The first Markov talks about the pattern of new arrivals, the second Markov talks about the pattern of service times, and the 1, that means we only have one server. Here's the picture.

The number inside each circle is the number of customers in the system being served right now or waiting to be served. The picture extends off to the right forever. Traditionally, we use λ to represent the arrival rate of customers per unit time, and μ, the next letter in the Greek alphabet, to represent the average departure rate due to customers being served per unit time. dt is some tiny fraction of one unit of time, which in the limit goes to 0. When someone shows up, the number of people in the system increases by 1. When someone leaves, the number decreases by 1. I've suppressed the probabilities on the loops, where nothing happened during the short time interval dt, but they're easy to recover. Remember that all of the arrows leaving a node have to always add up the total probability of one.

Well, a steady state vector for this system is going to tell us how often, in the long run, the system will be in each of these states. It's easiest to find the steady state vector if we forget about probabilities for the moment and think of the diagram as a rather elaborate water fountain. Each state, each circle, is a basin containing a certain amount of water, and the arrows tell you what fraction of that water will move along that arrow to a different basin during one transition. Dump 1 gallon water in this fountain wherever you want, and let it run until things steady out, with the water level in each basin stabilized. The number of gallons in each of the basins will give you the steady state vector of the system. In this stable, steady state configuration, there are P0 gallons of water in basin 0, P_1 gallons in basin 1, P_2 gallons in basin 2, and so on. Obviously, if you add all these Ps, you have to get 1, since we started out with one gallon.

OK, now focus just on the arrows joining basin 0 to basin 1. In the steady state, there are P_0 gallons of water in basin 0, and the fraction that moved to basin 1 during a transition is $\lambda\ dt$. So in one transition, P_0 times ($\lambda\ dt$) gallons move from basin 0 to basin 1. At the same time, there are P_1 gallons of water in basin 1, and the fraction moving to basin 0 during the transition is, according to the arrow, $\mu\ dt$. So P_1 times ($\mu\ dt$) gallons move from basin

1 to basin 0. If we're talking steady state, the transition doesn't change the amount of water in any basin. This means that what flowed from 0 to 1 must be exactly the same as the number of gallons that flowed from 1 to 0. That is, $P_0 \times (\lambda\, dt) = P_1\, (\mu\, dt)$. Solve this for P_1 and you get $P_1 = (^\lambda/_\mu)P_0$. You can apply exactly the same reasoning to each other pair of adjacent basins, and you get parallel results, so, $P_2 = (^\lambda/_\mu)P_1$, $P_3 = (^\lambda/_\mu)P_2$, and so on forever. In general, $P_{i+}1 = (^\lambda/_\mu)\, P_i$.

Now, remember that all of these P_i are actually probabilities, and they have to add up to one, $P_0 + P_1 + P_2$, and so on, equals 1. The sum goes forever. It just says that the chance of no customers, plus the chance of one customer, plus the chance of two customers, and so on forever, has to add up to 100%.

And we can use this to find all of the P_i. Like this. Start with our equation, and make a second copy of it, but multiply the second copy through by $^\lambda/_\mu$, like this. Again, remember that these sums go on forever. OK, now recall that our Markov work said that when you multiply a P_i times $(^\lambda/_\mu)$, you just get the next P in the list, $P_{i+}1$. So let's rewrite that lower formula. Now subtract the lower formula from the upper one. On the left side, everything cancels out, except for P_0! Remember, these sums go on forever, so there is no last term left over on the bottom there. On the right, we get $1 - ^\lambda/_\mu$. So, $P_0 = 1 - ^\lambda/_\mu$ in this model, and each P_i after that is $^\lambda/_\mu$ times the P that came before it.

So if I tell you the arrival rate λ and the service rate μ, you'll know all of the P_i. And these numbers have a lot of practical significance. For example, P_0 is the fraction of the time that the system has no customers in it; no one's being served; no one's waiting in line. From the point of view of the people providing the service, that is not a good situation. They're paying to supply a server, but the server has nothing to do. From the point of view of the service provider, then, a small value of P_0 is desirable. You don't want to have your worker goofing off.

But turn it around and look from the customer's perspective. In a single server system, you like it when there is no one in the system when you arrive in the line; that means you are served immediately. If the server is busy 90% of the time, then 90% of the time you have to wait for the server to get to you. If the customers are impatient, this kind of delay may be worse for business than having servers that aren't that busy. That's why luxury hotels have a lot more staff than budget motels.

Since the fraction of the time that the server is free in a M/M/1 model is $1 - \lambda/\mu$, as we just found, then the fraction of the time the server that the server is busy is 1 minus this, just λ/μ, the arrival rate divided by the service rate. With a little thought, you could probably have guessed this. If customers arrive at an average of 4 per hour and if the server can handle an average of 5 per hour, then on average the server is only busy $4/5$ of the time, since on average $1/5$ of the server's capacity is going to be unused. And that means that P_0, the chance that there are no customers in the system, is $1/5$, and so $1/5$ of the time, a newly arriving customer gets served without waiting.

But wait a second! If the server can handle customers faster than they're coming in, why does anybody ever have to wait? Imagine an assembly line where assemblies come down the line every 5 seconds, but it only takes 4 seconds to add your part to the assembly. You work for 4 seconds, rest for 1, work for 4, rest for 1, and so on, and no assembly ever has to wait for your service.

True, but remember that we're assuming that arrivals follow a Markov process, with each tiny time interval being equally likely to contain an arrival. What does this memoryless arrival process look like? Think about popcorn popping. Not right when it starts, or near the end, but the part in the middle. You know what it sounds like, like this.

My friends, every Markov arrival process, which is also called a Poisson process, is going to sound just like popcorn popping, except that it may be either sped up or slowed down. It turns out to be a good model of a lot of different arrival processes. Telephone calls at a switchboard, requests for a document on a web server, people arriving at a grocery store, cars driving past a mile marker on the highway. And it's an oddity of the mathematics that, if each little time interval has the same, independent chance of containing an arrival, then the observed arrival pattern tends to include clusters of arrivals interspersed with relatively dead intervals—which solves the mystery of why popcorn popping sounds the way that it does. It's not peer pressure. A kernel doesn't say, oh, all of my friends are popping! I've got to! It's just a Poisson process at work.

Now, let's look again at our waiting at the line, where people arrive at an average of four per hour. That means that there's a $^1/_{15}$ chance of an arrival during any minute. Think of really slow popcorn. There are going to be some times with long gaps between one arrival and the next, and the server is generally going to be standing around with nothing to do. But sometimes we'll get clusters of several people arriving by closely separated times, and then some of them are probably going to have to wait. And remember we've got similar uncertainty in service times. The server, on average, handles five customers per hour, or 12 minutes per customer, but that's only an average. The actual times are spread out, like the time between consecutive pops of popcorn, sometimes very short, occasionally quite long.

We can use our equations for our Ps to find out how much effect this variability adds to our system. We said if λ is four customers arriving per hour, on average, and μ is five customers per hour, the service rate, on average, then the server is busy $^4/_5$ of the time, so $P_0 = {}^1/_5$. But recall that for this simple model, each P is the P before it, multiplied by $^\lambda/_\mu$, and here, that's $^4/_5$. So we get this. Each bar is $^4/_5$ as tall as the one to its left.

To tell you the truth, I still find this kind of surprising. If the arrival rate is only 80% of the service capacity, I think that most of us would imagine that lines wouldn't generally be much of a problem. But you can see here that a single server system, with these kind of random arrivals and service times, is going to have a considerable line a lot of the time. Usually, the system will have three or more people in it, and more than 10% of the time, it will have 10 or more people in it. By the system, I mean the people waiting in line along with any people currently being served. The average number of customers in this system, in the long run, turns out to be 4.

For my money, though, the single queuing statistic of greatest importance is the average time that a customer spends in the system, from the moment that that customer joins the system, until the moment that his or her service is completed. This is called W, which stands for the word wait, since it's the average waiting time. You can use the values of the P_i for the M/M/1 model to find the value of W, but the work involves infinite series. When the smoke clears, though, it turns out that the answer has a particularly simple form, $W = \frac{1}{(\mu - \lambda)}$, which I think is just a lovely equation.

Applying it to our current example, where customers arrive at a rate of 4 per hour and the server can handle 5 per hour, this means that a newly arriving customer will spend an average $\frac{1}{(5-4)}$, or 1 hour in the system. Yeah. That's right. An hour. If customers arrived evenly spaced in time and service times were constant, everyone would be served immediately, and every one would spend 12 minutes in the system. But the random nature of arrivals and service has increased this by 400%! And that's only the average; some customers spend much longer than an hour.

Now, this is very simple model, but it provides some pretty important insights into how queuing systems behave in general. Random variations in arrival times and service times can considerably degrade system performance. A system can be made much more efficient if, for example, arrivals can be scheduled, such as making appointments. But in much of Western life, we enjoy the flexibility of requesting service whenever the whim hits us. Arrivals, then, look pretty much like a Markov process, and the problems that we're seeing here manifest. And, of course, you can't schedule arrivals in some situations, like, when computers break down or people need emergency medical attention.

Another important point, the problem only gets worse as the system is pushed closer and closer to its capacity. In fiscally challenging times, many organizations try to do more with less, so they may add to the load on a queuing system without increasing its capacity. Let's get a feeling for the effect of this with our current queuing example. Suppose we decide to increase the average load on our queuing system by only 20%, from 4 customers per hour to 4.8 customers per hour. People who have never studied queues tend to expect this to degrade system performance by about 20%. People may have to wait 20% longer for service, for example. Is it true? Well, let's use our M/M/1 equation for the average time in system; $W = \frac{1}{(\mu - \lambda)}$, which is now $\frac{1}{(5 - 4.8)}$. That's 5 hours. This is staggering. A customer is actually in service for only 12 minutes. But to get that service, the average time in the system is now 5 hours! A 20% increase in load resulted in a 400% increase in the average time a customer spends in the system!

As the load on a queuing system approaches its capacity, the performance of the queuing system degrades with mind-blowing speed, assuming that there's no limit on how long the line is allowed to be. If you ever saw the *I Love Lucy* episode, maybe you can recall the scene in the candy factory, classic comedy: two women wrapping candies that are brought to them on a steadily moving conveyor belt. All is going wonderfully, until the spacing between the candies shrinks, and shrinks, and shrinks; λ, the arrival wait, is increasing, and μ, the service rate, is staying the same. You might not be able to predict that Lucy ends up shoving the candies into her mouth, her clothes, her hat, but you can certainly predict disaster.

In most queuing systems, the problem is just as severe, if not so amusing. When the arrival rate grows bigger than the capacity of the servers, the overload work backs up further and further, and the system never reaches a steady state. Our equation for W actually falls apart entirely in this case, and gives a negative answer. Mathematically, it's because we passed the radius of convergence for the infinite series that describes this. More simply, we got a nonsense answer to a nonsense question. The system has no steady state.

So what do you do when you have too much work? The obvious answer is to add more workers. On the candy conveyor line, we just saw that two servers, which clearly that wasn't enough. Or you can limit the jobs you accept. If I have traditional call waiting, for example, I can be serving one person and having one person waiting to talk to me, but if another person calls, they are not permitted to wait. They get a busy signal. There are a lot of ways that you can modify a queuing system. You can introduce a priority discipline, like in an emergency room, where more important calling units are taken first.

The thing is, each of these different choices changes the equations that describe the model. Most of them are derived in a way similar to the way that we analyzed the Markov diagram for the M/M/1, but with more complicated math. There are tons of queuing theory models, and you can often look up the equations that fit your situation. But some queuing systems are so complicated that it's just easier to simulate the model of the system and work with that. We'll be discussing simulation in our next lecture. Once you've got your model, though what do you do with it? Queuing theory isn't an optimization technique. It doesn't tell you the best thing to do. But it does allow you to consider the consequences of your various possible choices and then you can see what options look most attractive.

Suppose we're talking about a tool crib in a factory. The people waiting are the workers who need the tools from the crib. Suppose these are skilled workers who cost the company $20 an hour. The workers in the crib are less skilled, and cost $12 per hour. In a lot of the queuing problems with people, we have to worry about the fact that people impatient; they're customers, and losing their good will and patronage is bad for business. In this problem, though, the people in line are employees. We're not worried about their impatience; we're worried that we're paying them $20 an hour to stand in line!

Let's stick with our earlier M/M/1 model and assume an average of 4.8 workers per hour that want the tools, and a server able to handle 5 workers per hour. Recall the average time in the system for a worker was a whopping 5 hours. What is this costing us? Well, each skilled worker that joins the queue stays an average of 5 hours, and they cost us $20/hour. That's $100 per worker that gets in line, and on average, 4.8 get in line per hour. So on average, we're losing $480 worth of skilled labor each hour. Add the $12 for the tool crib worker and that's $492 per hour. That's obviously not acceptable.

But we can work out the same kind of calculation for 2 or more workers, using somewhat more complicated equations. And here are the results. As you can see, adding just one more server reduces the average waiting time from 5 hours to 15.5 minutes, and that includes the 12 minutes of actually being served. Hourly cost drops from $492 to just under $49, a 90% cost decrease. More than two tool-crib attendants aren't worth it, though, the additional reduction in waiting times doesn't pay for their cost. So, two tool-crib workers is the way to go.

Playing around with queuing models can reveal a lot of non-intuitive solutions for practical problems. For example, imagine two secretaries, each with a boss, and each with enough work to be busy $^3/_4$ of the time. It turns out that if the two secretaries are reorganized into a pool, handling the work of the two bosses on a first-come, first-served basis, the average time to complete a job is cut almost in half, even though neither secretary is working harder than they were before. The secretarial pool helps to address the clustering problem of Poisson arrivals.

I actually used this idea to help a telecommunications company give better customer service to clients who called the [company's] regional offices with questions. If a call wasn't region specific, we could reroute it to an operator who was free in another regional office—simple and effective. The firm ended up using fewer operators, but giving better service.

But when the calling units are people, the solution isn't always found in mathematics. Psychological factors play an important role. The Houston airport was receiving complaints from customers about long waits at baggage claim. They added more baggage handlers and got the average time from a disembarkation to collecting bags to 8 minutes, well within industry standards. But they still got complaints. On closer examination, they realized that the passengers were spending one minute walking from the plane to baggage claim and seven minutes waiting at baggage claim for their bags. The airport moved the baggage claim farther from the planes, so it took passengers six minutes to walk there. Now they only waited two minutes for their bags, and the complaints ceased. Now I think about that every time I take that hike at an airport.

And waiting feels longer when you're unoccupied. That's why there are often mirrors near elevators. That's why the Disney parks always give you something to look at as you wait for a ride. In fact, the Disney parks are the masters of the psychological aspects of managing queues. The single-file queues move at a brisk pace, making customers feel like they're making good progress. The times posted, like 30 minutes from this point, always overestimate the actual average time from that point. Having mentally committed to a 30-minute wait, guests feel they're ahead of the game when they're boarding the ride after only 23 minutes. And there's a chance you might be able to get your picture taken with Chip and Dale.

Waiting in line is boring. But the study of waiting lines is anything but. Understanding human psychology can make queues less painful to the people in them. And even better, understanding the mathematics of queues can help us design systems to give better traffic flow, better health care, faster emergency response, and more efficient production and service.

Monte Carlo Simulation for a Better Job Bid
Lecture 23

In this lecture, you will learn about simulation. With simulation, we numerically model the relationships between the important parts of the problem: the decisions that we control and the random events that we don't. It's often called a Monte Carlo simulation, a name that comes from the famous casino in Monte Carlo, Monaco. We generate random values for the random variables in our problem and then see how a proposed strategy works when faced with these random events. The technique is remarkably powerful and versatile, making it, like linear programming, one of the most widely used techniques in operations research.

Simulation

- Suppose that you're a construction contractor who is bidding on a construction job. It's a competitive bid, so your success is going to be dependent on your ability to determine, with reasonable accuracy, what completing the job is going to cost. You have enough experience with jobs like this that you can model its different activities pretty well.

- Preparing a bid will cost you $5000, win or lose. If you actually land the job, you'll have other costs: materials, labor, and possible late fees if your work gets behind schedule. But, of course, with the winning bid, you'll get paid for your efforts.

- So, what should you bid? Bid too high and you have little chance of getting the job, not to mention losing the $5000 prep fee. Bid too low and you'll get the job, but you may be sorry that you did. Your profit is going to be whatever you bid minus your expenses. We're going to use simulation to get a handle on what those expenses might be.

- To begin, we're not going to worry about how to handle the randomness—we'll just put in average values, and then make them stochastic in a second pass through the model. So, let's work out the cost of completing the job by using a Calc spreadsheet.

- The prep cost is a fixed $5000. The materials cost is going to vary from job to job, but for now we'll use the average of similar jobs in the past, $60,000.

- Next comes labor cost and late penalty—but the labor cost depends on how many weeks of labor we need, and the penalty depends on how long the whole job takes. Again, let's stick in fixed values for these random quantities for now: 9 weeks of labor and 2 weeks of delay. There are a few extra columns, too, for things that we'll need to figure out what the total cost of the job will be.

	A	B	C	D	E	F	G	H
1	Simulation 1: Actual Job Cost							
2								
3	Prep cost	Materials cost	Labor weeks needed	Labor cost	Weeks delay	Job complete in week #	Completion penalty	Total cost
4	$5000	$60,000	9		2			

Figure 23.1

- Your laborers get paid by the week: $700 per person for each week of work. You currently have 4 workers. So, your labor cost will be given by the following formula: 2800 times the number of labor weeks needed.

	A	B	C	D	E	F	G	H
1	Simulation 1: Actual Job Cost							
2								
3	Prep cost	Materials cost	Labor weeks needed	Labor cost	Weeks delay	Job complete in week #	Completion penalty	Total cost
4	$5000	$60,000	9	=2800*C4	2			

Figure 23.2

- Labor weeks needed was in cell C4, so the formula for labor cost is 2800 times C4. Every formula in a spreadsheet starts with an equal sign, and the asterisk means multiplication. (If you're using Excel rather than Calc, all semicolons become commas in Excel.) When we enter this in the spreadsheet, it computes the value: $25,200.

- When is the job done? To get completion time, take the labor time and add the delay time.

	A	B	C	D	E	F	G	H
1	**Simulation 1: Actual Job Cost**							
2								
3	Prep cost	Materials cost	Labor weeks needed	Labor cost	Weeks delay	Job complete in week #	Completion penalty	Total cost
4	$5000	$60,000	9	$25,200	2	=C4+E4		

Figure 23.3

- With our current numbers, that's 11. Next, we have to deal with the completion penalty. Part of the contract for this job is that it needs to be finished in 12 weeks or less. Finish it late and you owe a penalty of $12,000 for every week that you're over the limit. Again, we need a formula. One way to get it is an IF statement, as shown in **Figure 23.4**.

	F	G	H
1			
2			
3	Job complete in week #	Completion penalty	Total cost
4	11	=IF(F4<=12; 0; 12000*(F4−12))	

Figure 23.4

- IF statements always say the following: If the first thing is true, then write the second thing; otherwise, write the third thing. So, in this case, it says: If cell F4—the job completion time—is less than or equal to 12, then write 0 (no penalty); otherwise, write the penalty of 12000*(F4−12), because every week over 12 costs you $12,000.

- When we enter this formula, we get a late penalty of 0 this because our job was done in week 11. All that's left is to add the costs.

	A	B	C	D	E	F	G	H
1	Simulation 1: Actual Job Cost							
2								
3	Prep cost	Materials cost	Labor weeks needed	Labor cost	Weeks delay	Job complete in week #	Completion penalty	Total cost
4	$5000	$60,000	9	$25,200	2	11	$ -	=A4+B4+D4+G4

Figure 23.5

- This comes out to $90,200. And that's our model for how much the job will actually cost.

	A	B	C	D	E	F	G	H
1	Simulation 1: Actual Job Cost							
2								
3	Prep cost	Materials cost	Labor weeks needed	Labor cost	Weeks delay	Job complete in week #	Completion penalty	Total cost
4	$5000	$60,000	9	$25,200	2	11	$ -	$90,200

Figure 23.6

- But this model is deterministic. We still have to put in the randomness—for materials cost, labor time needed, and delay time.

- The materials for the construction on a job like this average $60,000, but there's a fair amount of variation in this. Looking at similar jobs in the past, you decide that the cost of materials is approximately as shown in **Figure 23.7**.

- This picture is called the probability density function of this **random variable**. It shows that costs that are close to $60,000 are common and that the farther you get from $60,000, the less common that cost is. In statistics, a bell-shaped curve like this is called a normal distribution. This one has a mean of $60,000. Its spread, as measured by standard deviation, is $4000.

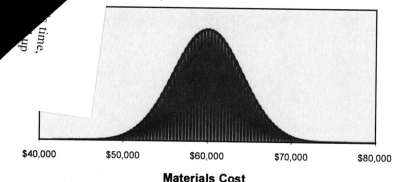

$40,000	$50,000	$60,000	$70,000	$80,000

Materials Cost

Figure 23.7

- We need to generate random materials costs for our model. If you're using a simulation add-in to your spreadsheet (such as Analytic Solver, Crystal Ball, or @Risk), there are straightforward ways to get this, but for Excel and Calc, we are going to introduce a formula that gets the job done. For our normally distributed materials cost with a mean of $60,000 and a standard deviation of $4000, we can generate values by using the following.

  =NORMINV(RAND(); 60000; 4000)

- Every time you enter something new anywhere in the spreadsheet, the sheet calculates a new value of RAND in this cell, and that means you're going to get a new value for your materials cost.

- Labor costs are a fixed $2800 per week, but the number of weeks required to complete the project is uncertain. The following is a table of possible times and their probabilities. The table deals with a discrete random variable. It can take on only a finite set of values.

Labor completion time with 4 workers	
Weeks	Prob.
7	0.1
8	0.3
9	0.3
10	0.1
11	0.05
12	0.07
13	0.08

Figure 23.8

- How do we get this to work in our spreadsheet? If we're not using an add-in, we need to enhance this table a bit first. We'll paste it in the spreadsheet and, to its left, add a range of values for each row, as shown in **Figure 23.9**.

	A	B	C	D
13			Labor completion	
14			time with 4 workers	
15	RAND() from...	RAND() up to...	Weeks	Prob.
16	0	0.1	7	0.1
17	0.1	0.4	8	0.3
18	0.4	0.7	9	0.3
19	0.7	0.8	10	0.1
20	0.8	0.85	11	0.05
21	0.85	0.92	12	0.07
22	0.92	1	13	0.08

Figure 23.9

- The =RAND() function always returns a value between 0 and 1, and all values in that range are equally likely.

- To get Excel or Calc to follow this rule, we use a command called VLOOKUP ("vertical lookup"). It uses the table we just made. The command will look like the following.

	A	B	C	D
13			Labor completion	
14			time with 4 workers	
15	RAND() from...	RAND() up to...	Weeks	Prob.
16	0	0.1	7	0.1
17	0.1	0.4	8	0.3
18	0.4	0.7	9	0.3
19	0.7	0.8	10	0.1
20	0.8	0.85	11	0.05
21	0.85	0.92	12	0.07
22	0.92	1	13	0.08
23		Labor time	=VLOOKUP(RAND(); A16:D22; 3)	

Figure 23.10

- The first argument of VLOOKUP is the number to look up, which is going to be random. The second argument tells us where in the spreadsheet this 4-column table is, by specifying its upper-left and lower-right corners.

- The final argument specifies which column of that table contains the information that we want. In this case, it's the third column of the table, which contains the number of weeks of labor.

- This is the formula that goes in the weeks of labor cell in our spreadsheet. In a parallel fashion, we'll set up a table that tells us how many weeks of delay we have. The following is our simulation.

	A	B	C	D	E	F	G	H	I
8	**Simulation 1: Actual Job Cost**								
9									
10	Prep cost	Materials cost	Labor weeks needed	Labor cost	Weeks delay	Job complete in week #	Completion penalty	Total cost	
11	$5000	$60,000	9	$25,200	2	11	$ -	$90,200	
12									
13			**Labor completion**						
14			**time with**				**Weeks of delay**		
			4 workers						
15	RAND() from...	RAND() up to...	Weeks	Prob.		RAND() from...	RAND() up to...	Weeks	Prob.
16	0	0.1	7	0.1		0	0.51	1	0.51
17	0.1	0.4	8	0.3		0.51	0.77	2	0.26
18	0.4	0.7	9	0.3		0.77	0.9	3	0.13
19	0.7	0.8	10	0.1		0.9	0.96	4	0.06
20	0.8	0.85	11	0.05		0.96	0.99	5	0.03
21	0.85	0.92	12	0.07		0.99	1	6	0.01
22	0.92	1	13	0.08					

Figure 23.11

- Every time the spreadsheet recalculates, the cost will change, because we'll have different random numbers. You can force the spreadsheet to do this by pressing the F9 key. By doing this, it becomes apparent that it's not rare to have a late penalty and that they can be expensive. We also get some idea of what variation we can expect in job cost.

- What we want the spreadsheet to do is to repeat the simulation many times—just as we can do by pressing the F9 key—but to record the result of each trial. Once again, simulation add-ins will do this automatically.

- The results give the values for 3 quantities for the first 15 simulation runs. We commanded the spreadsheet to repeat our situation 1000 times and to record the values of these three quantities in each trial.

Trial	Job duration	Penalty	Total cost
1	13	$ 12,000	$ 121,064
2	10	$ -	$ 86,783
3	13	$ 12,000	$ 105,525
4	13	$ 12,000	$ 100,551
5	14	$ 24,000	$ 130,529
6	10	$ -	$ 79,923
7	15	$ 36,000	$ 134,784
8	10	$ -	$ 81,127
9	13	$ 12,000	$ 110,290
10	9	$ -	$ 90,962
11	15	$ 36,000	$ 129,195
12	10	$ -	$ 91,488
13	11	$ -	$ 93,666
14	15	$ 36,000	$ 143,968
15	10	$ -	$ 89,033

Figure 23.12

- We can use the functions MAX, AVERAGE, and MIN to find the biggest, average, and smallest values of each variable. The following is what we get for the three variables we are tracking.

Summary of 1000 trials			
	Job duration	Penalty	Total cost
Maximum	19	$ 84,000	$ 121,064
Mean	11.249	$ 6,960	$ 86,783
Minimum	8	$ -	$ 105,525

Figure 23.13

- These numbers will fluctuate a bit every time you recalculate the spreadsheet; every recalculation generates new random numbers. But the results over 1000 runs are pretty stable. Total costs average about $98,000, but the range is quite wide—from around $75,000 to over $190,000.

- Even though the average cost of a job is just over $98,000, over 5% of the time the cost is $130,000 or more, and costs of $120,000 or more occur about 1 time in 8. If cash flow is an issue for your company, this is something to keep in mind, because a bid of $120,000 on a job like this, even if accepted, will actually lose you money 1 time in 8.

- Why is this happening? Anything over 12 weeks starts the late fee clock, and at $12,000 per week, those fees can quickly eat up your profits. In fact, we're paying on average almost $7000 in late fees, but it can go as high as $72,000.

What-If Analysis

- One of the wonderful things of the analytic approach in general and of simulation in particular is what-if analysis. A simulation can allow you to explore the impacts of changes in your situation. In this case, it's going to help us get a grip on our bottom line.

- In order to prepare for this, we have to modify our simulation model. One of the precepts of good modeling, particularly in a spreadsheet environment, is that if there is a constant in your problem—a number—it deserves its own cell in your spreadsheet. Then, whenever you want to use that number in a formula, the formula should refer to that cell rather than containing the number

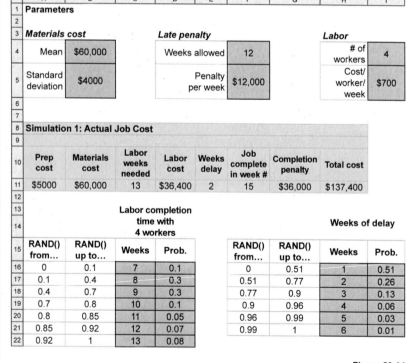

	A	B	C	D	E	F	G	H	I
1	**Parameters**								
2									
3	*Materials cost*			*Late penalty*				*Labor*	
4	Mean	$60,000		Weeks allowed		12		# of workers	4
5	Standard deviation	$4000		Penalty per week		$12,000		Cost/ worker/ week	$700
6									
7									
8	**Simulation 1: Actual Job Cost**								
9									
10	Prep cost	Materials cost	Labor weeks needed	Labor cost	Weeks delay	Job complete in week #	Completion penalty	Total cost	
11	$5000	$60,000	13	$36,400	2	15	$36,000	$137,400	
12									
13			**Labor completion time with 4 workers**				**Weeks of delay**		
14									
15	RAND() from...	RAND() up to...	Weeks	Prob.		RAND() from...	RAND() up to...	Weeks	Prob.
16	0	0.1	7	0.1		0	0.51	1	0.51
17	0.1	0.4	8	0.3		0.51	0.77	2	0.26
18	0.4	0.7	9	0.3		0.77	0.9	3	0.13
19	0.7	0.8	10	0.1		0.9	0.96	4	0.06
20	0.8	0.85	11	0.05		0.96	0.99	5	0.03
21	0.85	0.92	12	0.07		0.99	1	6	0.01
22	0.92	1	13	0.08					

Figure 23.14

itself. This allows you to change one number in your spreadsheet and have the effects of that change percolate throughout everything on the sheet. (See **Figure 23.14.**)

- This is essentially the same simulation we had before, but now everywhere we had a number, such as 12, we replace it with the name of the cell that contains that "12" in our parameters section. We ended up with six parameters.

- There is one change to be made in our simulation. We've now made the number of workers an explicit parameter in the problem, but the table for the labor time is written with the assumption of 4 laborers. How long does the job take when we change the number of laborers?

- We can answer this by looking at the labor requirement not in weeks, but in man-weeks. A man-week is the amount of work that one man—or woman—can do in a week. So, if 4 people working on a job can do it in 5 weeks, it takes 4×5, or 20, man-weeks. It could have been done by 5 people working for 4 weeks instead or 10 people working for 2 weeks.

- To compute the time to complete the job, we'll first find the number of man-hours implied by our 4-worker table. That just means multiplying the time given in the table by 4. Then, we'll divide that by our actual number of workers to find out how many weeks they'd need for the job. Finally, we'll round that up to the next whole number, because workers get paid by the week, or fractions thereof.

- The old formula for labor time was the following.

    =VLOOKUP(RAND(); A16:D22; 3)

- The new one is as follows.

    = ROUNDUP(4 * VLOOKUP(RAND(); A16:D22; 3) / I4; 0)

- Start with the old formula for how many weeks it takes 4 workers to finish the job. Multiply that by 4 to get the man-weeks for the job. Then, divide that by how many workers we have. That's in cell I4. Finally, round the result up to the next whole number. The zero at the end tells Calc to round to zero decimal places—that is, to a whole number.

- We'll also modify the labor cost to say that even if the job is done in less than 5 weeks, the workers still have to be paid for 5 weeks of labor. In the following, I4 is the number of labors, I5 is what we pay each per week, and C11 is the number of weeks of labor that we need. If it's less than 5, we still pay for 5.

  =MAX(C11,5)*I4*I5

- More workers should be able to get the job done faster, cutting down on late penalties. On the other hand, you have to pay them. But because the job is completed faster, you don't have to pay them for as long. The following shows what happens to total cost as we change the number of workers.

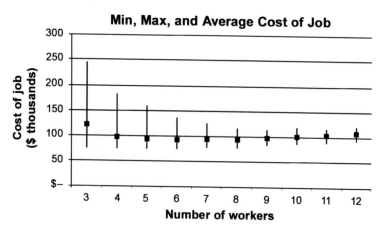

Figure 23.15

- The small squares show the average cost of a job with the specified number of workers, and we see that we can save some money, on average, by hiring more. Six seems to be the best number, saving us about $4000 on average, even though we need to pay more workers.

- The vertical bars show the maximum and minimum costs in the 1000 simulated trials. With 8 workers—twice the original complement—the average price is about $2000 dollars cheaper than its original value, but the job price never rose above $115,000. The job is 99.9% likely to be completed on time—a fact you can use in negotiating with the customer. In fact, it's usually done by the sixth week.

Important Term

random variable: A measurable quantity, often numeric, associated with the outcome of a stochastic experiment.

Suggested Reading

Ragsdale, *Spreadsheet Modeling & Decision Analysis*.

Online Instructions for Creating Simulation Data Tables

Calc (kludgy and tedious): http://aaronjackson.net/enblog/?p=1137.

Excel (much easier): http://www.youtube.com/watch?v=tpIhQuxQeNs.

Questions and Comments

1. It's possible to model a queuing system by either using the queuing equations discussed in the previous lecture or simulation techniques, as discussed in this one. What are the relative advantages and disadvantages of the two approaches?

Answer:

The queuing models have fairly strong assumptions. If these assumptions are violated by a substantial amount, the model is inappropriate. Our queuing work also focused on long-term behavior only, and our equations, for the large part, reported only expected values. Simulation allows us to handle different model assumptions, to investigate transient behavior as well as steady state behavior, and to see the range of possible outcomes—not just the mean.

On the other hand, if the queuing equations apply to a question at hand, then they are faster to apply than a simulation model. Because they do not involve taking random samples, they are also not subject to sampling error in their answers. Finally, their explicit functional form allows one to use calculus or similar tools to investigate how the system's behavior changes as its parameters change.

2. Our simulation in this lecture is a nice demonstration of what is sometimes called "the flaw of averages." People without much experience in dealing with variability often assume that you can find the average value of an output by computing its value for the average values of its inputs. In our model, the average number of weeks of labor required for the job is 9.23, and the average delay time is 1.87 weeks. That means that the average time to job completion is 11.1 weeks, which is under our 12-week deadline. Such people conclude that they don't have to worry about late penalties—when the reality, as seen in the lecture, is that the average late fee is almost $7000.

Monte Carlo Simulation for a Better Job Bid
Lecture 23—Transcript

We've been addressing randomness more directly in our recent lectures, including stochastic processes to help us to predict what to expect. For a lot of our techniques, we've considered the best solution to be the one that gives the best expected payoff, the best average result.

In some cases, that's as it should be. When you face the same situation over and over, and all that you really care about is the total payoff at the end, expected payoff is exactly what you want. A realtor doesn't care how many houses he or she sells on a particular day; the important thing is the total for the month, or the year. In cases like this, statistics says that the long-run average payoff that you actually collect is likely to be very close to the expected payoff, never mind the ups-and-downs along the way. But, when a situation is faced only once, or only a handful of times—ah, now you have to worry about the downside risk, and so far, this is something that we've given limited attention to.

In queuing theory, our big-time equations computed expected values, like average time spent in line. But there's also variability to consider. It may be fine if a restaurant serves you, on average, in 6 minutes, but not if most people get lightning-fast service, while one person in twenty actually waits two hours. That's why we touched on the probabilities of having to wait, or of the line being longer than a given size.

We actually did a bit better in decision theory. We talked about how to adjust the payoffs on the tree to reflect feelings of risk love and risk aversion. And we could also look at the possible outcomes of our strategy and find the probabilities of each. Still, even in decision theory, our strategy was chosen based on what was best, on average.

There's another issue, too. While models like the queuing and Markov models that we've recently studied can provide us with good insights on how some systems behave, they also tend to have strong requirements in their assumptions. It's hard to use Markov analysis on processes that aren't memoryless. Most queuing models are even more restrictive, requiring that the number of servers is constant over time, and that all servers work at the same mean rate, and so on. Small deviations from these assumptions won't invalidate the model, but gross deviations will. And real-life situations often show such gross deviations. So what can we do?

We can use simulation. In simulation, we numerically model the relationships among the important parts of the problem, the decisions that we control and the random events that we don't. It's often called a Monte Carlo simulation, a name that comes from the famous casino in Monte Carlo, Monaco. We generate random variables for the values for the random variables in our problem, and then see how a proposed strategy works when faced with these random events. The technique is remarkably powerful and versatile, making it, like linear programming, one of the most widely used techniques in operations research.

Ever have a flight delayed due to weather? The capacity of the U.S. air transport system is hard pressed by demand to begin with. From queuing theory's perspective, we know that means long waiting lines and frequent late flights. But throw in a major weather event that disrupts the system, and things can get much, much worse. In 2005, the FAA had only a vague idea of the cost of such disruptions, and no real strategy at all for handling the problem. So, they decided to take a closer look, in conjunction with Metron Aviation and the Volpe Transportation Center. They designed simulation models for the air traffic system, including the impact of storms. Based on what they found, they came up with a new approach to managing air traffic and built it into an expert system. This system could make recommendations in real time, as traffic blockages occurred.

Computerized expert systems may suggest changes in flight paths, but that doesn't mean that the air traffic controllers are going to pay any attention to them. The success of a project required buy-in of the controllers, and given that the error on their part can result in the loss of hundreds of lives, they were rightly skeptical.

So, the research team went on to develop an interactive simulation that allowed controllers to work with the expert system in a simulated environment. Simulation is wonderful for handling situations that are too dangerous or too costly to do live. The interaction of controllers with expert system allowed refinements to the system, resulted in rules for its use and deployment, and generated buy-in from the controllers.

Bottom line? In the first two years of the deployment, the savings to the aircraft companies and the flying public totaled almost $190 million. Compare this to the cost of the study, which was about $5 million. The projected savings are estimated at about $2.8 billion over a 10-year period. That's a considerable chunk of change in anyone's ledger.

Now, it's doubtful you face a problem on that scale. But the versatility of simulation and the increasing availability of simulation software make it a powerful tool for any operations researcher. So how do you do a simulation? Well, as always, the better you understand a problem, the better you can model it. We need a structural model. What factors in the situation are key, and how do they interact? Once we understand how the pieces fit together, we need to understand the nature of the random events in the problem. This usually requires data, lots of data. For the FAA study, how often do storms occur? How large are they? How long do they last? How long does a storm like this delay a flight like that? And what's the distribution of flight-delay times when storms aren't present? And so on. All of that data goes into mathematically describing the possible outcomes of random events and how likely each outcome is. That gives us a probability distribution for each random event.

Of course, we can't just generate one set of random events and see what happens. That would be like visiting Time Square on New Year's Eve at 11 PM, and figuring you have the place pretty much sussed. If you visit it on a Tuesday afternoon in June, you're going to get a very different impression.

It's the basic lesson of statistics; to draw decent conclusion about a variable quantity, you have to look at a sufficiently large sample. So, we want to run our simulation over and over, each run having its own randomly generated values for its chance events.

This kind of repetitive work is heaven for a computer. And assuming we've built a good simulation, the compiled output gives us a pretty clear idea of the range of outcomes that we might expect, not only what's possible, but also how likely. And that can lead to better strategies for handling the situation. And then we can implement those strategies in our model, rerun it, and see if our new ideas do, in fact, work.

Simulation isn't an optimization technique; it doesn't automatically figure out what the best strategy is. But in spite of that, a well-designed model can allow you to play what-if games, to modify the model in order to see how that modification affects the possible results. This glimpse into worlds other than our own can be incredibly useful in identifying both risks and opportunities in managing uncertainty.

Let me highlight this with a business application, and at the same time, demonstrate how simulation is actually done. I'm going to continue to work in the spreadsheet environment; it's easily accessible, and it allows you to see how the pieces fit together as we build the model. But I'll be honest: The bare-bones spreadsheet environment is not the perfect place to do simulation. When our simulation is built, we'll want to run it many times and summarize the results in useful ways, like charts and tables. Ideally, you'd like to be able to interact with these charts and tables, drilling down to specific information of interest. These tools are possible in a spreadsheet, but for this lecture, we'll create them; most aren't built-in.

That said, there are also third-party add-ins that give Excel this functionality. Three well-known ones are Analytic Solver, Crystal Ball, and @Risk, and they greatly simplify the variable generation, data-gathering, and data-reporting steps. If you are interested in this material, give them a free trial and decide whether they might we worth purchasing. If you're doing more advanced simulation, you may even want to learn one of the dozens of simulation programming languages that exist, such as SIMSCRIPT or ACSL. Anyway, a basic spreadsheet program has enough to give us the basic feel for what's going on.

So, here's our problem. You are a construction contractor, and you're bidding on a construction job. It's a competitive bid, so your success is going to be dependent on your being able to determine, with reasonable accuracy, what completing the job is going to cost you. You have enough experience with jobs like this that you can model its different activities pretty well.

Preparing a bid will cost you $5000, win or lose. If you actually land the job, you'll have other costs: materials, labor, and possible late fees if your work gets behind schedule. But of course, with the winning bid, you'll get paid for your efforts. So, what should you bid? Bid too high and you have little chance of getting the job; bye-bye, $5000 prep fee. Bid too low, and you'll get the job, but you may be sorry you did. Your profit is going to be whatever you bid is minus your expenses. We're going to use simulation to get a handle on what those expenses might be.

To begin, I'm not going to worry about how to handle the randomness. I'll just put in average values, and then will make them stochastic on the second pass through the model. So let's work out the cost of completing the job. Prep cost. Fixed. $5000. The materials cost, well, that's actually going to vary from job to job, but for now, I'll use the average figure for similar jobs in the past, $60,000. Next comes labor cost and late penalty, but the labor cost depends on how many weeks of labor we need, and the penalty depends on how long the whole job takes. Again, let's stick in some fixed values for now, and we'll make them random later. Nine weeks for labor, two weeks for delay.

You can see I've created a few extra columns, too; things that we need to figure out what the total cost of the job will be. OK. Your laborers get paid by the week; $700 for each week that they work. You currently have four workers. So your labor cost will be given by the formula, 2800 times the number of weeks of labor needed, like this. Labor weeks needed was in cell C4, so the formula for labor cost is 2800 times C4. As you know, every formula in a spreadsheet starts with an equal sign; the asterisk means multiplication.

When I enter this into the spreadsheet, it computes the value, $25,200. OK. When is the job done? Well, take the labor time and add the delay time, and that's completion time, like this. With our current numbers, that turns out to be 11. And now, the completion penalty, part of the contract for this job is that it needs to be finished in 12 weeks or less. Finish it late, and you owe a penalty of $12,000 for every week that you're over the limit. Again, we need a formula. One way to get it is an IF statement, like this. IF statements always say, if the first thing is true, then write the second thing; otherwise write the third thing. So here it says this; if the cell F4, the job completion time, is less than or equal to 12, then, write a 0; no penalty. Otherwise, write the penalty of 12,000 times the quantity (F4 − 12), because every week over 12 is going to costs you $12,000. By the way, the separator that you're seeing here is the semicolon, which is what Calc always uses. If you're working in Excel, type exactly the same thing, but use commas instead of semicolons. Excel always uses commas.

OK. We enter this formula, and we get a late penalty of zero this time, because our work was done in week 11. All that's left is to add up the costs, like this, which comes out to $90,200. And that's our model for how much the job will actually cost. Well, almost. We still have to put in the randomness for materials cost, for labor time needed, and for delay time. I'm up for it. Let's get to work.

The materials for the construction on a job average $60,000, we said, but there's a fair variation in this. Looking at similar jobs in the past, you decide that the cost of materials is approximately like this. This picture is called a probability density function, or PDF, of this random variable. To interpret a PDF, imagine that it's the silhouette of a pile of sand. Every grain of sand in that pile represents the cost of one job. Lots of grains are piled up on or close to $60,000 in this PDF. That means that costs that are close to $60,000 are common. The further you get from $60 K, the less common that cost is.

In statistics, a bell-shaped curve like this is called a normal distribution. This one has a mean of $60,000, which works out for any distribution to be the balance point of the pillar of sand. You can imagine it sitting on a teeter-totter. Its spread, on the other hand, is measured by its standard deviation, and here it's $4000. For normal curves, the standard deviation is easy to find. Take a look at how I shaded this picture. In the shaded area, the curve looks like an umbrella. In that region, the curve is concave. Now look at the two light-colored tails. There, the curve is convex, like a bowl. The place where we switch from umbrella to bowl is called the point of inflection. For a normal curve, the point of inflection is always one standard deviation away from the mean. Here it happens at 60,000 ± 4000, so the standard deviation is $4000. Fair warning: This only works for normal curves. But for curves that are close to normal, about $2/_3$ of all of the observations, all of the grains of sand are within one standard deviation of the mean. About $1/_3$ of the sand in our picture would be out of the umbrella's shadow and would still get wet in a rainstorm.

Anyway, we need to generate random materials costs for our model. If you're using simulation add-in to your spreadsheet, there are straightforward ways to get this, but for Excel or Calc, I'm going to introduce a formula that gets the job done, although I'm not going to go into why it works. For our normally distributed materials cost with a mean of 60,000 and a standard deviation of 4000, we can generate values by using this. The RAND() generates a random number between 0 and 1, with all values in that range being equally likely. The NORMINV then uses that random number to generate a random variable for a normally distributed variable, with the specified mean and standard deviation; it essentially picks out one grain of sand. Every time you enter something new anywhere in the spreadsheet, the sheet recalculates the value of this RAND in this cell, and that means you're going to get a new value for your materials cost, a new grain of sand. And again, if you're in Excel rather than Calc, remember, all semicolons become commas in Excel.

OK, that's materials. How about labor? Labor costs are a fixed $2800 per week, but the number of weeks required to complete the job is uncertain. Here's a table of possible times, and their probabilities. Unlike our pile of sand PDF, this table talks about the discrete random variable. It can take on only a finite set of values. How do we get this work done in our spreadsheet? Well, I'm not going to use an add-in, so I need to enhance this table first, a bit. I'll paste it into the spreadsheet, and to its left, I'll add a range of values for each row, like this.

Here's the idea. The RAND function that I told you about a minute ago always returns a value between 0 and 1, and all the values in that range are equally likely. That means that 10% of the time, it'll give you a value between zero and 0.1, because that's 10% of the interval from 0 to 1. By the same logic, 30% of the time, it will give a value between 0.1 and 0.4; the range from 0.1 to 0.4 is 30% of the total distance from 0 to 1. So numbers in that range come up 30% of the time.

Here's another way of looking at it. RAND picks a random point from 0 to 1 on the line. You can see that range in the black numbers. Since the length between 0.1 and 0.4 is 30% of the total, the RAND value falls in this range 30% of the time, and we've assigned this range the meaning, 8 weeks of labor; you can see that in red. So, 8 weeks of labor happens 30% of the time, just as our table required. To get Excel or Calc to follow this rule, we use the command called VLOOKUP, vertical lookup. It uses the table we just made. Here's how the command would look. The first argument of VLOOKUP is the number to look up, which is going to be random. The second argument tells us where in the spreadsheet this four-column table is by specifying its upper-left and lower-right corners. We don't want to include headings, so our table starts in row 16. The final argument specifies which column of that table contains the information that we want. For us, it's the third column of the table, which contains the number of weeks of labor.

This is the formula that goes into our weeks of labor cell in our spreadsheet. And in parallel fashion, I'll set the table up to tell us how many weeks of delay we have. And, here's our simulation.

Every time the spreadsheet recalculates, the cost will change, because we'll have different random numbers. We can force the spreadsheet to do this by pressing the F9 key. Let's see it work. And we can see some things already, like the fact that it's not rare to have a late penalty and that they can be expensive. We're also getting some idea of what variation we can expect in job cost.

Let's be a bit more systematic about this. What we want the spreadsheet to do is to repeat the simulation many times, just as we've been doing when hitting the F9 key, but to record the result of each trial. Once again, simulation add-ins will do this automatically. But even without them, it can still be done quite easily in Excel. In Calc, it can be done, but it's more than a bit of a pain. Rather than get into the detail of it here, I'll provide a link to a sensible set of instructions in the bibliography.

Anyway, here's the result, or the beginning of it. We're looking at the values for three quantities for the first 15 simulation runs. I actually told the spreadsheet to repeat our situation 1000 times and to record the values of these three quantities in each trial. Now, no one wants to look at all of those values individually, so let's get some summary statistics. We can use the functions MAX, AVERAGE, and MIN to find the biggest, average, and smallest values for each variable. To find the maximum value of the of the cells from L5 to L1004, for example, you type, = MAX(L5:L1004), pretty straightforward.

Here's what we get for the three variables that I was tracking. These numbers will fluctuate a little bit every time you re-compute the spreadsheet; every re-computation is based on new random numbers. But the results over 1000 runs are pretty stable. And it looks like our total average cost is about $98,000, but that the range is quite wide, from around $75,000 to over $190,000. I can get a better sense of this by creating a frequency table and histogram.

The FREQUENCY function is built for making such a table. Let's look at the total cost in $10,000 increments, starting at $70,000. We start by putting the lower and upper cutoffs for each category in the spreadsheet, row by row, like this. Now, the Frequency column will contain how many observations fall in each category. We can get all this with a single command. First, we select all of the blank cells in the frequency column, then type this, =FREQUENCY(, and then select all of the 1000 values of total cost; then a semicolon, or a comma if you're working in Excel; and now we highlight the upper bounds of each category; then end parentheses. And here's the odd part. Like last time, don't hit enter. Hit control, shift, enter, all at the same time. It's an array formula; it fills in the entire column. Anyway, we get this. And now we can use Excel or Calc's chart capability to draw a bar chart.

The materials costs were normally distributed, but the total cost definitely isn't. Look at that long tail on the upper side of the graph; it's something that you ignore at your peril. Even though the average cost of a job is just over $98,000, over 5% of the time, the cost is 130,000 or more, and costs of 120,000 or more occur about 1 time in 8. If cash flow is an issue for your company, this is something to keep in mind, because a bid of $120,000 on a job like this, even if accepted, will actually lose you money one time in eight.

Why is this happening? Well, let's take a look at the time required to complete the job. We'll make a bar chart. And a real part of our problem begins to become clearer. Anything over 12 weeks starts the late-fee clock ticking, and at $12,000 a week, those fees can quickly eat up your profits. And we can see from the graph that we're paying late fees about 26% of the time. If you prefer, we can graph how much we're paying in late fees instead. And as we found before, we're paying an average of almost $7000 in late fees, but it can go has high as a whopping $72,000. So, we're a little wiser about what we can expect in costs, and we've identified the source of expense that at first sight may not have seemed so important. But the question, of course, is, what can we do about it?

This brings us to one of the wonderful things in the analytic approach in general, and of simulation in particular, what-if analysis. A simulation can allow you to explore the impacts of changes in your situation. In our case, it's going to help us get a grip on our bottom line. In order to prepare for this, I'm going to modify our simulation model in a way that I should have done from the start. One of the precepts of good modeling, particularly in a spreadsheet environment is: Thou shalt not hard code thy parameters. Putting it simply, if there's a constant in your problem, a number, then it deserves its own cell in your spreadsheet. Then, whenever you want to use that number in a formula, the formula should refer to that cell, rather than containing the number itself.

Why this extra complication? Well, one reason is that if the number happens to be an ugly one, like 2,520,598, then it's faster to refer to the cell than to type the number, and easier to avoid making a typo [in] one of the times that you enter it. But the big reason is that it allows you to change one number in your spreadsheet and then have the effects of that one change percolate through everything else on the sheet. So let me do a little revamp of our sheet, creating an area in the top for the parameters. It's essentially the same simulation we had before, but now everywhere that we had a number, like 12, we replace it with the name of the cell that contains that 12 in our parameters section. We ended up with six parameters.

There is one change to be made in our simulation, actually. Now that I've made the number of workers an explicit parameter in the problem, well, the table for labor time was written with the assumption of four laborers. How long does the job take now when I change the number of laborers answer this by looking at the labor requirement a little differently. N weeks, but in man-weeks. A man-week is the amount of work that one man, or woman, can do in one week. So, four people working on a job that takes five weeks, requires 4 × 5, or 20 man-weeks. It could have been done by 5 people working for four weeks, instead, or 10 people working for two weeks.

We should be a little careful here, since there could be synergies. A single worker can't accomplish some building tasks, like laying the ridge beam of a roof. And a swarm of workers would just get in one another's way, actually slowing progress rather than speeding it. It's also possible that workers are specialized in their abilities, or work at different speeds, and so on. One often runs into this kind of thing when modeling real-world situations. What do we do about it?

Well, the idea that each additional worker adds the same amount of oomph to the building process probably isn't perfectly true, but is it close enough to true? A model tries to extract the important aspects of the situation and to ignore the insignificant ones. Here, we'd want to seek domain knowledge. We'd want to talk to someone who knows the ins-and-outs of building projects. Since you are our experienced contractor in this scenario, you'd probably already know that this model is pretty accurate, as long as you have, say, at least 3 workers, but no more than 20. But you may also know that you're not going to be able to get workers if you can't promise them at least five weeks work.

With these caveats in mind, I'm going to compute the time to complete the job. I'll first find the number of man-hours implied by our four-worker table; that just means multiplying the time given in the table by 4. Then I'll divide that by our actual number of workers, to find out how many weeks that they'd need for the job. Finally, I'll round that number up to the next whole number, because the workers get paid by the week, or fractions thereof. The old formula for labor time was this. And the new one looks like this.

Start with the old formula for how many weeks it takes 4 workers to finish the job. Multiply that by 4 to get the man-weeks for the job. Then divide that by how many workers we have. That's in cell I4. Finally, round the result up

...oer. The 0 at the end tells Calc to round to 0 decimal ...vhole number. And remember, for Excel, all of those commas. We'll also modify the labor cost to say that, ...one in less than 5 weeks, the workers still have to be paid ...i labor, like this. I4 is the number of labors, I5 is what we ...:k, and C11 is the number of weeks of labor that we need. If ...we still pay for 5.

By the way, when you give a spreadsheet model to someone else, an end user, it's important that they know what quantities in the sheet they could enter new values for, and what quantities are being computed from those values. I follow the convention that any cell that you're allowed to change is blue-bordered with a grey interior. You'll see this in our Parameters section, but also in the probability tables for job delay and labor time needed. That means you could modify these distributions, too, if you wanted.

So, OK, let's play with our new, improved sheet. How can we get those costs down? Maybe we can negotiate with the customer on the deadline of 12 weeks. If they'd extend that to 13 weeks, how much would it help us? Well, change the 12 in the Weeks Allowed parameter box to 13, and find out. Well, it helps, some. The mean cost is now about $95,000, where before it was about $98,000. If the customer would accept a one-week extension before levying fees, it would save you about $3000 on average. Maybe you could offer them $1000 off of your quoted price for each such an extension. Of course, they might not go for it. Worse still, it really hasn't addressed a lot of your downside risk. In the 1000 simulation runs, you still could end up paying up to $60,000 in penalties. Any better ideas?

Materials don't look promising. You'd either need to find cheaper materials or a more accommodating supplier. Same for delay times, whether supplier delays or acts of God. But how about workers? More workers should be able to get the job done faster, cutting down on late penalties. On the other hand, you have to pay them. But since the job is completed faster, you don't have to pay them as long. Here's what happens to total cost as we change the number of workers. The red dots show the average cost of a job with the specified number of workers, and we see that we can save some money, on average, by hiring more; 6 seems to be about the best number, saving us around $4000 on average, even though we need to pay more workers. But perhaps more importantly, look at the vertical bars that show the maximum and minimum costs in the 1000 simulated trials.

With eight workers, twice your original complement, the average price is about $2000 cheaper than the original value, but the job price never rose above $115,000. The job is literally 99.9% likely to be completed on time, a fact that you can use in negotiating with the customer. In fact, the job is usually done by week 6. So, we can cut our original average cost. Virtually guarantee the job will be done on time. Be able to bid even if the customer raises the late penalty, and greatly reduce the uncertainty in our own return, allowing us to bid more intelligently and to predict our profit margin more accurately. And that is the power of simulation.

In the next and final lecture to this series, we'll return to this simulation, coupling the stochastic work that we've done in recent lectures with the optimization work from the middle part of the course. The result will be stochastic optimization, with the goal of finding best answers to problems that include random elements, where you get to define for yourself what constitutes best. I think you'll be amazed by the power and flexibility of the machinery that we can now bring to bear.

Stochastic Optimization and Risk
Lecture 24

This final lecture revisits the simulation from the previous lecture, coupling the stochastic work that we've done in recent lectures with the optimization work from the middle of the course. The result is stochastic optimization, with the goal of finding best answers to problems that include random elements—where you get to define for yourself what constitutes "best." In this lecture, you will use your entire mathematical optimization toolkit to solve a difficult real-life problem.

Stochastic Optimization

- Getting a handle on our costs with the work we did in the previous lecture is only part of the problem. You still have to make your bid, and you have two competitors who are hoping to get the job for themselves. Let's call them Fred and Wilma. Up to this point, the three of you have each used teams of 4 laborers and the same suppliers, so you've faced the same cost curves. Given your new understanding of the economics of the kind of jobs you generally do, though, you now have an ace up your sleeve. Assuming that you get the job, you're going to use 8 workers. They'll stick with 4.

- To make a bid, each of you is going to estimate what you think the job will cost you and then add to that a certain percent markup, which, hopefully, will be your profit. For now, let's assume that each of you routinely decides on a 40% markup.

- The real question that we're trying to answer is how much you should bid. Your goal in bidding will be to maximize your average profit, but you may want to have a constraint limiting the chance of losing too much money.

- In rough terms, we need to run a simulation with a set of input values, look at the results, use that information to make changes to the input variables, examine the new results to see whether the changes were in the direction that we desired, and continue this way until we narrow in on an optimal solution. This is **stochastic optimization**.

- It's now possible to do problems like this in a spreadsheet environment with a surprisingly small amount of work—on the part of the human, at least. The kind of work that we'll be doing can be accomplished by augmenting Excel with add-ins, such as Analytic Solver, Crystal Ball, or @Risk. It's surprising how quickly these products can grind on difficult problems—that is, once you've formulated the problem. You still have to get to the superhighway.

- Let's modify the simulation that we wrote for costing the construction job to fit our new needs.

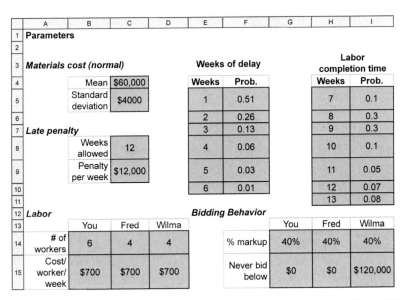

Figure 24.1

- We've rearranged things a bit, but the same information is there: the distribution of materials costs, the late penalty information, the probability distributions for delay time and labor requirements, and information about the workers and their cost. We've added similar labor parameters for Fred and Wilma, who are only using 4 workers each.

- In the lower-right corner, we've added a section about how each of you price the job, based on your estimate of what you think it would cost you to complete it. This section allows you to specify not only what markup you want on the job but also to set a minimum bid for any job in order to limit downside risk. Currently, only Wilma is using this option.

- When we first encountered this problem, we had to supplement the tables that you see in the upper-right corner with two more columns, specifying what ranges of random numbers corresponded to each output. Those RAND() columns are no longer necessary, nor is using RAND() or VLOOKUP. Any simulation add-in will include pull-down menus for all of the common distributions.

- The cost simulation itself looks very much like what we had before, only this time there are three copies of the original simulation, one for each contractor.

Actual Job Cost (You)

Prep cost	Materials cost	Labor weeks needed	Labor cost	Weeks delay	Job complete in week #	Completion penalty	Total cost
$5000	$60,059	4	$28,000	1	5	$ -	$93,059

Actual Job Cost (Fred)

Prep cost	Materials cost	Labor weeks needed	Labor cost	Weeks delay	Job complete in week #	Completion penalty	Total cost
$5000	$60,059	8	$22,400	1	9	$ -	$87,459

Actual Job Cost (Wilma)

Prep cost	Materials cost	Labor weeks needed	Labor cost	Weeks delay	Job complete in week #	Completion penalty	Total cost
$5000	$60,059	8	$22,400	1	9	$ -	$87,459

Figure 24.2

- We want to simulate the bidding. But a person's bid is based on his or her estimate of the actual job cost. We have to model the accuracy of those estimates. With an add-in, we can take our historic data and fit a distribution to it.

- Suppose that we look at the percent error in the last 100 jobs that you bid and find that their distribution looks like the following bar graph. You've been close to right on average—the balance point of the bars is very close to 0% error—but there's some spread, high and low. The spreadsheet, with the add-in, can automatically compare this data to a host of different theoretical probability distributions and assess how well each distribution fits the data. We simply highlight the data, select "fit," and instantly get a set of distributions to choose from, ranked from best to worst, according to the application of a statistical goodness-of-fit test. For our data, the best is shown as a curve over the bars.

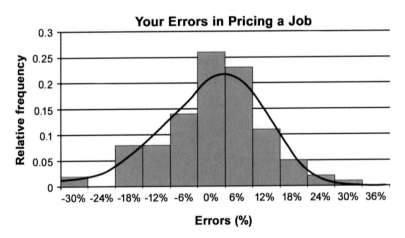

Figure 24.3

- The software picked a shifted Weibull distribution, which provides the graph and a variety of statistical charts and measures to assess the goodness of the fit between the distribution and the data. In this case, everything checks out.

- Once we approve this choice for the distribution, we click the cell in our simulation that's going to contain your estimate error. Thereafter, each time the sheet updates, it will automatically compute a new error for your estimate, drawn from this Weibull distribution.

- Because we don't have data on the accuracy of the bids of your competitors, we'll run with the assumption that their errors are distributed the same way that yours are. But your estimates are independent, so each contractor get his or her own random variable.

- The new part of the model will look like the following.

Simulation 2

Job cost estimates errors			Bids			Winner	Your profit
You	Fred	Wilma	You	Fred	Wilma	Fred	-$5000
10.80%	−6.21%	−12.73%	$144,357	$114,842	$120,000		

Figure 24.4

- The first three columns contain those shifted Weibull distributions that capture the errors in pricing. In the trial shown, you overestimated the cost of the project while Fred and Wilma priced it below the actual cost.

- Now we can compute what each person actually bids. Let's walk through the logic for Wilma. The same reasoning will apply for you and Fred. For the current trial, Wilma's actual cost ends up at about $87,500—but, of course, she doesn't know that. What happens?

- Her error shows that she underpriced the job by about 13%, which means that her estimate is only 87% of the actual job cost, which is about $76,300. Now she applies her markup. She wants a 40% markup, according to our parameters section, so she multiplies her estimate by 140%, or 1.4. This gives her bid, about $107,000.

- However, Wilma had a minimum bid of $120,000 as a hedge against downside risk. She'll use either the bid she just computed or $120,000, whichever is larger. In this case, because $107,000 is less than $120,000, she bids $120,000, as shown in the table.

- In general, Wilma's bid is calculated by the following formula.

$$= \text{MAX}\Big(\text{min. bid,} \underbrace{\big(\text{actual cost}\big)*\big(1+\text{percent error on estimate}\big)*\big(1+\text{markup percentage}\big)}\Big)$$
$$\text{Wilma's intended bid}$$

- Her intended bid is the job cost adjusted for her error and then increased by her markup percentage. And she'll bid that or her minimum bid value, whichever one's higher. An equivalent formula is entered for you and for Fred. (See **Figure 24.4**.)

- What happened in the trial? Fred was the low bidder at about $115,000, so he gets the job—and actually makes about $27,000 in the process. Because you didn't get the job, you are out the $5000 prep fee. But this is only one trial. Every time we hit the F4 key, we get another simulation.

- Next, we would simulate the situation thousands of times, and then collect and summarize the results. But with an add-in, we can just tell the spreadsheet what cells we want to keep track of, and then clicking on them automatically provides us with interactive graphics. It also computes many summary statistics on this distribution, such as the mean cost of the job, which is about $97,000, and your mean profit, which is about $8300.

- But there's a problem with the number for your mean profit: You're going for a 40% markup, and 40% on a job that costs nearly $100,000 is $40,000 profit. But your average profit isn't 40%; it's only about 9%.

- The following profit graph shows a huge spike at −$5000. This is your payoff when you don't get the job, which is happening about 61% of the time. Your average profit includes all of those times when you lose $5000.

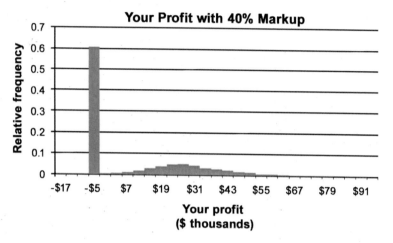

Figure 24.5

- Interestingly, that's not all that's going on. To see the rest of it, let's temporarily reduce everyone's prep fee to 0 and make it so that people are going to use a 0% markup and bid exactly what they think the job is worth.

- If people never made a mistake in estimating the job's cost, the whole thing would be a breakeven proposition. So, get the job or not, we should break even on average. However, when we change the sheet's parameters to give a prep fee of 0 and a 0% markup for each contractor, the sheet reports that you don't break even. On average, you lose over $2000. If you check Fred and Wilma, you'll see similar results.

Simulation 2

Job cost estimates errors			Bids			Winner	You
You	Fred	Wilma	You	Fred	Wilma	You	-$15,062.
−13.79%	17.97%	2.14%	$94,154	$169,667	$146,895		

Average profit over all trials $(2,127.19)

Figure 24.6

- This is called the **winner's curse**, and it arises anytime people are bidding on something whose true value they don't quite know. It's true that your bid, while sometimes high and sometimes low, is right on average. In the trial on the screen, you got the job, because you had the low bid. But you had the low bid because you estimated the job at almost 14% under its actual cost. As a result, you won. And in winning, you lost—over $15,000. So, you have to shade your bid, accounting for the winner's curse.

- Let's put back the prep fees and restore our 40% markups for each bidder.

Bidding Behavior

	You	Fred	Wilma
% markup	40%	40%	40%
Never bid below	$0	$0	$120,000

Figure 24.7

- Your current problem is that you aren't making enough money because you aren't getting the job often enough. What if you slash your markup and play with your minimum bid value? Given that the other two contractors are behaving as we've described, what's the best thing for you to do?

- One approach would be to plug in a bunch of different values in place of our 40% markup and see what happens to our average profit as a result. We could do this manually, but add-in programs make it possible to do this very easily. You specify the values that you want and the spreadsheet reports the results.

525

d that increasing the markup decreases the downside
once you reach a 20% markup, there's a limited effect.
% of the cases lose you more than $5000, your prep fee.
arkup of 20% or more is pretty good insurance against
rything.

er your markup, the higher your profit—if you get the job.
And once in a while, both of your competitors are going to grossly
overestimate the cost of the job, and you can swoop in with your
high and profitable bid. But that doesn't mean that you'll want such
a markup.

The Solution

- If we have more than one variable to look at, it's quite a bit more difficult to implement than what we just did. With stochastic optimization, we can discover what the best combination of variable values for our contracting business is.

- Using the add-in, we specify that our objective is to maximize our expected profit. We tell it that our decision variables are our markup percentage and our minimum bid. We add a constraint, specifying that we want to lose more than $5000 at most 1% of the time. We also let the optimization decide on how many workers we should use. That will be an integer variable.

- We really should go further, such as specifying generous upper and lower bounds on each of our decision variables and specifying that we want to use an evolutionary solver. The software has gotten smart enough, though, that if we don't specify these things, it will probably make those decisions on its own. We set our original values for our variables to exactly match Fred's, in which case our average profit is just under $7000.

- Then, we tell the solver to run—and to get a really good feel for the range of possibilities, we have it conduct each simulation 5000 times. In less than three minutes, it's done, with the suggestion that you use a markup of 28%, never bid under $28,500, and, surprisingly, use 6 workers, not 8.

Important Terms

stochastic optimization: Finding a (near) optimal solution to a problem involving chance.

winner's curse: In an auction in which the true value of the item up for bid is unknown, the phenomenon that the winner often pays more for the item than it is worth.

Questions and Comments

1. Consider a private value auction, where each bidder has his or her own valuation on what the item is worth to him or her. Suppose that each bidder knows his or her own valuation but not the valuation of others. Would shading one's bid below the true valuation still make sense?

Answer:

Yes. If you bid exactly what you believe the item to be worth, you would always gain zero, regardless of whether you won the bid. Only by bidding something less than you think the item is worth can you capture a surplus. Of course, the more you shade, the more chance you have of losing the bid.

An interesting variant is the second price auction, where the high bid wins the item but only has to pay as much as the second-highest bid. One can show without too much work that in such an auction, each bidder should bid what the item is worth to him or her.

2. A tempting approach to determining good strategies is to look at those competitors who have had the most success in the past and to emulate what they're doing. Why might a direct application of this strategy be a bad idea?

Answer:

It's necessary to see how those who succeed differ from those who fail. For a concrete example, consider people being given $100 and being allowed, if they wish, to use some or all of it for a single bet in roulette. If you look over a large number of cases, you'll find that all of the most successful people bet all of the money, and they bet it on a single number. But people following this same strategy are also more likely to lose $100 than anyone else. In fact, the best strategy on average is not to bet.

It's interesting to keep this in mind when thinking about the financial collapse of the early 21st century. Until the meltdown, the most successful people were the ones taking extreme risks with high leverage. That didn't make the policy a good one.

Stochastic Optimization and Risk
Lecture 24—Transcript

In the first lecture in this series, I said that applying operations research to a problem was often like taking a long trip by car. If your goal is getting there, rather than sight-seeing, you use superhighways; they're fast and reliable and let you cover a lot of distance. In OR, the superhighways are the mathematical algorithms that we apply to a formulation of a problem.

Just like with real road systems, mathematical superhighways are something relatively new. It wasn't that many years ago that, to have any hope of applying those useful algorithms, you had to know a lot of mathematics. Being bright and industrious let you get to the point where you understood the algorithms. But even then, applying them to real-world problems was often a formidable task. The calculations required were often just too wild and wooly. Modern computing, cheap, and readily available, changed all that.

The result is that I didn't need to teach you to grind through the simplex method, or gradient descent, or genetic algorithms, or equations for finding multiple regression coefficients and confidence intervals. With nothing more than a PC and a spreadsheet, you can handle all that, and more. And with spreadsheet add-ins, you can go much further still. It's really quite remarkable, to be honest.

But, of course, there's a catch. The only things allowed on the operations research superhighway are mathematical formulations, and the only things coming off of that system are information about those formulations. So, maybe you can chill out while you're on the highway, but every trip begins with formulation. Every trip ends with interpretation. And to date, that's work that no computer can do. That's why we've focused on these aspects of solving problems throughout this series. We've assembled quite a toolkit, with each tool suited to its own kind of challenge. Problems that won't yield to one technique may yield to another, or perhaps to a combination of several. We learn analytics one technique at a time, but when applying it in the field, the solution often involves attacking it on more than one front.

So in this final lecture, I'd like to demonstrate what it looks like to bring our entire toolkit to bear. And in addition to using some techniques we've already discussed, I'm going to introduce a powerful hybrid of our stochastic work of the most recent sections of the course and our optimization work from the middle lectures. This combination is called, sensibly enough, stochastic optimization, and it really is the best of both worlds.

To get started, let's go back to the example from last lecture. You were a building contractor who was going to be submitting a sealed bid on a construction job. Low bid wins the job. The first step was figuring out how much the job was going to cost. The problem was stochastic because the materials cost, the weeks of labor required, and the weeks of delay in construction were all random variables. As a result, we ran our simulation thousands of times with different values for the random variables, and compiled the results. That gave us the frequency distributions for the quantities of interest.

Then we played some what-if games. We saw, for example, that you could lower your average cost by increasing your number of workers from four to six or eight. Six was a little cheaper than eight on the average, but eight greatly reduced the risk of running up huge late fees on a job that took more than 12 weeks. When you compare the eight-worker peak, there in blue, to the original four-worker peak, you can see that it's farther to the right, indicating that it has a higher modal cost. But remember that the mean of a distribution is the balance point of the pile of sand in the picture. The blue eight-worker curve balances out at about $94.5 thousand, while the red four-worker curve balances to the right about $2000 higher, due to that long tail on the right.

But getting a handle on costs is only part of the problem. You still have to make your bid, and you have two competitors who are hoping to get the job for themselves. Let's call them Fred and Wilma. Up to this point, the three of you have each used teams of four laborers and the same suppliers, so you've faced the same cost curves. Given your understanding of the economics in the kind of work that you generally do, though, you now have an ace up your sleeve. Assuming you get the job, you're going to use eight workers. They'll stick with four.

To make a bid, each of you is going to estimate what you think the job will cost you, then add to that a certain percent markup, which, hopefully, will be your profit. For now, let's assume that each of you routinely decides on a 40% markup. You might be happy with that, or you might want to change it. In fact, the real question that we're trying to answer is how much you should bid. Your goal in bidding will be to maximize your average profit, but you may want to have a constraint limiting the chance of losing too much money. Hmmm, a constraint and an objective and a decision variable of what markup percentage you want—that all sounds quite a lot like an optimization problem. Let's see how far the optimization approach can take us.

The size of your markup determines your bid, and indirectly your profit. Is this relationship linear between markup and profit? Well, if you get the job, each 1% increase in your markup adds the same number of dollars to your bid, and hence to your profit, so yes. Linear. Or, rather, no, because that increase in markup also reduces the chance that you get the job in the first place. From our work with expected values, we know that finding your average profit is going to involve multiplying the job's profit by the probability that you get the job at all. Since both the profit and the probability depend on the markup, we've got a nonlinear objective. Too bad.

So, we'll need a nonlinear technique. Which one? We need to survey the mathematical landscape of our profit function in terms of our variable, percent markup. What does it look like? If it's all gentle rolling hills, we can use a steepest descent algorithm. We'll have to watch out for local minima, but steepest descent algorithms tend to be fast. We can tell the spreadsheet to use a bunch of different start points for its search, and then take the best one.

But the landscape isn't all rolling hills. Look at what happens when your bid is very close to the lower bid of your two competitors. Bid one cent less than anybody else, and you get the job and make what you make in the construction business. Two cents more than that though, and you lose the job, getting nothing but the bill for the $5000 prep fee. So you're not going to be able to use a walk-downhill approach, because there is a cliff in the profit landscape, right where your bid is tied for the lowest bid.

Which means that we need to use an evolutionary algorithm. They don't like a lot of constraints, but we don't have many, just a limit on losing a lot of money. Evolutionary algorithms are going to want us to put upper and lower limits on our variables, but that should be easy enough. Fantastic—except that we've gotten so caught up with this optimization analysis that we've been ignoring the elephant in the room. This problem is stochastic, stochastic all over the place. We want to solve an optimization problem, but we've never done that when dealing with randomness at the same time. There's no chance that we can come up with a strategy that's guaranteed to give you the best profit here. The best we can hope for is to maximize, say, expected profit, or, median profit. Similarly, we can't absolutely assure that we don't lose more than a certain amount of money if we make a bad bid. We can only assure that such a screw up doesn't happen more often than we're willing to accept.

So, what can we do? Well, in rough terms, we need to run a simulation with a set of input values, look at the results, use that information to make changes to the input values, examine the new results to see whether the changes were in the direction desired, and continue this way as we narrow in on an optimal solution, in two words, stochastic optimization.

If that sounds like a heck of a lot of computation to you, you're right. Amazingly, though, it's now possible to do problems like this in a spreadsheet environment with a surprisingly small amount of work, on the part of the human, at least. But to do so, we've got to cross the border that separates what current-day spreadsheets can do on their own from what they can do with third-party add-ins. The kind of work that I'm going to demonstrate today can be accomplished by augmenting Excel with add-ins like Analytic Solver, Crystal Ball, or @Risk. I'm not going to demonstrate those products explicitly, that's not my focus, but I'll show you the kinds of results that they allow you to obtain, and give you a sense of what it takes to use them. I will say that these products improve each year, and that I am sometimes flabbergasted by how fast they can grind on hard problems, using only a PC. Once, that is, that you've formulated the problem. You still have to get to the superhighway.

So let's take the simulation that we wrote for costing the construction job and modify it for our new needs. First, you'll recall, we had a section for Parameters, the constants of the problem. Here it is now. I've already arranged things a bit, but the same old information is there. The distribution of materials costs, the late penalty information, the probability distributions for delay time and labor requirements, and information about your workers and their cost. I've added similar labor parameters for Fred and Wilma, who are only using four workers each. In the lower-right corner, I've added a section about how each of you price a job, based on your estimate of what you think it would cost you to complete it. You'll see that I've let you specify not only what markup you want on the job, but also to set a minimum bid on any job in order to limit downside risk. Right now, only Wilma is using that option. So I added a bit to the parameters section. I've also taken away a bit. When we first encountered this problem, I had to supplement the tables that you see in the upper right-hand corner with two more columns, specifying the ranges of the random numbers that corresponded to each output. It looked like this.

Those RAND() columns are no longer necessary, nor is using RAND() or VLOOKUP at all, for that matter. Any simulation add-in will include pull-down menus for all of the common distributions. Tables like our weeks of delay can be handled by a custom discrete choice from the pull-down menu, and all you have to do is to fill in the boxes saying, here's where the values are, and here are the probabilities that go with them, easy-peasy.

For our normally distributed materials cost, we click on the cell where materials cost is to appear. We choose normal from a pull-down menu. Then we enter the mean and standard deviation that we want, and we are done. There are dozens of distributions to choose from, exponential, beta, uniform, Poisson, binomial, triangular, some that even I haven't heard of. The software will show you a PDF of whatever distribution that you've specified, then allow you to modify it in various ways to better match your real-world variable.

OK, how about the cost simulation itself? Well, it looks very much like what we had before, only in triplicate. I've made three copies of the original simulation, one for each contractor. That way, I can change the numbers of laborers, weekly labor cost and prep fees for one contractor without affecting the others. Fred and Wilma currently have identical values of these parameters. All three of you face the same size job, the same delay times, and the same materials cost, so I generated one random value for each of these and applied it to all three contractors. Some of your values look different because of your eight-worker decision, a shorter job completion time, but a higher labor cost, since you may recall that you had to guarantee your workers at least five weeks pay. Note that, on this particular job that you see, Fred and Wilma's four-worker cost is actually cheaper than your eight-worker one.

But now comes the new stuff. We want to simulate the bidding. But a person's bid is based on his or her estimate of the actual job cost. We have to model the accuracy of those estimates. How do we get a handle on that? Well, we could guess, of course. That's more or less what we did earlier, when we modeled materials prices as the normal curve, using the mean and standard deviation from historical data. Normal curves are easy, and a lot of real-world quantities follow a normal distribution. But maybe we can do better.

With an add-in, we in fact, can. We can take our historic data and fit a distribution to it. Suppose we look at the percent error in the last 100 jobs that you bid, and find that the distribution looks like this. As you can see you've been close to right on average; the balance point of these bars is very close to 0% error, but there's some spread, high and low. This set of 100 is only a sample, though. It's unlikely, for example, that you never underestimate a job by around 24%, where there's a gap in the histogram off on the left. But, the spreadsheet, with the add-in, can automatically compare this data to a host of different theoretical probability distributions and assess how well each distribution fits the data. We simply highlight the data, say fit, and instantly get a set of distributions to choose from, ranked from best to worst, according to the application of a statistical goodness-of-fit test. For our data, the best looks like this.

The software picked a shifted Weibull distribution, shown here in red. It provides the graph and a variety of statistical charts and measures to assess the goodness of fit between the distribution and the data. In this case, everything checks out. Once I approve this choice for the distribution, I click the cell in my simulation that's going to contain the estimate error. Thereafter, each time the sheet updates, it will automatically compute a new error for your estimate, drawn from this Weibull distribution. Since I don't have any data on your competitors, I'll run with the assumption that their errors are distributed the same way that yours are. You all have about the same amount of experience in the business, after all. But your estimates are independent, so each contractor get was going to get his or her own random variable. So, here's what the new part of the model is going to look like. The first three columns contain those shifted Weibull distributions that capture the errors in pricing. In the trial shown, you overestimated the cost of the project, while Fred and Wilma priced it below the actual cost.

Now we can compute what each person actually bids. Let's walk through the logic for Wilma. The same logic will apply for you and Fred. For the current trial, Wilma's actual cost ends up being about $87,500, but of course, she doesn't know that. What happens? Well, her error shows that she underpriced the job by about 13%; 100% − 13% is 87%, so her estimate is only 87% of the actual job cost, which is about $76,300. Now she applies her markup. She wants a 40% markup according to our parameters section, so she multiplies her estimate by 140%, or 1.4; this gives her bid, of about $107,000. Well, almost. Because remember that Wilma had a minimum bid of $120,000 as a hedge against downside risk. She'll either bid what she just computed, or, $120,000, whichever is larger. In this case, since $107,000 is less than $120,000, she bids $120,000, as shown in the table.

If you follow the logic we just laid out, generally, you'll see that Wilma's bid will be this. Her intended bid is the job cost, adjusted for her error, then increased by her markup percentage. And she'll bid that or her minimum bid value, whichever one's higher. OK, done. And an equivalent formula is entered for you and Fred. What happens in the trial shown? Fred was the low bidder at about $115,000, so he gets the job, and actually makes a about $27,000 in the process. By the way, both the winner cell and the profit cell are just IF statements. For example, the profit cell says, IF the winner is you, then write (bid − actual cost), which would be your profit; otherwise write loss of the prep fee.

So, good for Fred. Not so good for you. Since you didn't get the job, you are out the $5000 prep fee—this time. Remember, that's only one trial. Every time we hit the F9 key, we get another simulation. Like this. This time, you overestimated your cost by almost 11%, but Wilma overestimated by even more. Fred underestimated his true cost, but this was a long job, so he accrued late penalties that you didn't, so your bid still comes in low. You make a profit of over $57,000. Sweet.

Well, you know what comes next. We simulate the situation thousands of times, then collect and summarize of those results. Except, with an add-in, we don't need to. We can just tell the spreadsheet what cells we want to keep track of, and then [clicking] on them automatically provides us with interactive graphics. It also computes a lot of summary statistics on this distribution, like the mean cost of the job, which is about $97,000, and your mean profit, which is about $8300, which is, wait a minute. You're going for a 40% markup, right? Forty percent on a job that costs nearly $100,000 is $40,000 profit, but your average profit isn't 40%; it's only about 8000, 9%. What gives?

Well, let's check out the profit graph and see what we can see. Oh, and there it is. See that huge spike at −$5000? That's your payoff when you don't get the job, which you can see is happening about 61% of the time. Your average profit includes all of those times when you lose $5000. Interestingly, though, that's not all that's going on. To see the rest of it, let's temporarily reduce everyone's prep fee to 0. I'm also going to make this the most cutthroat market imaginable, so the people are going to use a 0% markup; they're going to bid exactly what they think the job is worth. Now, admittedly, this isn't smart, but what do you think's going to happen?

Well, if people never made a mistake in estimating the job's cost, the whole thing would be a breakeven proposition. If you don't get the job, you get 0, since we're waiving the prep fee. If you do get the job, you bid it at cost, and so your profit is still 0. But of course, people do make mistakes in estimating job value, but we saw on average that, their estimate is right; overbidding and underbidding balance out almost perfectly. So, get the job or not, we should break even on average. But if so, then what's the deal with this?

When I changed the sheet's parameters to give a prep fee of 0 and a 0% markup for each contractor, the power of what-if analysis, again, the sheet reports that you don't break even. On average, you lose over $2000. If you check Fred and Wilma, you'll see similar results. Why?

It's called the winner's curse, and it arises any time that people are bidding on something whose true value they don't quite know. Yeah, it's true that your bid, while sometimes high and sometimes low, is right on average. In the trial on the screen, you got the job, since you had the low bid. But you had the low bid because you estimated the job at almost 14% under its actual cost! As a result, you won, and in winning, you lost, over $15,000. Hardly cause for celebration.

And if you think about it, the amount that you bid doesn't matter at all, unless no one bids below you—which means that you only get the job if nobody else believes that they can do the job for what you think you can do it for. And if your evaluation of the cost of the job is below both of the other two bidders, there's a pretty good chance that you've underestimated the actual cost of the job, and in winning, you're going to lose money. So even if you aren't looking to make a profit, you have to shade your bid, accounting for the winner's curse. How much each bidder should shade is an interesting and complicated question in game theory.

But, we're not doing charity work. Let's put back in the prep fees and restore our 40% markups for each bidder. Your current problem is that you aren't making enough money because you aren't getting the job often enough. What if you slash your markup? And while we're at it, I'm going to allow you to play with your minimum bid value, too. Given the other contractors are behaving as we've described, what's the best thing for you to do? Well, one approach would be to plug in a bunch of different values for that 40% markup and see what happens to our average profit as a result. We could do this manually, as we did earlier in the course, but all of the add-in programs that we mentioned make it possible to do this very easily. You specify the values that you want, and the spreadsheet reports the results, including some very nice charts. Like this.

I love this. This is a trend chart, showing what happens to profit as we vary the percentage markup by 5% increments from 0% to 45%. The blue dots in the middle are the mean profits. On the left, with 0% markup, you get the job about 97% of the time, but when you do, you lose about $100 on average due to the winner's curse. You also lose the prep fee in the 3% of the jobs where you don't clinch the contract, so you lose on average about $250. But the expected profits increase as we increase our markup from 0% to somewhere around 25%. Then, average profit starts to drop again as we price ourselves out of the market. So for a best mean profit, we want a markup of about 25%.

Now the colored bands, which give us a sense of risk. Imagine this as a single piece of green felt, with a piece of gray felt lying over top of it and partially concealing it. The gray band is the 75% confidence band, meaning that 75% of our trial runs gave profit within it. So the profits for a 0% markup were 75% likely to lie between −$11,000 and +$11,000. The corresponding green band lies beneath the gray one, so when the markup is 0%, it runs from about −$17,000 to +$16,000. This is the 90% confidence band, that is, 90% of the time, the profit at 0% markup lies in this range; 5% of the time it may be above $16,000, 5% of the time it may be below −$17,000, but the rest of the time, it remains in the confidence band. Once you understand this, you can see that increasing the markup decreases the downside risk, but that once you reach 20% markup, there's a limited effect. At most, 5% of the cases lose you more than $5000, your prep fee. So any markup of 20% or more is pretty good insurance against losing your shirt.

On the other hand, the bands at the top of the chart continue to climb. Why? Well, the higher your markup, the higher your profit, if you get the job. And every once in a while, both of your competitors are going to grossly overestimate the cost of the job, and you can swoop in with your high and profitable bid. But that doesn't mean that you'll want such a markup. For example, even though, with a 45% markup, you have a 5% chance of making more than $45,000 or so, your average profit has dropped. Such a high markup costs you more in lost jobs than it's worth, on average.

Well, what we just did is very nice, but if we have more than one variable to look at? It's quite a bit harder to implement. What's the best combination of variable values for our contracting business? We can find out with stochastic optimization. Using the add-in, we specify that our objective is to maximize our expected profit. We tell it that our decision variables are the markup percentage and the minimum bid. We add a constraint, specifying that we want to lose no more than $5000 at most 1% of the time. What the heck, let's go for broke, and also let the optimization decide on how many workers we should use. That will be an integer variable.

We really should go farther, like specifying generous upper and lower bounds on each of the variables, and specifying that we want to use an evolutionary Solver. But the software has gotten pretty smart these days, and even if we don't specify these things, it will probably make those decisions on its own. I set our original values for our variables to exactly match Fred's, in which case our average profit is just under $7000.

Then I told Solver to run, and to get a really good feel for the range of possibilities, I had it conduct each simulation 5000 times. Evolutionary Solvers trade away speed for flexibility, but in less than 30 seconds, the Solver had set up the problem and gotten to work. Within a minute, it had improved the original average profit. In less than three minutes, it was done, with the suggestion that you use a markup of 28%, never bid under $28,500, and surprisingly use 6 workers, not 8.

Since you'd never want to bid under $28,500 anyway, the spreadsheet setting that as a cutoff for a minimum bid was its way of saying that a minimum bid is a stupid idea in this problem. This keeps the chance of losing more than $5000 to about 0.7% and gives an average profit of just under $13,000. And you win over $2/_3$ of the bids.

That means, of course, that you're still losing $5000 on $1/_3$ of the time, but that's the price of doing business. The only way to reduce that is to lower your markup and win more jobs, and that reduces average profit. The kind of production possibilities frontier work that we did in multiobjective programming can let us explore that possibility. It shows, for example, that lowering your markup to about 22.5% will get you an extra 5% of the contracts, but it reduces average profit by about $500. You can decide if it's worth it. And that's our solution.

In recent years, the kind of interface offered by the add-ins I've mentioned has been gaining increasing prominence, and that's good, because that kind of intuitive, interactive, drill-down environment is an ideal place to do analysis in data and in analytics. The software does the cranking, you do the interpretation—the last part of your mathematical cross-country trip, from the superhighway to your final destination.

For example, now that we've optimized our work-force size, markup, and minimum bid, we can see what random factors have the most impact on our bottom line. The software will compute the correlation of your profit to each of the random variables over the thousands of trial runs, then present the result to you in a tornado diagram.

You can see why it's called a tornado diagram, the bars are always listed from the longest to the shortest, making a rough tornado. In general, some of the bars are pointing left, indicating there's a negative correlation. Here, all the bars show positive correlations, meaning that an increase in any of these factors tends to increase your profit.

Number one on the list is Fred's error. The more he overestimates the job cost, the more you like it, the more he underestimates it, the less you like it, and for obvious reasons. A low bid from Fred is never good news for you, since it makes it easier for him to steal your job. Interestingly, Wilma's error is far less important. Since she never bids below $120,000, her mistakes are less likely to result in you losing a job.

If you look at the two bars between Fred and Wilma, they show that you actually like long jobs with substantial delays. Why? Comparative advantage. With your larger work force, you're less affected by late penalties, so you can undercut the competition. So keep an eye out for this kind of job. Materials cost helps you far less, since all three of you face the same cost here. A big materials cost means that your markup gives you more money for the job at 28%, but that's a weak effect.

And finally, there's virtually no correlation between whether your estimate is high or low and how much you make. Screwing up your estimate is bad if your estimate is too high or too low, so the net effect is a wash. In a similar way, I could specify parameters of interest, like cost of labor per week or the size of the weekly penalty, and could obtain tornado charts showing how your bottom line depends on these factors. Worth doing, if the factor is something you might be able to do something about.

Of course, in the end, the problems that we've been considering are about real-life applications, a fact that we always need to keep in mind. If you implement the strategy we just developed for your construction business, you should do quite well, for a time. But if you model Fred's situation, you'll find that his mean profit is now negative. You know that situation isn't going to last. He might get out of the business. Or he might slash his markup margin to make himself competitive with you. If he comes sniffing around your job site, he might also get the idea that hiring more labor more than pays for itself. And now you'll have a new problem, of how to avoid a price war. The field of game theory, a subject which I developed in a another course, might be able to help us here, but the tools and analysis we've developed in this course are what can clue us in to the fact that the issue is going to come up. In the long run, the decisions that we make can change what our competitors do.

Like so many aspects of life, business is an ever-evolving situation. But analytics can help you to stay one step ahead of the competition. And I hope that, now that you've reached the end of this course, you feel you have a better idea of how you can make predictive analytics, optimization, and computational power work for you.

I've had a great time in this course. I hope I've managed to convey not only the usefulness of analytics, but the beauty and elegance of some of its techniques. We've opened some doors to amazing mathematical worlds, and you can walk through those doors any time you like and take a closer look at whatever you've seen that happens to have captured your fancy.

And me? I keep doing that, too. But for the last 24 lectures, I've been the kid in the candy shop. For someone who loves to teach, there are few satisfactions as rich as the company of someone who loves to learn, and that, clearly, describes you. So thank you. Thanks for sharing the trip.

Entering Linear Programs into a Spreadsheet

The best way to learn to do anything in Microsoft Excel or OpenOffice Calc is to do it yourself. Linear programming can be done in either application with almost identical effort and virtually identical steps. For this guide, it is recommended that you boot up one of these programs and work through the sample problem step by step.

The assumption of this guide is that you know the basics of your spreadsheet program—how to refer to cells by row and column label, how to highlight cells by holding the left mouse button and dragging the mouse, how to copy one cell into another cell or a range of other cells, how to represent a range of cells (such as A2:A5 for all cells between A2 and A5), and a few other basics. If you don't know how to do something in Excel or Calc, you can get help by pressing the F1 key while in the spreadsheet or (sometimes the best option) by using an Internet search engine to look for exactly what you need.

A Sample Problem

The following LEGO problem is a product mix problem with vertical integration, including a make-or-buy decision. But it can also be modeled it as a transshipment problem (raw materials and third-party packets "shipped" to packets available "shipped" to products), making it similar to problems from Lecture 8.

The linear program is about taking plastic and metal to make two kinds of intermediate products called "construction packets" and "accessory packets," and then combining these to make two products for sale: kit 1 and kit 2. You can buy construction packets from a third party, too, if you want to.

The LEGO Problem

The LEGO company sells building toys and is currently selling two different kinds of building kits, kit 1 and kit 2. LEGO has found that it can save money by "modularizing" the kits that it sells. LEGO manufactures a specific collection of pieces, bags them together, and calls the result a "packet." A kit is made simply by collecting together the packets needed for that particular kit. LEGO makes two kinds of packets: the construction packet and the accessory packet. Kit 1 consists of 4 construction packets and 1 accessory packet, and it sells for $13. Kit 2 consists of 3 construction packets and 2 accessory packets, and it sells for $16. Because of limited demand for LEGO products in the marketplace, the total number of kits sold will not exceed 3000. (You may assume that LEGO will construct only those kits that it sells.)

To make a construction packet, LEGO uses 50 grams of plastic and incurs a cost of 20 cents. To make an accessory packet, LEGO uses 25 grams of plastic and 20 grams of steel. An accessory packet costs 35 cents to produce. (Neither of these packets may be sold directly.) LEGO may also buy construction packets from a third-party supplier for 30 cents each. LEGO has available 100,000 grams of plastic and 50,000 grams of steel. How should LEGO conduct its purchasing and production if it wishes to maximize its profit?

The "measurable quantity" formulation of the program is as follows.

> Maximize # of $ of profit
>
> subject to
>
> # of grams of plastic used ≤ # of grams of plastic available
> # of grams of steel used ≤ # of grams of steel available
> # of construction packets used ≤ # of construction packets available
> # of accessory packets used ≤ # of accessory packets available
> # of kits sold ≤ # of kits demanded

The details of this problem don't matter, but the problem allows us to work through most of the issues that will come up when you put a linear program into your spreadsheet. The following is the program.

Maximize
$13KIT1 + 16KIT2 - 0.20CMAKE - 0.35AMAKE - 0.30CBUY$

subject to

$50CMAKE + 25AMAKE \leq 100,000$
$20AMAKE \leq 50,000$
$4KIT1 + 3KIT2 \leq CMAKE + CBUY$
$KIT1 + 2KIT2 \leq AMAKE$
$KIT1 + KIT2 \leq 3000$

all variables ≥ 0

The Setup

First, we put the problem in standard form. That means that all constant terms go on the right side of the constraints, all variable terms go on the left side of the constraints, and any parentheses in the problem are "multiplied out." Here, the third and fourth constraints have variables on their right-hand sides. We'll subtract them off to the left. This leaves us with the following.

Maximize
$-0.20CMAKE - 0.35AMAKE - 0.30CBUY + 13KIT1 + 16KIT2$

subject to

$50CMAKE + 25AMAKE \leq 100,000$
$20AMAKE \leq 50,000$
$4KIT1 + 3KIT2 - CMAKE - CBUY \leq 0$
$KIT1 + 2KIT2 - AMAKE \leq 0$
$KIT1 + KIT2 \leq 3000$

all variables ≥ 0

The program is now in standard form. Next, we're going to represent this program in a kind of shorthand form by creating the following tableau. The tableau contains everything that the program tells us, except for the fact that the program is a MAX and that all decision variables must be nonnegative. These are easy to remember, so don't worry about them.

The Coefficient Tableau for the LEGO Problem

KIT1	KIT2	CMAKE	AMAKE	CBUY		RHS
13	16	−0.2	−0.35	−0.3	=	PROFIT
0	0	50	25	0	≤	100,000
0	0	0	20	0	≤	50,000
4	3	−1	0	−1	≤	0
1	2	0	−1	0	≤	0
1	1	0	0	0	≤	3000

Figure A.1

Do you see how the tableau works? We've essentially recorded only the numbers that "fill in the blanks" in the tableau. To recover a line from the program at the top of the page, you just proceed across a row of the tableau and multiply each number in that row by the variable that heads its column. (RHS, which stands for "right-hand side," isn't a variable, so no multiplication is done for the RHS of the constraints.) So, for example, the second row of the tableau says $0KIT1 + 0KIT2 + 50CMAKE + 25AMAKE + 0CBUY \leq 100000$, which is precisely the first constraint in our program. The top row is the objective. Note that we did record the direction of each inequality. They were all ≤ for the constraints here, but some could be = or ≥.

In the spreadsheet program, we're going to record the information in the tableau almost exactly as we did there—with a few additions. We'll have a row of cells below our first row (the variable names) where we'll eventually place numerical values for all of the decision variables. When we finish, these cells will hold the optimal solution. We'll also include a column labeled LHS (which stands for "left-hand side"). The cells in this column will hold the computed numerical values of the left-hand side of each of our constraints, as well as the computed value of the objective function. Finally, we'll also include a column to the left of our program, to remind us what each constraint says.

545

The following spreadsheet includes borders and shading. Why did we do this? We shaded all of the cells containing the numbers that we typed in. That way, it will be easy later to know what numbers we can change to new values (to make up a new problem) and what numbers we can't (because they're really formulas). We've put double-lined boxes around the cells that will hold our decision variable values. For now, we've put the values 1, 2, 5, 4, and 10 in those cells, just as placeholders. You can type any numbers you want there—later, Solver will find the best values.

	A	B	C	D	E	F	G	H	I
1	Variables								
2		KIT1	KIT2	CMAKE	AMAKE	CBUY		Make up any	
3		1	2	5	4	10		values you like	
4								in these cells!	
5	Maximize	13	16	−0.2	−0.35	−0.3		=	PROFIT
6									
7	Constraints						LHS		RHS
8	Plastic	0	0	50	25	0		≤	100,000
9	Steel	0	0	0	20	0		≤	50,000
10	Construction	4	3	−1	0	−1		≤	0
11	Accessory	1	2	0	−1	0		≤	0
12	Demand	1	1	0	0	0		≤	3000

Figure A.2

Now, we're going to write the mathematical relationships to tie all of these numbers together. You know how the numbers in the table are supposed to combine to produce meaningful calculations. For example, look at constraint 1 in row 8 of the spreadsheet, which says that we can't use more than the 100,000 grams of plastic that we have on hand. How much *are* we using, with our made-up decision variable values we put in row 3 of the spreadsheet? $0KIT1 + 0KIT2 + 50CMAKE + 25AMAKE + 0CBUY$ becomes $0(1) + 0(2) + 50(5) + 25(4) + 0(10) = 350$ grams.

Excel and Calc both have a built-in function that does exactly this, without all of the typing: SUMPRODUCT. Let's do it. Click in cell G8. For Excel, type = SUMPRODUCT(B3:F3,B8:F8), and then press return. SUMPRODUCT just multiplies each number in the one range by the corresponding one in the other range and then adds the results, exactly as we desire. If you are using

Calc, you'll use a semicolon where Excel uses a comma—for example, SUMPRODUCT(B3:F3;B8:F8). When you press enter, you should see our 350 grams of plastic used appear in cell G8.

We're going to do this for each of the constraints, combining the variable values in row 3 with the constraint coefficients in each row. We could type out the equation for each row, but we're going to enter it a little differently than this. Enter the formula as it is entered in the following spreadsheet (again, if you are using Calc, replace the comma with a semicolon). We've just edited the formula that was typed into G8 by inserting 4 dollar signs.

	A	B	C	D	E	F	G	H	I
1	Variables								
2		KIT1	KIT2	CMAKE	AMAKE	CBUY			
3		1	2	5	4	10			
4									
5	Maximize	13	16	−0.2	−0.35	−0.3		=	PROFIT
6									
7	Constraints						LHS		RHS
8	Plastic	0	0	50	25	0	=SUMPRODUCT($B$3:$F$3,B8:G8)	≤	100,000
9	Steel	0	0	0	20	0		≤	50,000
10	Construction	4	3	−1	0	−1		≤	0
11	Accessory	1	2	0	−1	0		≤	0
12	Demand	1	1	0	0	0		≤	3000

Figure A.3

Those dollar signs did not affect the calculation of 350 grams in G8, but they will be important. Once you've typed the formula and pressed return, move your cursor to the lower-right corner of cell G8, where a small black square is visible. Your cursor will turn to a skinny cross. When it does, press and hold the left mouse button, and then drag downward until you have highlighted cells G9 to G12. Let go of the mouse button. You've just entered the formulas for the left-hand sides of all of the other constraints. The dollar signs told the spreadsheet that you *didn't* want the row 3 values used in the formula to move when you dragged the formula downward. Do a fast double-click on cell G12 and you'll understand. It uses data from row 12 and row 3, as it should. Let's finish things off by entering the formula for the objective in cell G5. It's = SUMPRODUCT(B3:F3,B5:F5). As always, in Calc, that comma is going to be replaced by a semicolon.

The Solution

Before you can use Solver to solve the problem, you need to tell Excel or Calc that you want to use it. The steps to do this vary a bit, depending on your spreadsheet program, but you only have to do this once. In some versions of Excel, you do this by clicking on the "File" tab on the top of the sheet, choosing "Options" from the left-hand menu, and then choosing "Add-ins." At the bottom of the window that opens, choose the button labeled "Go" next to "Excel add-ins." When the window pops up, check the option for "Solver add-in." Click "OK." If all went well, you should now see "Solver" listed on your "Data" tab. Calc comes with the linear solver already installed, and a beta version of a nonlinear and evolutionary solver is available online. If you have a different version of Excel, you can search the Internet for instructions appropriate for your version.

To use Solver, you need to tell it where the decision variables for your program appear in the spreadsheet, what the program constraints are, and where the objective function value is. You're also going to have to tell it some other bits of information, such as the fact that all of your decision variables in this problem are nonnegative. Let's do this in the spreadsheet where you've created our linear program. **Figure A.4** shows you what it looks like in Calc. Excel's requestor is almost identical. The biggest difference is that you'll find "Solver" under the "Data" tab in most versions of Excel and under the "Tools" menu in Calc.

Most of this is pretty self-explanatory, once you know the names that the spreadsheet program uses for the parts of a linear program. Our objective value (profit) is in cell G5, which we put in our "Target cell" box. In Excel, it would be labeled as "Set objective." Below this, we specify the cells holding our decision variables, B3 through F3. Excel calls these "variable cells," while in Calc they are the "changing cells." We can either type this in the box, or we can click in the "variable cells" box to activate it, and then click and drag in the spreadsheet to highlight cells B3 through F3.

This is a MAX program, so we need to click on the radio button for "Maximum"—but it's already selected for us. We do, however, need to enter the constraints. The way that we've entered our program into the spreadsheet makes entering the constraints especially easy. We've computed the LHS of every constraint. We just have to make sure that it bears the right relation to the RHS of each constraint. For our program, by coincidence, every LHS is supposed to be ≤ to the corresponding RHS. Enter the first constraint in the first row under "Limiting conditions" in Calc, or by clicking on the "Add" button next to the "Subject to the constraints" box in Excel. You'll then need to provide three bits of information: what cell represents the LHS of the constraint in question, what cell represents the RHS of the constraint, and what symbol (≤, =, or ≥) belongs between them. You can enter constraints one at a time, each time moving to a new row in Calc or pressing "Add" in Excel, but here we used a shortcut to do all constraints at once. We could do this because they were all ≤ constraints. Consecutive constraints with the same relational operator can be done in one row.

If you're using Excel, make sure that at the bottom of the requestor, "Solver," is set to "Simplex LP" for its solving method and the box for "assume variables nonnegative" is checked. In Calc, you'll automatically be using the simplex method, but you'll need to click on the "Options" button and tell Calc that your variables are nonnegative, and then click "OK."

Click on the "Solve" button, and Solver should now solve your problem. When it's done, it will present a window letting you know that a solution has been found. In Excel only (not yet in Calc), it's also possible to include a sensitivity report (a topic considered in Lecture 12).

Common problems arise from either 1) forgetting to impose the nonnegativity constraints (put a check mark in the appropriate box in Solver to do this) or 2) wanting to highlight an entire row of the spreadsheet (and not just the LHS *cell*) as the left-hand side of a constraint. In this example, we entered all of our constraints at once, but if you are unsure of what you are doing, enter them one at a time. Each constraint will have a single cell on its LHS and a single cell on its RHS.

Like most things with computers, the spreadsheet solution of linear programs is much easier than it seems at first. Working through this example should help you get past many of the more common obstacles.

Entering Linear Programs into a Spreadsheet

Glossary

100% rule: A rule that often allows the determination of whether changes in the constant terms of multiple constraints in a linear program can be made without altering shadow prices. Alternatively, a rule that often allows the determination of whether changes in multiple objective function coefficients in a linear program can be made without altering the optimal values of the decision variables.

absorbing state: In a Markov system, a state that cannot be left once it is entered.

adjusted r^2: A measure of the goodness of fit of a multiple regression model. The value of r^2 is adjusted downward for each additional independent variable, to help address the problem of choosing a model that overfits the data.

affinity analysis: Umbrella term for the data mining techniques that seek to determine "what goes with what."

alternative optima: Two or more best solutions to the same problem.

analytic hierarchy process (AHP): A procedure for identifying the relative importance of factors in a project by decomposing the project into activities. These activities may themselves be broken down in a similar way through as many levels as desired. Pairwise comparisons are made between items at the same level of this decomposition, and AHP combines these comparisons into a ranking of the activities.

Apriori method: A method for generating frequently occurring subsets of items for an association rules application.

association rules: A data mining task of creating rules of the following form: "If an individual has *this* collection of traits, predict *that* individual also has that set of traits."

auxiliary variable: Also called a dependent variable. A variable whose value is completely determined by the values of the other variables in the problem. Formulations may be simplified by using auxiliary variables but must include an equality constraint for each one, which essentially defines how its value is computed.

binary variable: Also called a 0/1 variable or a Boolean variable. A variable that can only assume the values of 0 or 1.

binding: A constraint is binding on a solution if it has zero slack in that solution.

Bayesian analysis: A technique for revising the probabilities of outcomes in light of new information.

branch and bound: A technique for finding optimal solutions to integer linear programs by repeatedly replacing a linear integer program with two variants, each of which differs from the original by the inclusion of one additional constraint.

Calc: The OpenOffice suite's equivalent to Excel. It's freely downloadable but lacks some of the features of Excel.

calling units: General term applied to those individuals or objects that make use of a queuing system.

centroid: In data mining, the centroid of a cluster is the point for which each variable is the average of that variable among points in that cluster.

classification: Using the information available about an individual to predict to which of a number of categories that individual belongs.

classification tree: A data mining classification technique. The algorithm repeatedly splits a subset of the individuals into two groups in a way that reduces total "impurity."

cluster: A collection of points considered together because of their proximity to one another.

coefficient: The number multiplied by a variable is its coefficient.

coefficient of determination: See r^2.

confidence (data mining): For an association rule, the probability that the consequent is true given that the antecedent is true.

confidence (statistics): Reported when generating a confidence interval for a specified population parameter. The confidence of the interval is the probability that the series of calculations that generate the interval, when applied to a random sample of specified size, will result in an interval that contains the population parameter.

confidence interval: An interval of values generated from a sample that hopefully contains the actual value of the population parameter of interest. See **confidence (statistics)**.

conservation constraint: An equality constraint in a linear program that generally requires that counting the same thing in two different ways must give the same answer. It can often be interpreted as "total in = total out."

constraint: A rule that must be obeyed by any acceptable solution.

convex: A region is convex if a line segment drawn between two points of the region always stays within the region. A square is convex; a star is not. One can intuitively think of a function being convex if someone walking on the surface of the function has "line of sight" to every other point on the function's surface. Hence, a single valley is convex, but two valleys separated by a ridge are not.

correlation: Intuitively, a measure of the extent to which the variation in two variable quantities are linked. Its value ranges from −1 to 1. The closer the correlation is to 1, the closer the variables, when plotted in a scatter plot, would be to a straight line. For plots where the slope of the trend line would be positive, the correlation is positive. It is negative for plots where the slope of the trend line would be negative.

covariance: Similar to the correlation, the covariance of two quantities is their correlation multiplied by the standard deviations of two variables.

cyclic component: The component of a time series forecast that attempts to capture cyclic variation. This differs from seasonal variation by showing nonconstant duration or intensity or unpredictable onset.

data envelopment analysis (DEA): A technique for evaluating the relative efficiency of decision-making units by comparing them to virtual producers. See **efficient**.

decision-making unit (DMU): The usual term given to a producer in data envelopment analysis.

decision theory: A collection of techniques for determining what decisions are optimal in a situation that involves stochastic elements, especially those situations that evolve over time.

decision tree: A structure consisting of branching decision nodes and chance nodes used to analyze sequential decision making in the face of uncertainty.

decision variable: In a mathematical model, any measurable quantity over which you have direct control. A solution to a model is a specification of the values of all of its decision variables.

dendrogram: A graph detailing the sequence of cluster mergers that eventually ends in a single cluster.

derivative: The derivative of a function is itself a function, one that essentially specifies the slope of the original function at each point at which it is defined. For functions of more than one variable, the concept of a derivative is captured by the vector quantity of the gradient.

deterministic: Involving no random elements. For a deterministic problem, the same inputs always generate the same outputs. Contrast to **stochastic**.

discriminant analysis: A data mining technique for separating a set of data into two categories with a hyperplane so as to separate a two-category output variable.

distance, average: In cluster analysis, the definition of the distance between two clusters as being the average of the distances between each point in one cluster and each point in the other.

distance, complete linkage: In cluster analysis, the definition of the distance between two clusters as being the maximum of the distance between a point in one cluster and a point in the other.

distance, Euclidean: Our normal concept of distance, applied in an arbitrary number of dimensions.

distance, single linkage: In cluster analysis, the definition of the distance between two clusters as being the minimum of the distance between a point in one cluster and a point in the other.

distance, statistical: Also called Mahalanobis distance. A distance measure that takes into account the variance and covariance of the various variables used to define it.

dual price: See **shadow price**.

dummy variable: A variable that can take on values of only 0 or 1. Dummy variables are useful in forecasting and data mining for representing two-category variables. A categorical variable with k categories can be represented by a collection of $k-1$ dummy variables.

e: A natural constant, approximately 2.71828. Like the more familiar π, e appears frequently in many branches of mathematics.

efficient: In data envelopment analysis, a producer is inefficient if a virtual producer could create the same outputs as the producer under scrutiny while using less of each input. A producer that is not inefficient is efficient. See **weakly efficient**.

ergodic: A Markov system is ergodic if there exists some n such that, for any two pure states i and j, it is possible to move from state i to state j in exactly n transitions.

error: In a forecasting model, the component of the model that captures the variation in output value not captured by the rest of the model. For regression, this means the difference between the actual output value and the value forecast by the true regression line.

event: A statement about the outcome of an experiment that is either true or false for any given trial of the experiment.

evolutionary algorithm: See **genetic algorithm**.

Excel: The Microsoft Office suite's spreadsheet program.

expected value: The average (or mean) value of a numeric random variable.

exponential growth/decay: Mathematically, a relationship of the form $y = ab^x$ for appropriate constants a and b. Such relations hold when the rate of change of a quantity is proportional to its current value.

exponential smoothing: A time series forecasting technique that forms the basis of many more-complicated models. Exponential smoothing can be thought of as a variant of the weighted moving average.

feasible: Possible. In a mathematical program, feasible solutions are those that satisfy all of the constraints.

feasible region: The set of decision variable combinations that correspond to feasible solutions to the problem. In a two-variable linear program, the feasible region, if it exists, will be one contiguous convex region bordered by straight lines.

hyperplane: The higher-dimensional equivalent of a plane. Think of a flat wall. Don't worry—most people can't visualize them.

genetic algorithm: An optimization technique involving a stochastic search of possible solutions in a way that somewhat mimics DNA recombination.

Gini index: A measure of purity for classification trees. The purity of a rectangle is obtained by computing the fraction of observations in that rectangle that belong to each output category, squaring these fractions, and then subtracting their sum from 1.

gradient: A vector quantity that indicates the direction of "straight uphill" for a function of one or more variables. It is the higher-dimensional generalization of the derivative of a one-variable function.

heat map: A data visualization technique in which colors or shades are used to convey information about a variable in a chart, graph, or map.

heteroscedastic: A collection of random variables is heteroscedastic if their standard deviations are not all equal.

homoscedastic: A collection of random variables is homoscedastic if they all have the same standard deviation. Linear regression generally assumes that the error terms in the forecast are homoscedastic.

independent: Two events are independent if knowing the outcome of one does not alter the probability of the other. In symbols, A and B are independent if and only if $P(A) = P(A \mid B)$. Two events that are not independent are dependent.

infeasible program: A program that has no feasible solutions. Its constraints cannot all be satisfied by any set of decision variable values.

INFORMS: The Institute for Operations Research and the Management Sciences.

integer program: A program in which all of the decision variables are required to be integers—that is, positive or negative whole numbers or zero. If only some of the variables have this requirement, the program is termed "mixed integer."

irreducible: A Markov system is irreducible if it is possible to follow a set of transitions that passes through every state in the system and eventually returns to where it began.

k-nearest neighbors: A classification/prediction technique in data mining in which the value assigned to a new observation is computed by the output values of the k training set observations closest to it.

limited resource constraint: The most common kind of linear programming constraint. It says that you can't use more of something than what is available.

linear expression: An algebraic expression consisting of the sum or difference of a collection of terms, each of which is either simply a number or a number times a variable. Linear expressions graph as "flat" objects—straight lines, planes, or higher-dimensional analogs called hyperplanes.

linear program: A model in which the objective function and the two sides of each constraint are all linear expressions. Linear programs are especially easy to solve.

linear regression: A method of finding the best linear relationship between a set of input variables and a single continuous output variable. If there is only one input variable, the technique is called simple; with more than one, it is called multiple.

logarithm: The inverse function to an exponential. If $y = a^x$ for some positive constant a, then $x = \log_a y$. The most common choice for a is the natural constant e. $\log_e x$ is also written $\ln x$.

LP relaxation (of an integer program): The linear program that results from suspending the integer requirements of an integer linear program.

manifold, two-dimensional: A surface that on a very small scale looks like a plane. If you imagine a surface that doesn't have rips, sharp ridges, or cliffs, you're probably imagining a 2-manifold.

Markov analysis: A set of techniques for analyzing the behavior of systems that can be viewed as occupying one of a number of states, with a particular probability of transitioning from one state to another at various points in time.

Markov diagram: A graphical representation of Markov process. Each pure state of the system is represented by a node, and each possible transition from state to state is indicated with an arrow connecting those nodes.

mean absolute deviation (MAD): A measure of forecast accuracy, MAD is the average amount by which the forecast differs from the actual value.

mean absolute percentage error (MAPE): A measure of forecast accuracy, MAPE is the average percentage by which the forecast differs from the actual value.

mean squared error (MSE): A measure of forecast accuracy that is similar to variance in its calculation, MSE is the average of the squares of all of the residuals for a forecast.

measurable quantity: The quantities represented by numbers in a mathematical model. In most cases, a measurable quantity can be defined as "the number of (unit of measure)" of something, such as the number of hours spent on a project.

memoryless: A process where the probability of reaching a particular state after the passage of so much time depends only on the current state and not on the events that led to that state. Rolling a die is memoryless.

minimum performance constraint: The second-most-common kind of linear programming constraint. It says that the amount obtained of something is at least as much as is required. Also called a quota constraint.

model: A simplified representation of a situation that captures the key elements of the situation and the relationships among those elements.

Monte Carlo: A kind of simulation, named after the famous Moroccan casino. The simulation is run a large number of times with different values for the random quantities in each run.

Moore's law: Formulated by Intel founder Gordon Moore in 1965, it is the prediction that the number of transistors on an integrated circuit doubles roughly every two years. To date, it's been remarkably accurate.

multicollinearity: The problem in multiple regression arising when two or more input variables are highly correlated, leading to unreliable estimation of the model coefficients.

multiobjective programming: Any optimization problem with more than one objective.

multiperiod planning: Problems in which the same pattern of decisions must be made repeatedly over time.

naïve forecast: The forecast for future times of whatever just happened.

nonbinding: A constraint is nonbinding on a solution if it has positive slack in that solution. (Negative slack means that the constraint is violated.)

nonnegativity constraint: A constraint requiring a particular decision variable to be at least zero. Also called a trivial constraint.

nontrivial constraint: Any constraint other than a nonnegativity constraint.

normalizing: Also called standardizing. Linearly rescaling a variable to make its mean 0 and its standard deviation 1. This is done by taking each of the variable's values, subtracting the mean of the variable, and then dividing the result by the variable's standard deviation.

objective (function): The mathematical expression that represents the goal of an optimization problem. The better the objective function value, the better the solution. Objective function values are usually either maximized or minimized in an optimal solution.

objective function coefficient range (OFCR): The range of values over which a coefficient in the objective function of a linear program may vary without changing the optimal values of the decision variables.

objective function line (OFL): In a graphically solved linear program, a line of constant objective function value. OFLs act like contour lines on a topographic map. In problems with more than two variables, OFLs are replaced with planes or hyperplanes of constant objective function value.

operations research: The general term for the application of quantitative techniques to find good or optimal solutions to real-world problems. Often called operational research in the United Kingdom. When applied to business problems, it may be referred to as management science, business analytics, or quantitative management.

optimization: Finding the best answer to a given problem. The best answer is termed "optimal."

optimum: The best answer. The best answer among all possible solutions is a global optimum. An answer that is the best of all points in its immediate vicinity is a local optimum. Thus, in considering the heights of points in a mountain range, each mountain peak is a local maximum, but the top of the tallest mountain is the global maximum.

overfitting: In forecasting or prediction, choosing a more complex model that performs better on the training data but does more poorly on new data because the model is responding to spurious patterns (noise) in the training data, rather than capturing a true relationship between inputs and the output. (3, 6)

oversampling: In data mining, the choice of representing a class in a data set at a relative frequency higher than its occurrence in the population. Oversampling is appropriate if the class of interest forms only a small proportion of the entire population.

parameters: The values of the constants relevant to a problem, such as the quantity available of a resource. Models are often created so that the parameters can be changed to reflect different situations.

Pareto optimal: A solution is Pareto optimal if one cannot do better at accomplishing one goal without doing worse on at least one other goal.

penalty: A modifier to the objective of a problem that reflects that the proposed solution is deficient in completely satisfying a constraint. Used in genetic algorithms and soft constraints.

Poisson process: A process that generates successes at various points along some interval. The defining characteristics of a Poisson process are that each tiny interval have the same minuscule chance of containing a success and that a success in any one such interval is independent of a success in any other. Arrival processes are often modeled as Poisson.

polynomial: A mathematical expression that consists of the sum of one or more terms, each of which consists of a constant times a series of variables raised to powers. The power of each variable in each term must be a nonnegative integer. Thus, $3x^2 + 2xy + z - 2$ is a polynomial.

population: The set of all individuals of interest for purposes of the current study, or the set of all relevant variable values associated with that group.

posterior probability distribution: See **prior probability distribution**.

power law: A relationship between variables x and y of the form $y = ax^b$ for appropriate constants a and b.

prediction: In data mining, using the information available about an individual to estimate the value that it takes on some continuous output variable. Some people use the term to include classification, as well.

prediction interval: The prediction interval is an interval with a specified probability of containing the value of the output variable that will be observed, given a specified set of inputs. Compare to **confidence interval**.

principal components analysis (PCA): A technique for reducing the number of variables in a data set by identifying a collection of linear combinations of those variables that capture most of the variation of a larger set of the original variables. These new variables can then be used in place of the larger set.

prior probability distribution: Bayesian inference modifies the probabilities of events in light of new information. The probability distribution in effect before new information is provided is called the prior probability distribution. The new distribution, accounting for the new information, is called the posterior probability distribution.

probability: A measure of the likeliness of the occurrence of uncertain events. Probabilities lie between 0 and 1, with more-likely events having higher probability. If all possible outcomes of an experiment are equally likely, the classical probability or a priori probability of an event can be defined as the fraction of those outcomes in which the event occurs. In other cases, probability may be empirically defined from the fraction of historical cases in which the event occurred.

probability, conditional: The probability that some event occurs, given that some other collection of events occurred. The probability that A occurs given that B occurs is written $P(A \mid B)$.

probability, joint: The probability that all events in some collection occur.

program: Mathematically, an optimization model. The term corresponds to the idea of a program as a schedule, as in television programming, because many early optimization models were scheduling problems.

queue: Waiting line. The study of waiting lines is called queuing theory. Technically, the queue does not include any calling units that are currently in service, only those waiting for service to begin.

quota constraint: The second-most-common kind of linear programming constraint. It says that the amount obtained of something is at least as much as is required. Also called a minimum performance constraint.

r^2: The coefficient of determination, a measure of how well a forecasting model explains the variation in the output variable in terms of the model's inputs. Intuitively, it reports what fraction of the total variation in the output variable is explained by the model.

random variable: A measurable quantity, often numeric, associated with the outcome of a stochastic experiment.

random walk: A sequence in which each term is obtained from the previous one by the addition of a random term. The simplest example would be that x_n equals the number of heads obtained in x coin flips.

redundant: A constraint is redundant if satisfying the other constraints in the problem automatically satisfies the redundant constraint. If $x \leq 5$ and $y \leq 5$, then $x + y \leq 20$ is redundant.

regression: A mathematical technique that posits the form of a function connecting inputs to outputs and then estimates the coefficients of that function from data. The regression is linear if the hypothesized relation is linear, polynomial if the hypothesized relation is polynomial, etc.

regression line: The true regression line is the linear relationship posited to exist between the values of the input variables and the mean value of the output variable for that set of inputs. The estimated regression line is the approximation to this line found by considering only the points in the available sample.

regression tree: A classification tree with a continuous output.

residual: Given a data point in a forecasting problem, the amount by which the actual output for that data point exceeds its predicted value. Compare to **error**.

right-hand-side range (RHS range): The range of values over which the constant term in a constraint may vary without changing the shadow price of any constraint.

risk: Variability in the desirability of the possible outcomes of a situation involving uncertainty. Decision makers may be risk neutral, meaning that they choose whatever course of action gives the highest expected payoff. Alternatively, they may be risk averse or risk loving. Risk-loving individuals are inclined to accept a gamble even if, on average, it does not benefit them. Risk-averse individuals are inclined to refuse a gamble even if, on average, it does benefit them.

rollback: The procedure of evaluating a decision tree by computing expected values at each chance node and choosing the most-attractive options at each decision node.

saddle point: A point on a manifold where the derivative is zero in any direction but is neither a maximum nor a minimum. The middle of a saddle has this characteristic, although technically a terrace point is also a saddle point.

sample: A subset of a population.

seasonal component: The component of a time series forecast that captures the seasonality of the data—that is, its regular, periodic variation. Some sources use the term for such variation only if the period length is at least one year.

sensitivity: The probability that an observation is classified as having a trait of interest, given that it does indeed possess that trait. Contrast to **specificity**.

sensitivity analysis: An investigation of how an optimal solution changes as parameters of the problem are varied from some set of original values. Sensitivity analysis can be done with most optimization techniques, such as linear programming or decision theory.

shadow price: Also called dual price. The amount that the optimal objective function value increases when the right-hand side of a constraint is increased by 1. Shadow prices may change when one leaves the right-hand-side range of a constraint.

simple moving average: A forecast for a period in a time series made by averaging together the values for a specified number of nearby time periods. For predictive models, this will mean the n observations immediately preceding the current time period.

simplex method: An algorithm invented by George Dantzig for finding the optimal solution to any linear program that has one. For most linear programs, the simplex method finds the optimal solution with surprising speed.

simulation, digital: A powerful technique for what-if analysis, allowing the development of a complex model and an analysis of its possible outcomes. Most often this model involves randomness and is therefore done as a Monte Carlo simulation.

soft constraint: A constraint that need not be completely satisfied. Generally, if it is not, a penalty is assessed, with a larger penalty applied to a larger violation of the constraint.

solver: Generally, a piece of software for solving a class of problems. Specifically, the add-in in Excel or Calc capable of finding optimal solutions to linear and nonlinear programs.

slack: For any inequality constraint, the slack is the side of the constraint claiming to be bigger minus the side claiming to be smaller. Hence, the slack in $4x + y \geq z$ is $4x + y - z$. The slack in $8x \leq 12$ is $12 - 8x$. A constraint that is satisfied always has a slack of 0 or more.

specificity: The probability that an observation is classed as not having the trait of interest, given that it does not, in fact, possess that trait. Contrast to **sensitivity**.

standard deviation: A measure of dispersion, it is the square root of the variance.

standard error: Not an "error" in the traditional sense. The standard error is the estimated value of the standard deviation of a statistic. For example, the standard error of the mean for samples of size 50 would be found by generating every sample of size 50 from the population, finding the mean of each sample, and then computing the standard deviation of all of those sample means.

standard form: A linear constraint in standard form has all of its variable terms on the left side of the constraint and its constant term on the right side. $2x + 4y \leq 7$ is in standard form.

state vector: A string of values specifying the probability that a Markov system is in each of the possible pure states.

stationary time series: A time series that does not show a long-term trend or change in variance.

steady state vector: In Markov analysis, a state vector that is unchanged by the application of the transition matrix. A Markov system that reaches a steady state vector will never move from it.

steepest descent: Also called gradient descent. A collection of techniques for zeroing in on the maximum or minimum of a function. One first determines the direction in which the objective function is changing most rapidly in the desirable direction and then jumps from the current solution to one (nearly) in that direction.

stochastic: Involving random elements. Identical inputs may generate differing outputs. Contrast to **deterministic**.

stochastic optimization: Finding a (near) optimal solution to a problem involving chance.

SUMPRODUCT: A function in Calc and Excel that takes two row or column arrays of equal length, multiplies corresponding elements, and then adds the results. In vector calculus, this is the dot product or scalar product of two vectors.

SWOT: Stands for "strengths, weaknesses, opportunities, threats." A strategic management technique for identifying the direction that an organization should take by examining its inherent strengths and weaknesses and its comparative advantages and disadvantages in relation to its competitors.

system, queuing: All of the calling units either currently under service or in the queue are considered to be in the system.

tabu search: Also called taboo search. A search procedure of a part of the feasible region in which a list is kept of recently examined solutions. The procedure avoids returning to solutions already found unsatisfactory.

testing set: See **training, validation, and testing sets**.

time series: A data set consisting of one value of the output variable for each point in time. The points in time are usually evenly spaced. Alternatively, a forecasting technique used on such data.

tornado diagram: A chart summarizing how the value of a quantity of interest, such as the objective, depends on the random variables in a stochastic optimization problem.

training, validation, and testing sets: In data mining, where data is plentiful, model quality is usually assessed by using part of the data as a training set, to create one or more models, and then assessing their quality with the validation set. If more than one model is being assessed, the best is chosen based on validation results, and the final choice is then evaluated on a third set of data, the testing set.

transformation of variable: The technique of replacing a variable in a problem by some specified function of that variable. Common in forecasting, transforming a variable by replacing it with its logarithm, square root, or the like can make a nonlinear relationship into a linear one as well as address problems of heteroscedasticity.

transient state: In a Markov system, a state whose long-run probability approaches zero regardless of initial conditions.

transition matrix: In Markov analysis, a matrix whose entries p_{ij} give the probabilities of moving to a specified state j given that the system is currently in state i.

transshipment problem: A classic kind of linear programming problem. Units are shipped from supply points through transshipment points and on to demand points. The goal is often to minimize the total cost of the operation. A variety of constraints can be added to this problem.

unbounded program: A program in which one can find a feasible solution that makes the objective function value arbitrarily good. Intuitively, you can find possible answers to an unbounded program that are "infinitely good."

utility: In essence, "happiness points." Utilitarian models in economics posit that decision makers act so as to maximize their personal utility. A utility measure satisfies a number of sensible rules.

validation set: See **training, validation, and testing sets**.

variance: A commonly used statistical measure of the dispersion, or spread, of data. For a population, the variance is computed by deviation of each observation from the population mean, squaring those differences, and averaging the squares. For a sample, the same calculation is performed, but the result is multiplied by $n/(n-1)$, where n is the sample size.

virtual producer: In data envelopment analysis, a hypothetical producer created by taking a linear combination of actual producers.

weak efficiency: In data envelopment analysis, a producer is weakly efficient if it is efficient but some virtual producer creates its same outputs while using no more of any input and using less of at least one input.

weighted moving average: A forecast for a period in a time series made by averaging together the values for a specified number of nearby time periods, with each period's importance reflected by its weight. For predictive models, this will mean a weighted average of the n observations immediately preceding the current time period.

winner's curse: In an auction in which the true value of the item up for bid is unknown, the phenomenon that the winner often pays more for the item than it is worth.

Zipf's law: A relationship between the relative frequency of a word in a language and its position in a list of words in that language, sorted by frequency. The nth word in the list has a relative frequency proportional to $1/n$.

Bibliography

Anderson, Chris K. "Setting Prices as Priceline." *Interfaces* 4 (2009): 307–315. A discussion of how Priceline uses operations research to determine optimal prices and inventory allocations for Kimpton Hotels. A nice example of dynamic programming.

Anderson, David, Dennis Sweeney, Thomas Williams, Jeffrey Cam, and James Cochran. *Quantitative Methods for Business*. Mason, OH: South-Western Cengage Learning, 2013. Anderson, Sweeney, and Williams give a nice introduction into the skills and tools needed for business analytics. The book, now in its 10th edition, has stood the test of time.

Berry, Michael J., and Gordon Linoff. *Data Mining Techniques: For Marketing, Sales, and Customer Support*. Charlottesville: John Wiley and Sons, 1997. This is a quick read for a manager interested in an overview of the data mining techniques that exist and what they can and cannot do. A good introduction.

Bollapragada, Srinvas, Hong Cheng, Mary Phillips, Marc Garbiras, Michael Scholes, Tim Gibbs, and Mark Humphreville. "NBC's Optimization Systems Increase Revenues and Productivity." *Interfaces* 1 (2002): 42–60. An in-depth but clear discussion of the NBC problem discussed in the lecture.

Boyles, Stephen. "Notes on Nonlinear Programming." University of Texas. Last modified Fall 2009. Accessed March 30, 2014. A treatment of nonlinear programming that is a little more advanced than the treatment in the lecture but still accessible to the mathematically competent layperson. Many more advanced resources are readily available online for the interested student.

Budiansky, Stephen. *Blackett's War: The Men Who Defeated the Nazi U-Boats and Brought Science to the Art of Warfare*. New York City: Knopf, 2013. An informative historical perspective on the early days of operational research and its importance in World War II. The discussion is primarily from a human rather than mathematical perspective.

Butchers, E. Rod, Paul R. Day, Andrew P. Goldie, Stephen Miller, Jeff Meyer, David M. Ryan, Amanda C. Scott, and Chris A. Wallace. "Optimized Crew Scheduling at Air New Zealand." *Interfaces* 1 (2001): 30–56. A more in-depth treatment of the Air New Zealand application discussed in the lecture. The article discusses the extent and complexity of the challenge at a level quite accessible to the intelligent layperson.

Caixeta-Filho, José Vicente, Jan Maarten van Swaay-Neto, and Antonio de Pádua Wagemaker. "Optimization of the Production Planning and Trade of Lily Flowers at Jan de Wit Company." *Interfaces* 1 (2002): 35–46. Like most *Interfaces* articles, this one provides considerable detail into the problem and solution while remaining comprehensible to the layperson. The model included at the end of the article shows the complexity of the real-world model.

Cook, Wade, and Joe Zhu. *Data Envelopment Analysis: Balanced Benchmarking.* 2013. Like the lecture, this book presents its examples in spreadsheets, but its treatment is far more extensive. A thorough treatment of data envelopment analysis as it is usually applied.

Cornuejols, Gerard, Michael Trick, and Matthew Saltzman. "A Tutorial on Integer Programming." Clemson University. Last modified 1995. Accessed March 29, 2014. http://www.math.clemson.edu/~mjs/courses/mthsc.440/integer. A convenient and understandable discussion of a number of the classic problems in integer and mixed-integer programming.

Cox, Jeff, and Eliyahu Goldratt. *The Goal: A Process of Ongoing Improvement.* Great Barrington: North River Press Publishing Corporation, 2012. In his struggles to save his job and his factory, Eliyahu Goldratt's protagonist discovers many key ideas of operations management. A fast, informative, and engaging read, especially for those interested in the production sector.

Dudley, Jay, and Thomas Buckley. "How Gerber Used a Decision Tree in Strategic Decision-Making." *Graziadio Business Review* 3 (1999). http://gbr.pepperdine.edu/2010/08/how-gerber-used-a-decision-tree-in-strategic-decision-making/. Accessed March 30, 2014. Additional information on the Gerber phthalates story.

Dunham, Margaret. *Data Mining: Introductory and Advanced Topics*. Upper Saddle River: Prentice Hall, 2003. A more advanced data mining text. The latter sections deal with web mining, special mining, and temporal mining.

Elsner, Ralf, Manfred Krafft, and Arnd Huchzermeier. "Optimizing Rhenania's Mail-Order Business through Dynamic Multilevel Modeling (DMLM)." *Interfaces* 1 (2003): 50–66. The full story of Rhenania's recovery.

Gass, Saul, and Assad Arjang. *An Annotated Timeline of Operations Research: An Informal History*. Boston: Springer Science Business Media, Inc., 2005. https://uqu.edu.sa/files2/tiny_mce/plugins/filemanager/ files/4290078/operation_research/operation research 1.pdf. Accessed March 29, 2014. This comprehensive chronology starts with operations research's precursors in the 16th century and traces operations research developments and the people responsible for them through the year 2004. Given the scope of the material, discussion of any particular topic or individual is necessarily brief, but the book lends itself to browsing and gives a sense of the extent of the subject.

Grinstead, Charles M., and J. Laurie Snell. *Introduction to Probability*. 2nd rev. ed. American Mathematical Society, 2006. http://www.dartmouth. edu/~chance/teaching_aids/books_articles/probability_book/book.html. Accessed March 30, 2014. A wonderful, free book on probability theory. Engaging examples, clear explanations, and an interesting interplay between probability calculations and empirical investigation.

Horner, Peter, and Barry List. "Armed with O.R." *ORMS Today* (August 2010): 24–29. An interesting interview with Admiral Mike Mullen, Chairman of the Joint Chiefs of Staff, and the usefulness of his master's degree in operations research to his career.

Hyndman, Rob, and George Athanasopoulos. *Forecasting: Principles and Practice*. OTexts, 2012. https://www.otexts.org/fpp. Accessed March 30, 2014. An excellent text that is available online for free. Hyndman and Athanasopoulos use the statistical language R to create their examples. R and the interface RStudio are also available online at no cost, giving the interested student access to a powerful statistical and forecasting tool. Furthermore, the text includes the R text needed to execute the models created in the book.

Miller, Jeff, and Patricia Hayden. *Statistical Analysis with the General Linear Model.* 2006. http://www.uv.es/~friasnav/librofactorial.pdf. Accessed March 30, 2014. An excellent and free book on regression and much more. Those who are interested in general linear models but have little background in the field couldn't hope for a better introduction. The text both develops a sensible intuition and presents examples that are worked out in detail.

Narisetty, Amar Kumar, Jean-Phillipe Richard, David Ramcharan, Debby Murphy, Gayle Minks, and Jim Fuller. "An Optimization Model for Empty Freight Car Assignment at Union Pacific Railroad." *Interfaces* 2 (2008): 89–102. An in-depth but accessible discussion of the Union Pacific scheduling problem introduced in the lecture. The article ends with a specification of the actual model used.

Piattelli-Palmarini, Massimo. *Inevitable Illusions: How Mistakes of Reason Rule Our Minds.* Charlottesville: John Wiley and Sons, 1994. A fascinating look at how bad most people are at applying basic reasoning to everyday situations. Many of the mental tunnels discussed have their roots in incorrect intuitions about probability.

Ragsdale, Cliff. *Spreadsheet Modeling & Decision Analysis: A Practical Introduction to Management Science.* Mason, OH: South-Western Cengage Learning, 2011. A very good first textbook for those interested in learning to actually do analytics. Ragsdale works in the spreadsheet environment, making it possible for the interested student to formulate and solve nontrivial problems without investing in expensive software.

Samuelson, Douglas. "Election 2012: The '13 Keys' to the White House." *ORMS Today* (June 2011) 26–28. A discussion of a result by quantitative historian Allan Lichtman, who used a technique akin to discriminant analysis to identify a set of 13 criteria that he uses to predict the winner of presidential elections. It has correctly predicted the winner of the popular vote in every election since 1984.